The Many Faces of Osteoarthritis

Vincent C. Hascall
Klaus E. Kuettner

Editors

Birkhäuser Verlag
Basel · Boston · Berlin

Editors

Vincent C. Hascall
Orthopaedic Research Center
Department of Biomedical Engineering
The Cleveland Clinic Foundation
9500 Euclid Avenue
Cleveland, OH 44195
USA

Klaus E. Kuettner
Department of Biochemistry
Rush Medical College at Rush Presbyterian-St. Luke's
Medical Center
1653 West Congress Parkway
Chicago, IL 60612-3864
USA

A CIP catalogue record for this book is available from the Library of Congress, Washington D.C., USA

Deutsche Bibliothek Cataloging-in-Publication Data
The many faces of osteoarthritis / Vincent C. Hascall ; Klaus E. Kuettner ed.. -
Basel ; Boston ; Berlin : Birkhäuser, 2002
   ISBN 3-7643-6581-1

ISBN 3-7643-6581-1 Birkhäuser Verlag, Basel – Boston – Berlin

The publisher and editor can give no guarantee for the information on drug dosage and administration contained in this publication. The respective user must check its accuracy by consulting other sources of reference in each individual case.
The use of registered names, trademarks etc. in this publication, even if not identified as such, does not imply that they are exempt from the relevant protective laws and regulations or free for general use.
This work is subject to copyright. All rights are reserved, whether the whole or part of the material is concerned, specifically the rights of translation, reprinting, re-use of illustrations, recitation, broadcasting, reproduction on microfilms or in other ways, and storage in data banks. For any kind of use, permission of the copyright owner must be obtained.

© 2002 Birkhäuser Verlag, P.O. Box 133, CH-4010 Basel, Switzerland
Member of the BertelsmannSpringer Publishing Group
Printed on acid-free paper produced from chlorine-free pulp. TCF ∞
Cover design: Micha Lotrovsky, CH-4106 Therwil, Switzerland
Cover illustration: Diffraction enhanced x-ray imaging (DEI) of human synovial joint cartilage. The top image is one of an intact human knee joint taken with the DEI technique at the National Synchrotron Light Source at Brookhaven National Laboratory. The bottom two pictures are a DE image of a portion of a talar dome (left) and its histological profile stained with Safranin-O/fast green (right). In both DE images, the articular cartilage is clearly visible. Furthermore, the contrast heterogeneity that gives the appearance of a lesion in the lower DE image is histologically validated. (The top image with friendly permission of Carol A. Muehleman and Matthias E. Aurich, the bottom pictures reprinted from *Osteoarthritis and Cartilage*, volume 10, Mollenhauer J, Aurich ME, Zhong Z, Muehleman C, Cole AA, Masnah M. Oltulu O, Kuettner KE, Margulis A, Chapman LD. Diffraction-enhanced x-ray imaging of articular cartilage. pp. 163–171 (2002) by permission of the publisher WB Saunders.)

Printed in Germany
ISBN 3-7643-6581-1

9 8 7 6 5 4 3 2 1

www.birkhauser.ch

Contents

List of contributors xiii

Preface xxv

The life of Klaus E. Kuettner xxix

Acknowledgements xxxix

The many phases of osteoarthritis

Jody Buckwalter
Introduction 3

Audrey McAlinden, Naoshi Fukui and Linda J. Sandell
Type IIA procollagen NH_2-propeptide functions as an antagonist of bone morphogenetic proteins 5

Matthew L. Warman
Genetics and osteoarthritis 17

Ada Cole, Hans Häuselmann, Johannes Flechtenmacher, Klaus Huch, Holger Koepp, Wolfgang Eger, Matthias E. Aurich, Bernd Rolauffs, Arkady Margulis, Carol Muehleman, Allan Valdellon and Klaus E. Kuettner
Metabolic differences between knee and ankle 27

Marc C. Hochberg
Prevention of lower limb osteoarthritis: Data from The Johns Hopkins Precursors Study 31

Thorvaldur Ingvarsson
The inheritance of hip osteoarthritis in Iceland 39

Lydia K. Wachsmuth, Beate Durchfeld-Meyer, Nadine I. Jahn, Nicole
Verzijl, Uwe H. Dietz, Nicole Gerwin, Manfred Keil, Hans-Ludwig Schmidts
and Ruth X. Raiss
Dynamics of matrix loss in the spontaneous osteoarthritic mouse
strain STR-1N 45

Discussion 51

Morphogenics, development and repair

A. Hari Reddi
Introduction 63

Frank P. Luyten, Cosimo De Bari and Francesco Dell'Accio
Identification and characterization of human cell populations capable
of forming stable hyaline cartilage *in vivo* 67

Holger E. Koepp, Johannes Flechtenmacher, Klaus Huch, Eugene J.-M.A.
Thonar, Gene A. Homandberg and Klaus E. Kuettner
Osteogenic protein-1 promotes proteoglycan synthesis and inhibits
cartilage degeneration mediated by fibronectin-fragments. 77

Susan Chubinskaya, David C. Rueger, Richard A. Berger and
Klaus E. Kuettner
Osteogenic protein-1 and its receptors in human articular cartilage 81

Véronique Lefebvre, Benoit de Crombrugghe and Richard R. Behringer
The transcription factors L-Sox5 and Sox6 are essential for
cartilage formation 91

Arthur Veis, Kevin Tompkins and Michel Goldberg
Amelogenin peptides have unique milieu-dependent roles in morphogenic
path determination 101

Discussion 107

Matrix molecules in cartilage and other tissues

Dick Heinegård
Introduction 115

Karl E. Kadler and David F. Holmes
Electron microscope studies of collagen fibril formation in cornea, skin
and tendon: Implications for collagen fibril assembly and structure in
other tissues 117

Tim Hardingham
Hyaluronics and aggrecanics 131

Anders Aspberg
Lectin domains in hyaluronan-binding proteoglycans 147

*Mats Paulsson, Andreas R. Klatt, Birgit Kobbe, D. Patric Nitsche and
Raimund Wagener*
The matrilins: A novel family of extracellular adaptor proteins 151

*Thomas M. Schmid, Jui-Lan Su, Kathie M. Lindley, Vitaliy Soloveychik,
Lawrence Madsen, Joel A. Block, Klaus E. Kuettner and Barbara
L. Schumacher*
Superficial zone protein (SZP) is an abundant glycoprotein in human
synovial fluid with lubricating properties 159

Frank Zaucke, Robert Dinser, Patrik Maurer and Mats Paulsson
Establishment of *in vitro* cell culture models for the investigation of the
pathogenesis of cartilage diseases 163

Discussion 167

Hyaluronan and joint biology

Vince Hascall
Introduction 177

Roger M. Mason, Peter J. Coleman, David Scott and J. Rodney Levick
Role of hyaluronan in regulating joint fluid flow 179

Endre A. Balazs and Charles Weiss
Elastoviscous hyaluronan in the synovium in health and disease 189

*Wannarat Yingsung, Lisheng Zhuo, Masahiko Yoneda, Naoki Ishiguro,
Hisashi Iwata and Koji Kimata*
The covalent complex formation of hyaluronan with heavy chains of
inter-α-trypsin inhibitor family is important for its functions 207

Janet Y. Lee, Ryan B. Rountree, David M. Kingsley and Andrew P. Spicer
In vivo investigation of hyaluronan and hyaluronan synthase-2 function
during cartilage and joint development 213

Cheryl B. Knudson, Kathleen T. Rousche, Richard S. Peterson,
Geraldine Chow and Warren Knudson
CD44 and cartilage matrix stabilization 219

Discussion 231

Skeletal turnover in health and arthritis

Robin Poole
Introduction 239

J. Frederick Woessner
Metalloproteinases and osteoarthritis 241

Haya ben-Zaken, Rosa Schneiderman, Hannah Kaufmann and
Alice Maroudas
Age-dependent changes in some physico-chemical properties of human
articular cartilage 247

Theodore R. Oegema, Jr.
The role of the bone/cartilage interface in osteoarthritis 253

Wim B. van den Berg, Peter M. van der Kraan, Alwin Scharstuhl,
Henk M. van Beuningen, Andrew Bakker, Peter L.E.M. van Lent,
Fons A.J. von de Loo
Dualistic role of TGFβ in osteoarthritis cartilage destruction and
osteophyte formation 261

Discussion 267

From biomarkers to surrogate outcome measures in osteoarthritis

Stefan Lohmander
Introduction 273

David R. Eyre, Lynne M. Atley and Jiann-Jui Wu
Collagen cross-links as markers of bone and cartilage degradation 275

Tore Saxne, Dick Heinegård and Bengt Månsson
Markers of joint tissue turnover in osteoarthritis 285

Thomas Aigner, Pia M. Gebhard and Alexander Zien
Gene expression profiling by the cDNA array technology: Molecular
portraying of chondrocytes 293

Ivan G. Otterness and Dolores Vázquez-Abad
Clinical evaluation of markers for osteoarthritis 297

*Wolfgang Eger, Stefan Söder, Dietmar Thomas, Thomas Aigner
and Günther Zeiler*
Joint degradation in rapidly destructive and hypertrophic osteoarthritis
of the hip 303

Tamayuki Shinomura, Kazuo Ito and Magnus Höök
Differential gene trap: A new strategy for identifying genes regulated
during cartilage differentiation 309

Discussion 313

Assessment of joint damage in osteoarthritis

Paul Dieppe
Introduction 323

Charles G. Peterfy
Imaging cartilage changes in osteoarthritis 329

*Carol Muehleman, Jürgen A. Mollenhauer, Matthias E. Aurich,
Klaus E. Kuettner, Zhong Zhong, Ada Cole and Dean Chapman*
Diffraction enhanced x-ray imaging of articular cartilage 351

Iain Watt
Bone changes in osteoarthritis 355

Joel A. Block
Radiographic joint space width (JSW): A marker of disease progression in
osteoarthritis of the hip 363

Gabriella Cs-Szabo, Deborah Ragasa, Richard A. Berger and Klaus E. Kuettner
Small proteoglycans in knee and ankle cartilage 369

Discussion 373

Biomechanics and cartilage metabolism

Alan Grodzinsky
Introduction 383

Robert J. Wilkins, Bethan Hopewell and Jill P.G. Urban
Fixed charge density and cartilage biomechanics 387

Moonsoo Jin, Alan J. Grodzinsky, Thomas H. Wuerz, Gregory R. Emkey, Marcy Wong and Ernst B. Hunziker
Influence of tissue shear deformation on chondrocyte biosynthesis and matrix nano-electromechanics 397

Robert L. Sah
The biomechanical faces of articular cartilage in growth, aging, and osteoarthritis 409

John D. Kisiday, Moonsoo Jin, Bodo Kurz, Han-Hwa Hung, Carlos Semino, Shuguang Zhang and Alan J. Grodzinsky
Cartilage tissue engineering using a new self-assembling peptide gel scaffold .. 423

Matthias E. Aurich, Jürgen A. Mollenhauer, Klaus E. Kuettner and Ada A. Cole
Differential effects of IL-1β on human knee and ankle chondrocytes 429

Discussion 433

Biomechanics, motor control and osteoarthritis

Mark Grabiner
Introduction 441

Thomas P. Andriacchi
Dynamic function and imaging in the analysis of osteoarthritis at the knee 443

Eugene J.-M.A. Thonar, Debra E. Hurwitz, Thomas P. Andriacchi, Mary Ellen Lenz and Leena Sharma
Linking the biology of osteoarthritis to locomotion mechanics 453

Terese L. Chmielewski, Katherine S. Rudolph, Michael J. Axe,
G. Kelley Fitzgerald and Lynn Snyder-Mackler
Movement patterns of individuals with good potential to dynamically
stabilize their knees after acute ACL rupture 461

Lynn Snyder-Mackler and Terese Chmielewski
Neuromuscular control of the ACL deficient knee: Implications for the
development of osteoarthritis 465

Discussion 473

Photos from the conference 481

Index 489

List of contributors

Thomas Aigner, Cartilage Research, Department of Pathology, University of Erlangen-Nürnberg, Krankenhausstrasse 8-10, D-91054 Erlangen; e-mail: thomas.aigner@patho.imed.uni-erlangen.de

Thomas P. Andriacchi, Stanford University, Division of Biomechanical Engineering, Durand Building, Stanford, CA 94305-3030, USA; e-mail: tandriac@stanford.edu

Anders Aspberg, Connective Tissue Biology, Lund University, BMC, C12, 22184 Lund, Sweden; e-mail: anders.aspberg@medkem.lu.se

Lynne M. Atley, Department of Orthopaedics and Sports Medicine, Orthopaedic Research Laboratories, University of Washington, Box 356500, Seattle, WA 98195-6500, USA

Matthias E. Aurich, Department of Orthopedics, University of Jena, Waldkrankenhaus "Rudolf-Elle", Klosterlausnitzer Strasse 81, 07607 Eisenberg, Germany; and Department of Biochemistry, Rush Medical College, 1653 W. Congress Parkway, Chicago, IL 60612, USA; e-mail: rek_research.eisenberg@t-online.de

Michael J. Axe, Department of Physical Therapy, University of Delaware and First State Orthopaedics, Newark, DE, USA

Andrew Bakker, Rheumatology Research and Advanced Therapeutics, University Medical Center Nijmegen, 189, Geert Grooteplein 26-28, 6500 HB Nijmegen, The Netherlands

Endre A. Balazs, Matrix Biology Institute, 65 Railroad Avenue, Ridgefield, NJ 07657, USA; e-mail: eabalazs@matrixha.org

Richard R. Behringer, Department of Molecular Genetics, The University of Texas, M.D. Anderson Cancer Center, Houston, TX 77030, USA

Haya ben-Zaken, Department of Biomedical Engineering, Technion – Israel Institute of Technology, Haifa 32000, Israel

Richard A. Berger, Orthopedic Surgery, Rush Medical College at Rush-Presbyterian-St. Luke's Medical Center, 1653 W. Congress Parkway, Chicago, IL 60612, USA

Joel A. Block, Section of Rheumatology, Rush Medical College at Rush-Presbyterian-St. Luke's Medical Center, 1653 W. Congress Parkway, Chicago, IL 60612, USA; e-mail: jblock@rush.edu

Jody Buckwalter, Dept. of Orthopaedic Surgery, University of Iowa Hospital, 200 Hawkins Drive, Iowa City, IA 52242-1088, USA;
e-mail: joseph-buckwalter@uiowa.edu

Dean Chapman, Biological, Chemical and Physical Sciences Department, Illinois Institute of Technology, 3101 South Dearborn, Chicago, IL 60616, USA

Terese L. Chmielewski, Department of Physical Therapy and Interdisciplinary Graduate Program in Biomechanics and Movement Sciences, University of Delaware, Newark, DE 19716, USA; e-mail: tchm@udel.edu

Geraldine Chow, Department of Biochemistry, Rush Medical College, 1653 W. Congress Parkway, Chicago, IL 60612, USA

Susan Chubinskaya, Departments of Biochemistry and Section of Rheumatology, Rush Medical College at Rush-Presbyterian-St. Luke's Medical Center, 1653 W. Congress Parkway, Chicago, IL 60612, USA; e-mail: schubins@rush.edu

Ada Cole, Departments of Biochemistry and Anatomy, Rush Medical College, 1653 W. Congress Parkway, Chicago, IL 60612, USA; e-mail: ada_cole@rush.edu

Peter J. Coleman, Department of Physiology, St. George's Hospital Medical School, Cranmer Terrace, London, SW17 ORE, UK

Benoit de Crombrugghe, Department of Molecular Genetics, The University of Texas, M.D. Anderson Cancer Center, Houston, TX 77030, USA

Gabriella Cs-Szabo, Departments of Biochemistry and Orthopedic Surgery, Rush Medical College at Rush-Presbyterian-St. Luke's Medical Center, 1653 W. Congress Parkway, Chicago, IL 60612, USA; e-mail: gcaszabo@rush.edu

Cosimo De Bari, Laboratory for Skeletal Development and Joint Disorders, Department of Rheumatology, University Hospitals KU Leuven, Gasthuisberg, Herestraat 49, 3000 Leuven, Belgium

Francesco Dell'Accio, Laboratory for Skeletal Development and Joint Disorders, Department of Rheumatology, University Hospitals KU Leuven, Gasthuisberg, Herestraat 49, 3000 Leuven, Belgium

Paul Dieppe, MRC HSRC, Department of Social Medicine, University of Bristol, Canynge Hall, Whiteladies Road, Bristol BS8 2PR, UK;
e-mail: p.dieppe@bristol.ac.uk

Uwe H. Dietz, Thrombotic Diseases/Degenerative Joint Diseases, Aventis Pharma, 65926 Frankfurt, Germany

Robert Dinser, Institute for Biochemistry II, Medical Faculty, University of Köln, Joseph-Stelzmann-Strasse 52, 50931 Köln, Germany

Beate Durchfeld-Meyer, Drug Safety Evaluation, Aventis Pharma, 65926 Frankfurt, Germany

Wolfgang Eger, Department of Orthopaedic Surgery, Rummelsberg Hospital, Rummelsberg 71, 90592 Schwarzenbruck, Germany; and Department of Biochemistry, Rush Medical College, 1653 W. Congress Parkway, Chicago, IL 60612, USA;
e-mail: wolf.eger@t-online.de

Gregory R. Emkey, Continuum Electromechanics Group, Center for Biomedical Engineering, Massachusetts Institute of Technology, M.I.T. Room 38-377, Cambridge, MA 02139, USA

David R. Eyre, Department of Orthopaedics and Sports Medicine, Orthopaedic Research Laboratories, University of Washington, Box 356500, Seattle, WA 98195-6500, USA; e-mail: deyre@u.washington.edu

G. Kelley Fitzgerald, Department of Physical Therapy, University of Pittsburg, PA, USA

Johannes Flechtenmacher, Orthopaedic Clinic, Waldstrasse 67, 76133 Karlsruhe, Germany

Naoshi Fukui, Department of Orthopaedic Surgery, Washington University School of Medicine at Barnes-Jewish Hospital, St Louis, MO 63110, USA

Pia M. Gebhard, Cartilage Research, Department of Pathology, University of Erlangen-Nürnberg, Krankenhausstrasse 8-10, D-91054 Erlangen

Nicole Gerwin, Thrombotic Diseases/Degenerative Joint Diseases, Aventis Pharma, 65926 Frankfurt, Germany

Michel Goldberg, Laboratoire de Biologie et Physiopathologie Cranio-faciales-EA 2496, Université de Paris V, 1, rue Maurice Arnoux, 92120 Montrouge, France

Mark Grabiner, School of Kinesiology, University of Illinois at Chicago, 901 West Roosevelt Road, Chicago, IL 60608, USA

Alan J. Grodzinsky, Continuum Electromechanics Group, Center for Biomedical Engineering, Massachusetts Institute of Technology, M.I.T. Room 38-377, Cambridge, MA 02139, USA; e-mail: alg@mit.edu

Tim Hardingham, Wellcome Trust Centre for Cell-Matrix Research, School of Biological Sciences, University of Manchester, Stopford Building 2.205, Oxford Road, Manchester M13 9PT, UK; e-mail: timothy.e.hardingham@man.ac.uk

Hans J. Häuselmann, Laboratory of Experimental Cartilage Research, Center for Rheumatology and Bone Disease, Klinik Im Park, Hirslanden Group, Bellariastrasse 38, 8038 Zürich, Switzerland

Dick Heinegård, Lund University, Department of Cell and Molecular Biology, BMC, 22184 Lund, Sweden; e-mail: dick.heinegard@medkem.lu.se

Marc C. Hochberg, Division of Rheumatology and Clinical Immunology, Department of Medicine, University of Maryland School of Medicine, and Medical Service, Maryland Veterans Affairs Health Care System, Baltimore, MD 21201, USA; e-mail: mhochber@umaryland.edu

David F. Holmes, Wellcome Trust Centre for Cell-Matrix Research, School of Biological Science, University of Manchester, Stopford Building 2.205, Oxford Road, Manchester M13 9PT, UK

Gene A. Homandberg, Department of Biochemistry, Rush University, Rush-Presbyterian-St. Luke's Medical Center, 1653 W. Congress Parkway, Chicago, IL 60612, USA

Magnus Höök, Institute of Biosciences and Technology, Texas A & M University, Houston, TX 77030, USA

Bethan Hopewell, Physiology Laboratory, Oxford University, Parks Rd, Oxford OX1 3PT, UK

Klaus Huch, Department of Orthopedic Surgery, University of Ulm, Oberer Eselsberg 45, 89081 Ulm, Germany

Han-Hwa Hung, Center for Biomedical Engineering, Massachusetts Institute of Technology, Cambridge, MA 02139-4307, USA

Ernst B. Hunziker, M.E. Mueller Institute for Biomechanics, University of Bern, Switzerland

Debra E. Hurwitz, Department of Orthopaedic Surgery, Rush Medical College, Rush-Presbyterian-St. Luke's Medical Center, 1653 W. Congress Parkway, Chicago, IL 60612, USA

Thorvaldur Ingvarsson, University Hospital, 600, Akureyri, Iceland; e-mail: thi@fsa.is

Naoki Ishiguro, Department of Orthopedic Surgery, Nagoya University School of Medicine, Tsurumai, Nagoya, 466-8550, Japan

Kazuo Ito, Institute of Biosciences and Technology, Texas A & M University, Houston, TX 77030, USA

Hisashi Iwata, Department of Orthopedic Surgery, Nagoya University School of Medicine, Tsurumai, Nagoya, 466-8550, Japan

Nadine I. Jahn, Dept. of Pharmacy, University Bonn, 53113 Bonn, Germany

Moonsoo Jin, Continuum Electromechanics Group, Center for Biomedical Engineering, Massachusetts Institute of Technology, M.I.T. Room 38-377, Cambridge, MA 02139, USA; e-mail: msjin@mit.edu

Karl E. Kadler, Wellcome Trust Centre for Cell-Matrix Research, School of Biological Science, University of Manchester, Stopford Building 2.205, Oxford Road, Manchester M13 9PT, UK; e-mail: karl.kadler@man.ac.uk

Hannah Kaufmann, Department of Orthopaedics, Rambam Hospital, Haifa 32000, Israel

Manfred Keil, Drug Safety Evaluation, Aventis Pharma, 65926 Frankfurt, Germany

G. Kelley Fitzgerald, Department of Physical Therapy and Interdisciplinary Graduate Program in Biomechanics and Movement Sciences, University of Delaware, Newark, DE, USA

Koji Kimata, Institute for Molecular Science of Medicine, Aichi Medical University, Nagakute, Aichi, 480-1195, Japan; e-mail: kimata@amugw.aichi-med-u.ac.jp

David M. Kingsley, Department of Developmental Biology, Beckman Center, Stanford University, Palo Alto, CA 94305, USA

John D. Kisiday, Continuum Electromechanics Group, Division of Biological Engineering, Massachusetts Institute of Technology, Cambridge, MA 02139-4307, USA; e-mail: kisiday@mit.edu

Andreas R. Klatt, Institute for Biochemistry II, Medical Faculty, University of Köln, Joseph-Stelzmann-Strasse 52, 50931 Köln, Germany

Birgit Kobbe, Institute for Biochemistry II, Medical Faculty, University of Köln, Joseph-Stelzmann-Strasse 52, 50931 Köln, Germany

Holger E. Koepp, Department of Orthopedic Surgery, University of Ulm, Oberer Eselsberg 45, 89081 Ulm, Germany; e-mail: hjkoepp@t-online.de

Cheryl B. Knudson, Department of Biochemistry, Rush Medical College, 1653 W. Congress Parkway, Chicago, IL 60612, USA; e-mail: cknudson@rush.edu

Warren Knudson, Department of Biochemistry, Rush Medical College, 1653 W. Congress Parkway, Chicago, IL 60612, USA

Klaus E. Kuettner, Department of Biochemistry and Department of Orthopedics, Rush University, Rush-Presbyterian-St. Luke's Medical Center, 1653 W. Congress Parkway, Chicago, IL 60612, USA; e-mail: kkuettne@rush.edu

Bodo Kurz, Center for Biomedical Engineering, Massachusetts Institute of Technology, Cambridge, MA 02139-4307, USA

Janet Y. Lee, Department of Cell Biology and Human Anatomy, University of California, School of Medicine, Davis, CA 95616, USA; e-mail: jntlee@ucdavis.edu

Véronique Lefebvre, Department of Biomedical Engineering, Lerner Research Institute, Cleveland Clinic Foundation, Cleveland, OH 44195, USA;
e-mail: lefebvrv@bme.ri.ccf.org

Mary Ellen Lenz, Department of Biochemistry, Rush Medical College, Rush-Presbyterian-St. Luke's Medical Center, 1653 W. Congress Parkway, Chicago, IL 60612, USA

J. Rodney Levick, Department of Physiology, St. George's Hospital Medical School, Cranmer Terrace, London, SW17 0RE, UK

Kathie M. Lindley, Department of Protein Sciences, GlaxoSmithKline Inc., Five Moore Drive, Research Triangle Park, NC 27709, USA

Stefan Lohmander, Department of Orthopedics, University Hospital, 22185 Lund, Sweden; e-mail: stefan.lohmander@ort.lu.se

Frank P. Luyten, Laboratory for Skeletal Development and Joint Disorders, Department of Rheumatology, University Hospitals KU Leuven, Gasthuisberg, Herestraat 49, 3000 Leuven, Belgium; e-mail: frank.luyten@uz.kuleuven.ac.be

Lawrence Madsen, Department of Biochemistry, Rush Medical College at Rush-Presbyterian-St. Luke's Medical Center, 1653 W. Congress Parkway, Chicago, IL 60612, USA

Bengt Månsson, Department of Rheumatology and Department of Cell and Molecular Biology, Section for Connective Tissue Biology, Lund University, 22185 Lund, Sweden

Arkady Margulis, Department of Biochemistry, Rush Medical College, 1653 W. Congress Parkway, Chicago, IL 60612, USA

Alice Maroudas, Department of Biomedical Engineering, Technion – Israel Institute of Technology, Haifa 32000, Israel; e-mail: Alice@biomed.technion.ac.il

Roger M. Mason, Cell and Molecular Biology Section, Division of Biomedical Sciences, Faculty of Medicine, Imperial College, Sir Alexander Fleming Building, Exhibition Road, London, SW7 2AZ, UK; e-mail: roger.mason@ic.ac.uk

Patrik Maurer, Institute for Biochemistry II, Medical Faculty, University of Köln, Joseph-Stelzmann-Strasse 52, 50931 Köln, Germany

Audrey McAlinden, Department of Orthopaedic Surgery, Washington University School of Medicine at Barnes-Jewish Hospital, St Louis, MO 63110, USA

Jürgen A. Mollenhauer, Department of Orthopedics, University of Jena, Waldkrankenhaus "Rudolf-Elle", Klosterlausnitzer Strasse 81, 07607 Eisenberg, Germany; and Department of Biochemistry, Rush Medical College, 1653 W. Congress Parkway, Chicago, IL 60612, USA

Carol Muehleman, Departments of Anatomy and Biochemistry, Rush Medical College, 1653 W. Congress Parkway, Chicago, IL 60612, USA; e-mail: carol_muehleman@.rush.edu

D. Patric Nitsche, Institute for Biochemistry II, Medical Faculty, University of Köln, Joseph-Stelzmann-Strasse 52, 50931 Köln, Germany

Theodore R. Oegema, Jr., Department of Orthopedic Surgery and Biochemistry, University of Minnesota, 420 Delaware St., SE, Minneapolis, MN 55455, USA; e-mail: oegem001@umn.edu. Present address: Departments of Biochemistry and Orthopaedic Surgery, Rush Medical College, 1653 W. Congress Parkway, Chicago, IL 60612; e-mail: ted_oegema@rush.edu

Ivan G. Otterness, 241 Monument Street 5, Groton, CT 06340, USA; e-mail: iotterness@snet.net

Mats Paulsson, Institute for Biochemistry II, Medical Faculty, University of Köln, Joseph-Stelzmann-Strasse 52, 50931 Köln, Germany; e-mail: mats.paulsson@uni-koeln.de

Charles G. Peterfy, Synarc, Inc., 575 Market Street, 17th Floor, San Francisco, CA 94105, USA; e-mail: charles.peterfy@synarc.com

Richard S. Peterson, Department of Biochemistry, Rush Medical College at Rush-Presbyterian-St. Luke's Medical Center, 1653 W. Congress Parkway, Chicago, IL 60612, USA

Robin Poole, Shriners Hospital for Children, McGill University, 1529 Cedar Avenue, Montreal, Quebec H3G 1A6, Canada; e-mail: rpoole@shriners.mcgill.ca

Deborah Ragasa, Department of Biochemistry, Rush Medical College at Rush-Presbyterian-St. Luke's Medical Center, 1653 W. Congress Parkway, Chicago, IL 60612, USA

Ruth X. Raiss, Thrombotic Diseases/Degenerative Joint Diseases, Aventis Pharma, 65926 Frankfurt, Germany; e-mail: ruth.raiss@aventis.com

A. Hari Reddi, Center for Tissue Regeneration and Repair, Department of Orthopaedic Surgery, University of California, Davis School of Medicine, Sacramento, CA 95817, USA; e-mail: ahreddi@ucdavis.edu

Bernd Rolauffs, Klinik und Poliklinik für allg. Orthopädie, Albert-Schweitzer-Strasse 33, 48129 Münster, Germany; and Department of Biochemistry, Rush Medical College, 1653 W. Congress Parkway, Chicago, IL 60612, USA

Ryan B. Rountree, Department of Developmental Biology, Beckman Center, Stanford University, Palo Alto, CA 94305, USA

Kathleen T. Rousche, Department of Biochemistry, Rush Medical College at Rush-Presbyterian-St. Luke's Medical Center, 1653 W. Congress Parkway, Chicago, IL 60612, USA

Katherine S. Rudolph, Department of Physical Therapy and Interdisciplinary Graduate Program in Biomechanics and Movement Sciences, University of Delaware, Newark, DE, USA

David C. Rueger, Stryker Biotech, Hopkinton, MA 01748, USA

Robert L. Sah, Department of Bioengineering, Mail Code 0412, University of California-San Diego, 9500 Gilman Drive, La Jolla, CA 92093, USA; e-mail: rsah@ucsd.edu

Linda J. Sandell, Department of Orthopaedic Surgery, Washington University School of Medicine at Barnes-Jewish Hospital, St Louis, MO 63110, USA; e-mail: sandelll@msnotes.wustl.edu

Tore Saxne, Department of Rheumatology and Department of Cell and Molecular Biology, Section for Connective Tissue Biology, Lund University, 22185 Lund, Sweden; e-mail: tore.saxne@reum.lu.se

Alwin Scharstuhl, Rheumatology Research and Advanced Therapeutics, University Medical Center Nijmegen, 189, Geert Grooteplein 26-28, 6500 HB Nijmegen, The Netherlands

Thomas M. Schmid, Department of Biochemistry, Rush Medical College at Rush-Presbyterian-St. Luke's Medical Center, 1653 W. Congress Parkway, Chicago, IL 60612, USA; e-mail: tschmid@rush.edu

Hans-Ludwig Schmidts, Drug Safety Evaluation, Aventis Pharma, 65926 Frankfurt, Germany

Rosa Schneiderman, Department of Biomedical Engineering, Technion – Israel Institute of Technology, Haifa 32000, Israel

Barbara L. Schumacher, Department of Bioengineering, University of California at San Diego, 9500 Gillman Dr., La Jolla, CA 92093, USA

David Scott, Department of Physiology, St. George's Hospital Medical School, Cranmer Terrace, London, SW17 0RE, UK

Carlos Semino, Center for Biomedical Engineering, Massachusetts Institute of Technology, Cambridge, MA 02139-4307, USA

Leena Sharma, Northwestern University, Ward Building 3-315, 303 East Chicago Avenue, Chicago, IL 60611, USA

Tamayuki Shinomura, Department of Hard Tissue Engineering, Tokyo Medical and Dental University, Yushima, Bunkyo-ku, Tokyo 113-8549, Japan;
e-mail: t.shinomura.trg@tmd.ac.jp

Lynn Snyder-Mackler, Department of Physical Therapy and Interdisciplinary Graduate Program in Biomechanics and Movement Sciences, University of Delaware, Newark, DE 19716, USA; e-mail: smack@udel.edu

Stefan Söder, Cartilage Research Group, Institute of Pathology, University of Erlangen-Nürnberg, Krankenhausstrasse 8-10, 91054 Erlangen, Germany

Vitaliy Soloveychik, Department of Biochemistry, Rush Medical College at Rush-Presbyterian-St. Luke's Medical Center, 1653 W. Congress Parkway, Chicago, IL 60612, USA

Andrew P. Spicer, Center for Extracellular Matrix Biology, Texas A and M University, Health Science Center, 2121 W. Holcombe Blvd., Houston, TX 77030, USA

Jui-Lan Su, Department of Protein Sciences, GlaxoSmithKline Inc., Five Moore Drive, Research Triangle Park, NC 27709, USA

Dietmar Thomas, Department of Orthopaedic Surgery, Rummelsberg Hospital, Rummelsberg 71, 90592 Schwarzenbruck, Germany

Eugene, J.-M. A. Thonar, Department of Biochemistry, Orthopaedic Surgery and Internal Medicine, Rush Medical College, Rush-Presbyterian-St. Luke's Medical Center, 1653 W. Congress Parkway, Chicago, IL 60612, USA;
e-mail: Eugene_Thonar@rsh.net

Kevin Tompkins, Department of Cell and Molecular Biology, Northwestern University, 303 E. Chicago Avenue, Chicago, IL 60611, USA

Jill P.G. Urban, Physiology Laboratory, Oxford University, Parks Rd, Oxford OX1 3PT, UK; e-mail: jpgu@physiol.ox.ac.uk

Allan Valdellon, Regional Organ Bank of Illinois, 800 South Wells, Chicago, IL 60607, USA

Henk M. van Beuningen, Rheumatology Research and Advanced Therapeutics, University Medical Center Nijmegen, 189, Geert Grooteplein 26-28, 6500 HB Nijmegen, The Netherlands

Fons A.J. van de Loo, Rheumatology Research and Advanced Therapeutics, University Medical Center Nijmegen, 189, Geert Grooteplein 26-28, 6500 HB Nijmegen, The Netherlands

Wim B. van den Berg, Rheumatology Research and Advanced Therapeutics, University Medical Center Nijmegen, 189, Geert Grooteplein 26-28, 6500 HB Nijmegen, The Netherlands; e-mail: w.vandenberg@reuma.azn.nl

Peter M. van der Kraan, Rheumatology Research and Advanced Therapeutics, University Medical Center Nijmegen, 189, Geert Grooteplein 26-28, 6500 HB Nijmegen, The Netherlands

Peter L.E.M. van Lent, Rheumatology Research and Advanced Therapeutics, University Medical Center Nijmegen, 189, Geert Grooteplein 26-28, 6500 HB Nijmegen, The Netherlands

Dolores Vázquez-Abad, Pfizer Global Research and Development, New London, CT 06320, USA

Arthur Veis, Department of Cell and Molecular Biology, Northwestern University, 303 E. Chicago Avenue, Chicago, IL 60611, USA; e-mail: aveis@northwestern.edu

Nicole Verzijl, TNO Prevention and Health, 2301 CE Leiden, The Netherlands

Lydia K. Wachsmuth, Institute of Medical Physics, University Erlangen, 91054 Erlangen, Germany

Raimund Wagener, Institute for Biochemistry II, Medical Faculty, University of Köln, Joseph-Stelzmann-Strasse 52, 50931 Köln, Germany

Matthew L. Warman, Department of Genetics, Case Western Reserve University, 2109 Adelbert Road, Cleveland, OH 44106, USA; e-mail: mlw14@po.cwru.edu

Iain Watt, Department of Clinical Radiology, Bristol Royal Infirmary, Bristol BS2 8HW, UK; e-mail: iain.watt@ubht.swest.nhs.uk

Charles Weiss, Matrix Biology Institute, 65 Railroad Avenue, Ridgefield, NJ 07657, USA

Robert J. Wilkins, Physiology Laboratory, Oxford University, Parks Rd, Oxford OX1 3PT, UK

J. Frederick Woessner, Department of Biochemistry and Molecular Biology, University of Miami School of Medicine, P.O. Box 016960, Miami, FL 33101, USA; e-mail: fwoessne@med.miami.edu

Marcy Wong, M.E. Mueller Institute for Biomechanics, University of Bern, Switzerland

Jiann-Jiu Wu, Department of Orthopaedics and Sports Medicine, Orthopaedic Research Laboratories, University of Washington, Box 356500, Seattle, WA 98195-6500, USA

Thomas H. Wuerz, Continuum Electromechanics Group, Center for Biomedical Engineering, Massachusetts Institute of Technology, M.I.T. Room 38-377, Cambridge, MA 02139, USA

Wannarat Yingsung, Institute for Molecular Science of Medicine, Aichi Medical University, Nagakute, Aichi, 480-1195, Japan; present address: Department of Biochemistry, Faculty of Medicine, Chiang Mai University, Amphure Muang, Chiang Mai, 50200 Thailand

Masahiko Yoneda, Institute for Molecular Science of Medicine, Aichi Medical University, Nagakute, Aichi, 480-1195, Japan

Frank Zaucke, Institute for Biochemistry II, Medical Faculty, University of Köln, Joseph-Stelzmann-Strasse 52, 50931 Köln, Germany; e-mail: frank.zaucke@uni-koeln.de

Günther Zeiler, Department of Orthopaedic Surgery, Rummelsberg Hospital, Rummelsberg 71, 90592 Schwarzenbruck, Germany

Shuguang Zhang, Center for Biomedical Engineering, Massachusetts Institute of Technology, Cambridge, MA 02139-4307, USA

Zhong Zhong, NSLS Brookhaven National Laboratory, Upton, NY 11973, USA

Lisheng Zhuo, Institute for Molecular Science of Medicine, Aichi Medical University, Nagakute, Aichi, 480-1195, Japan

Alexander Zien, German National Research Center for Information Technology, Schloss Birlinghoven, 53754 Sankt Augustin, Germany

Preface

The conference "The Many Faces of Osteoarthritis", convened at Lake Tahoe, June 23–27, 2001, was held in my honor to acknowledge and to pay tribute to my contributions in the field and also to celebrate my birthday, which happened to fall in that time frame. The meeting was one of the happiest events in my professional life, and I am much indebted to the organizers and to all my colleagues and friends who contributed to its success. On a personal note, I was particularly pleased to meet again so many former trainees. The resulting book reflects the scientific presentations and discussions between a select group of invited investigators, all experts in this field. As will be evident, the meeting is a logical continuation of scientific workshops in the field, two of which were originally sponsored by Hoechst Pharmaceutical Company of Germany. This scientific meeting is now sponsored by two pharmaceutical companies, again Hoechst (now Aventis), and also Glaxo-SmithKline. The latter is a scientific partner for members of the Department of Biochemistry, which I have chaired since 1980 at Rush Medical College in Chicago.

In 1987, the Department of Biochemistry was awarded one of only three national grants in osteoarthritis as a Specialized Center of Research (SCOR) from the National Institute of Arthritis, Musculoskeletal and Skin Diseases of the National Institutes of Health. These Specialized Centers of Research support a cluster of individual, but interrelated, basic and clinical research projects. The SCOR at Rush entitled "Osteoarthritis: A Continuum (From Cartilage Metabolism to Early Detection and Treatment)" involved investigators from basic science and clinical departments, and I served as Program Director. In 1992, and again in 1997, Special Study Sections evaluated all SCOR grants in osteoarthritis and the program at Rush was approved each time for an additional five years, and is currently the only SCOR on osteoarthritis in the country.

To guarantee the best possible progress and research direction, the SCOR grant investigators are assessed and evaluated on an annual basis by a Scientific Advisory Committee composed of eight internationally renowned scientists (nicknamed "Scoriers") which not only came up with the idea of this meeting, but also put it together. In addition, they also evaluated the manuscripts for this book (not an easy task) and wrote a brief overview to reflect the general discussions during the meeting. The strength, however, lies in the fact that the meeting was set up in the familiar Gordon Research Conference style.

The conference "The Many Faces of Osteoarthritis" reflects the current state of knowledge and will help elucidate the etiopathology of osteoarthritis (OA), hopefully leading to early detection of the disease and novel treatment modalities. OA is the most common joint pathology and primarily affects the older population. Extended research, however, has shown substantial dissimilarities between OA and the aging processes. OA can be envisioned as a group of overlapping distinct diseases that may have different etiologies but show similar biologic, morphologic and clinical outcomes. The disease is characterized by unique pathological changes in some synovial joints, predominantly affecting the articular cartilage, but also the entire joint, including the synovial tissue and subchondral bone. It is only in later stages that it can be diagnosed indirectly by loss of articular cartilage as revealed in radiography, a method used to diagnose the disease in clinical practice and in epidemiological studies. Not all individuals with radiographic evidence of OA have clinical symptoms. However, the probability of symptoms increases with the severity of radiographic changes. It is hypothesized that both mechanical and biological events destabilize the normal coupling of synthesis and degradation of the matrix of the articular cartilage by its chondrocytes, with modulation also in the subchondral bone. The disease may be initiated by multiple factors including genetic, developmental, metabolic and especially traumatic ones that may have occurred much earlier in life.

In the early 1980s, when research in articular cartilage biochemistry was still in its infancy, I was asked, together with Dr. Vincent Hascall, to organize in September 1985 a Workshop Conference (sponsored by Hoechst-Werk Albert, Wiesbaden, Germany) entitled "Articular Cartilage Biochemistry." At this meeting, the results of ongoing research on the structure and metabolism, both of normal and osteoarthritic articular cartilage, and related arthritic disorders were presented. Through a detailed analysis of the matrix macromolecules, the biosynthesis and the normal and pathological metabolism of cartilage components by chondrocytes were reported. The questions asked were, for example: How do cells communicate in order to synchronize macromolecular synthesis and secretion with degradation in the different micro-regions of the extracellular matrix? What are the signals and the receptors in the intercellular transduction mechanisms? What are the critical interactions among intercellular macromolecules that infer tissue specificity upon extracellular processes? The participants of this meeting advanced the understanding of the mechanisms underlying cartilage degeneration in arthritic diseases. In 1986, the presentations and lively discussions of the workshop were published by Raven Press (New York) with Drs. Klaus Kuettner, Rudolf Schleyerbach and Vincent Hascall as co-editors.

The rapid growth of knowledge and methodologies in cartilage research resulted in an increase in diverse methods used to study cartilage biology and biochemistry. Therefore, in order to achieve a standardization of methods, I was asked by Dr. Alice Maroudas to conjointly organize an international seminar entitled "The Bat-Sheva Seminar on Methods Used in Research on Cartilaginous Tissues", which was held at the Nof Ginossar Kibbutz, Israel, in March 1989. This seminar was sponsored by the Bat-Sheva de Rothschild Foundation for the Advancement for Science in Israel. The proceedings of the meeting were published in a book by Academic Press (London and San Diego) in 1990 under the title "Methods in Cartilage Research" with Drs. Alice Maroudas and Klaus Kuettner as co-editors. The time was right for assembling the various methodologies into a single volume that reflected the sophistication of each aspect in this field and also provided a comprehensive source for investigators from other disciplines. The book describes for example: Qualitative and quantitation techniques for the tissue specimens, extraction methods, chondrocyte and explant cultures, and the tissue composition and organization. Furthermore, physical and mechanical properties as well as their relevance to physiological processes were delineated. Different approaches were described, compared, discussed and assessed. The major aim was to show investigators the various choices and possibilities in research and to discuss the appropriateness of study designs without ignoring their inherent shortcomings, limitations and difficulties.

Three years later another workshop conference (again sponsored by Hoechst-Werk Albert) entitled "Articular Cartilage and Osteoarthritis" highlighted current basic scientific and clinical research efforts to further advance the understanding of articular cartilage physiology and pathophysiology and also the etiopathology of osteoarthritis. Among the questions asked were: How does the microenvironment, containing collagens and proteoglycans, serve as a prerequisite for the maintenance of cell differentiation? What regulates the biosynthesis of proteoglycans and glycosaminoglycans? Does secretion and diffusion of recently identified "morphogens" play a role in remodeling and/or repair? What are the influences of mechanical stresses upon the biosynthesis of extracellular macromolecules? What is the role of non-collagenous macromolecules upon extracellular matrix specificity? How does a degradation enzyme contact a collagen fibril (in the case of collagenolysis) when the latter is normally encased in a matrix of other macromolecules? Clinical research findings on potential diagnostic markers of early OA and investigational, potential therapeutic interventions were presented. They provided invaluable sources of information, illuminating observations and promising new approaches for all orthopedic surgeons, rheumatologists and basic research scientists investigating joint diseases. This workshop symposium was afterwards published, again by Ravens Press (New York) in 1992, with Drs. Klaus Kuettner, Rudolf Schleyerbach, Jacques Peyron and Vincent Hascall as co-editors.

In April 1994, Dr. Victor Goldberg and I were asked to organize a workshop entitled "New Horizons in Osteoarthritis." Our task was to bring together an international, interdisciplinary group of leading scientists and clinicians to define the present knowledge and delineate future research directions on the etiopathogenesis of OA, as well as to develop new strategies in research for the understanding of the etiopathology of OA. This workshop was sponsored by the American Academy of Orthopedic Surgeons, the National Institute of Arthritis and Musculoskeletal and Skin Diseases, the National Institute on Aging, the National Arthritis Foundation, and the Orthopedic Research and Education Foundation. During this workshop, groups of investigators developed concepts and specific plans on various topics, which were presented, modified and endorsed by all participants. A book entitled "Osteoarthritic Disorders" was published by the American Academy of Orthopaedic Surgeons (Rosemont) in 1995, edited by Drs. Klaus Kuettner and Victor Goldberg. It contains invited summary manuscripts describing the current state of research in specific areas. Each summary is followed by a recapitulation of the extensive discussions. This covers the definition (including epidemiology) and classification of OA, cartilage changes in aging and changes in the osteoarthritic joint as an organ, role of mediators and inflammation in the degradative mechanisms, repair of cartilage, and the monitoring of preclinical and clinical progression and treatment of OA. At the time of the workshop the group identified gaps in factual knowledge and stated that more information was needed before questions addressed during the meeting could be answered. The book also contains selected reports, descriptions of the brainstorming discussions of future investigations, an overview, and consensus opinions and recommendations for new research directions in OA.

The group of investigators in the field of cartilage and osteoarthritis research is relatively small. Thus, most of the "Scoriers" were present at the four meeting/workshops/symposiums, mentioned above. Thus, the meeting in Lake Tahoe was a logical consequence of the numerous interactions that we had at these scientific "state-of-the-art" meetings. The remarkable growth of the field and the rapid increase in diversity of advancements, including the application of novel molecular biological approaches and new imaging techniques, was certainly reflected at this meeting at Lake Tahoe. The progress made since the first meeting about 16 years ago is enormous. Investigators now focus on receptors, transduction mechanisms, and specific cytokines that may regulate both the degradation and the repair/regeneration of the tissue. Still, significant research will be necessary in defining and clarifying questions about the basic etiopathology of OA. Clinical and epidemiological investigators will have to continue to interact with the basic scientists to identify the pathogenic characteristics of OA, in order to develop and assess new therapeutic interventions.

Special recognition is due to Dr. Hari Reddi, who was the local organizer of the meeting, and together with his administrative and well experienced staff (most of all Ms. Lana Rich), were responsible for inducing and maintaining an interaction between the scientists. Their time, energy and effort is highly appreciated. Special thanks go to Dr. Vincent Hascall, who was "volunteered" by his "Co-Scoriers" to be the "mastermind" of the scientific program of this meeting at Granlibakken/Lake Tahoe. His commitment and pursuit guaranteed its success. In this task he was invaluably supported by his secretary Ms. Kathy Vukovich. The meeting was dedicated, by Vince, to the friendship among scientists.

Spring 2002   Klaus E. Kuettner

The life of Klaus E. Kuettner

Introduction

Klaus E. Kuettner, Chairman of the Department of Biochemistry, Rush Medical School in Chicago, is one of a handful of especially gifted individuals in osteoarthritis research, who through the last two decades has raised the perception of this field in the eyes of important power structures, such as NIH, the Arthritis Foundation, multiple research journals, Gordon Conferences, ORS forums, Food and Drug Administration Agency, and European structures, including the World Health Organization. The latter has recognized these efforts by designating him to be their distinguished research center leader in osteoarthritis. On his 68th birthday, Klaus was honored at this symposium, and pending his resignation from his chairmanship, a glimpse of his life should interest the readership, including the many scientists whose careers he has influenced.

Childhood

Klaus was born in 1933 in Bunzlau, Silesia, an eastern province in Prussia, Germany. At the time, Bunzlau was a town of about 25,000 in which his family had lived for several generations. His great-grandfather was a potter; his grandfather started a company to use such building materials for making large commercial pipes; and his father expanded the company. Indeed, his father's factories manufactured stoneware fixtures for most of the German railways in the 1920–30s. From the earliest age on, Klaus felt an especially close relationship to his father, who encouraged him in scholastic endeavors and outdoor activities, thereby, inspiring him to pursue lofty goals. During World War II, the family, deeply opposed to the Third Reich, was able to survive. Toward the end of the war, the family escaped to Saxonia, central Germany and lived under Russian occupancy until 1949.

Training for science

After Klaus attended high school in Leipzig, the entire family escaped again to Minden, West Germany. There Klaus trained at The Pharmacy School and received his B.S.-equivalent in pharmacy in 1955. His interests widened from the initial goal of a career in the business of running a pharmacy, leading to his subsequent studies first at the University of Freiburg, where he gained an MS in pharmaceutical science in 1958. He then became a Ciba Fellow (Basel, Switzerland) and studied at the University of Berne (Switzerland), where he was awarded a Ph.D. degree in 1961. This predoctoral period set the stage for his future interests.

His Ph.D. thesis studied the release of histamine and other biological amines from mast cell granules, and their interactions with macroanions. Studies with ion-exchangers bearing sulfate and carboxylic groups made a good foundation for later studies of proteoglycans. His closeness to the Ciba group gave him valuable friendships with scientists and administrators of the Ciba company. It also gave him a glimpse of the pharmaceutical slant toward basic research, which helped later to initiate research and academic training opportunities with industrial partners.

In 1962, Klaus received a post-doctoral fellowship at Argonne National Laboratory in Chicago to study plutonium binding in cartilage. There, Dr. Arthur Lindenbaum was his assigned mentor. They were fascinated with the subject of metals binding to macroanions, such as glycosaminoglycans (GAGs), and at Argonne, they showed that plutonium bound selectively at the calcifying front in growth plate cartilage.

Starting on the academic ladder

Although Klaus had intended to return after 2 years to Ciba at Basel for further industrial biochemistry training, a turning point in his life occurred. Prof. Richard Winzler, Ph.D., offered him an instructorship in the Department of Biochemistry at the University of Illinois in Chicago with a secondary affiliation in the Department of Orthopaedics at Rush Presbyterian St. Luke's Medical Center in 1964 (at the time of the portrait in Fig. 1).

Lysozome and anti-invasion factor (AIF) research

Klaus' original research investigated the biochemical alterations in the matrix of the epiphyseal growth plate that occur during the differentiation, hypertrophy and calcification of cartilage, and its ultimate replacement by bone. He focused mainly on a study of the molecular arrangement of proteoglycans and lysozyme, and their role in regulating calcification.

Figure 1
Intense work ethic may be reflected in the pallid features and worried expression consistent with this stage of his career.

Twenty-one papers evolved, with many well-known scientists in the field collaborating in this work, notably Reuben Eisenstein and Nino Sorgente, and other long-term friends. Notably, collaboration with Julio C. Pita showed that mammalian lysozyme, but not egg white lysozyme, could reduce the size of proteoglycan aggregates in dilute solution by altering its conformation and interacting with link protein. Although naturally occurring overexpression and underexpression of this protein have failed to indicate gross interference with mineralization mechanisms, it remains likely that lysozyme plays an adjunctive role to reduce aggregate size to improve conditions for mineralization.

In the meantime, starting about 1975 and extending over the next 10 years, Klaus and his research colleagues opened the field of regulation of capillary invasion of growth cartilage. His dramatic presentation of this phenomenon was shown when growth cartilage was grafted to chick allantoic membrane with an active capillary bed. Chick capillaries failed to invade various intact cartilages, but did invade cartilage that had been extracted with 1.0 M NaCl or guanidine HCl. This provided an explanation for why bone tumors during their invasion circumvented growth plate and epiphyseal cartilages. This phenomenon was expanded into many papers showing the presence of anti-invasion factor(s) (AIF) in NaCl extracts of bone, aorta, articular cartilage, urinary bladder, and enamel matrix.

Indeed, important scientists across the USA (including the host of the current symposium, Hari Reddi) became involved collaboratively in multiple studies of AIF. By 1980, it was clear that the AIF consisted minimally of an endothelial cell antiproliferative agent, an anti-elastase, and an anti-MMP factor, all of low molecular weight. F. Suzuki's chondromodulin and Moses Folkman's TIMP-like factor, both of which inhibit angiogenesis in growth plates, now help explain how avascularity of hyaline cartilage is mediated! Although his work flourished along with good NIH RO1 support, life is what happens to one while making other plans.

A new career

By 1980, when the portrait in Figure 2 was taken, Klaus had reached the next turning point. A dilemma arose about whether to continue as the leader of a small team of workers solely focused on individual research projects as described above, or to take the Dean's unexpected offer to become Chairman of the Department of Biochemistry! After the first wave of euphoria, Klaus confronted local realities. The Department had only an extramural research budget of $ 33,000, five elderly professors, and only one person able to teach biochemistry courses. The challenge was irresistible, and Klaus sank his teeth into the severe problems of departmental remodeling and building.

His past studies on polyanions impressed on him the importance of proteoglycans for normal function of articular cartilage and their likely role in the pathophysiology of osteoarthritis. This area, then, became the central focus for research in his enlarging department. Because of a wide acquaintanceship with academicians in this arena, over the next 20 years, Klaus was unbelievably successful in attracting diverse talent to the department while continuing his own research, as is apparent to the attendees at this conference.

To elaborate on this a little further, in the 1970s, Klaus collaborated with Vince Hascall, then at the University of Michigan, and later at the National Institute of Dental Research, and later with Helen Muir and Tim Hardingham, at the Kennedy Institute, and with Dick Heinegard at the University of Lund. This established a dynamic network of friends and collaborators with cartilage research interests. This network continued to expand in the 1980s as Klaus took proactive roles in promoting the Orthopaedic Research Society, and in organizing Gordon Conferences on bones and teeth, and later on proteoglycans. This network proved most advantageous in his efforts to recruit an excellent and productive faculty.

Faculty recruitment

In this recruitment, Klaus was careful not to duplicate subdisciplines, but to find outstanding independent investigators whose productivity and goals cover the most important bases. Eugene Thonar came first and was pivotal for proteoglycan structure and markers. Next was Margaret Aydelotte for new cell culture methods and cellular heterogeneity; followed by Linda Sandell, for molecular biology of chon-

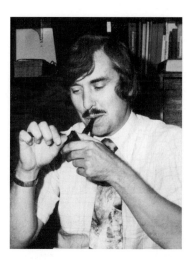

Figure 2
1980 – Now a debonair and confident academician.

drocytic matrix synthesis; Jim Kimura, for chondrocyte cell biology, Warren and Cheryl Knudson for hyaluronan cell biology; Tom Schmid for collagen chemistry; Gene Homandberg for fibronectin metabolism; Tibor Glant for autoimmune features of proteoglycans, Juergen Mollenhauer and H. Hauselmann in the 1990s for study of pericellular matrix metabolism in alginate beads; Koichi Masuda, who used these beads for tissue engineering of cartilage implants; Gabriella Cs-Szabo to study small proteoglycans, Susan Chubinskaya to explore the function of key anabolic mediators on articular cartilage, and Ada Cole to study profiles of MMPs and TIMP in knee and ankle cartilages. Jim Kimura and Linda Sandell very successfully made further career moves. Margaret Aydelotte retired in the late 1990s. A large number of scientists in other Departments (such as Richard Loeser, Joel Block, Carol Muehleman, and Jim H. Williams) were incorporated into the Department of Biochemistry programs to further their studies.

Klaus' philosophy has been to form a protective organization for this team of researchers, to offer a stable environment financially, and to make available educational and other types of opportunities.

Cartilage cells and matrix: the research program in osteoarthritis

Although the contributions from the team of independent researchers provide the bulk of the departmental output, Klaus retained certain areas of personal participation.

In the early 1980s, a novel method was developed in his laboratory for culturing adult articular chondrocytes for extended periods of time while maintaining phenotypic stability in agarose and later in alginate gels. These chondrocytes elaborated an extracellular matrix very similar or identical to that synthesized in vivo. The cells could be stimulated by modulators to degrade this new matrix, in a manner resembling the accelerated tissue breakdown characteristic of degenerative joint disorders, such as in osteoarthritis. This approach was named "chondrocytic chondrolysis." It was used to investigate the biochemical mechanisms of cartilage destruction and to test disease-modifying agents that may have therapeutic value for patients suffering from the arthritides. Two publications by M. Aydelotte and K. Kuettner (*Connec Tiss Res* 18: 205–222 and 223–234 (1988)), report biochemical and morphological studies defining differences in chondrocytes isolated primarily from the superficial layer and from the mid- to deep zones of bovine articular cartilage after the cells were isolated and resuspended in an agarose matrix. These studies, carried out with Barbara Schumacher, show the ability of mid- to deep zone chondrocytes to reassemble an extracellular matrix with retention of proteoglycans, whereas those from the superficial layer did not organize a similar matrix. They then exploited and refined an earlier observation by Bill Jourdian and his colleagues at the University of Michigan that chondrocytes can be maintained with phenotype stability in alginate cultures. The advantage of this matrix was that it could be dissolved in low salt, calcium-binding solvents without disrupting the interactions between the matrix macromolecules. This system was described in three papers (*Connect Tiss Res* 28: 143–159 (1992); *Matrix* 12: 116–129; *J Cell Science* 107: 17–27 (1994)).

Work in progress now utilizes the alginate system to define the metabolic differences between proteoglycans that are retained in close association with the cells (territorial) after dissolution of the alginate and proteoglycans, and those that are further removed (interterritorial). These studies have obvious consequences for expanded cell biological investigations, with special reference to characterizing separate cell populations.

A unique facility for access to human cartilage

Over 10 years ago, Klaus began collaboration with Steven Gitelis, M.D., Department of Orthopedic Surgery, who was then and currently remains the Medical Director of the Regional Organ Bank of Illinois. To the author's knowledge, this is the only such bank worldwide that can provide a quantity of weight-bearing articular cartilage of high quality for a thorough pursuit of biochemical properties so badly needed. To date, more than 2,000 human cartilages have been studied, and nearly every researcher in the Department now investigates human articular cartilages, an important attribute of Klaus' foresight.

Figure 3
2001 – Enjoying the sun and fun at Granlibakken.

NIH approval for the Rush OA research program

In the early 1980s, the Department of Biochemistry was awarded one of only three national grants in osteoarthritis to support a Specialized Center of Research (SCOR) from the National Institute of Arthritis, Musculoskeletal and Skin Diseases (NIAMS) of the National Institutes of Health (NIH). The coordinated effort of a SCOR grant addresses a specific health area and provides a multidisciplinary approach to the problem investigated. Originally, there were nine centers established: three in rheumatoid arthritis, three in osteoarthritis, and three in osteoporosis. The one at Rush entitled "Osteoarthritis: A Continuum (from Cartilage Metabolism to Early Detection and Treatment)" involved 15 investigators from five departments (Biochemistry, Orthopedic Surgery, Medicine [Rheumatology], Pathology, and Anatomy) with Klaus as program director (Fig. 3).

All SCOR grants in osteoarthritis were peer reviewed by a Special Study Section in 1992 and again in 1997. The renewal program at Rush received both times a favorable evaluation and was approved to continue its research objectives. Currently, this is the only ongoing SCOR for osteoarthritis.

The Research Educational Program

Klaus was convinced that to be completely successful, his department critically needed a Ph.D. doctoral training program. The timing (early 1980s) for this was unfortunate due to an overabundance of Ph.D. programs and training programs in the Chicago area at virtually every other medical school. Therefore, a novel and exciting program was necessary, and this was achieved through the focus on the single disease, osteoarthritis. The only other biochemistry program like it was that of Dennis Lowther at Monash University in Melbourne, Australia. Indeed, then, the Illinois Board of Higher Education approved the application for a Ph.D. degree teaching program at Rush Medical Center in 1984. This event led to an expansion of all phases of the Department's research and teaching programs.

In April 1991, Klaus' Department of Biochemistry was designated the first World Health Organization (WHO) Collaborating Center in the Field of Osteoarthritis (Rheumatology). One of the major "assignments" of this designation has been to train young, new, as well as experienced investigators in the area of cartilage/osteoarthritis research with the aim to transfer the knowledge to other laboratories and to the practice of medicine in the field of orthopaedic surgery/rheumatology. During the past 11 years, more than 60 young investigators (postdoctoral fellows) from around the globe have worked within the research laboratories of this department and collaborating departments. Over half of them held an M.D. (mostly in orthopedics) and after leaving, were able to incorporate their knowledge of basic biochemistry into the practice of academic-investigative medicine. At any one time, there are between five to ten postdoctoral fellows in the department.

At the time of this writing, there were 24 graduate students (Ph.D. candidates) in the department, and most of them chose research projects in cartilage biochemistry for their Ph.D. thesis work. In 1996, and again in 2001, the National Institute of Arthritis, Musculoskeletal and Skin Diseases (NIAMS) awarded the department a training grant entitled "Training in Cartilage Research," which supports the education of three graduate students and two postdoctoral fellows.

Industrial academic interactions

Klaus instituted a program for industry-academia interactions by providing a select group of graduate students to work on their thesis projects, with apprenticeships in large international pharmaceutical corporations. In this setting, the students learn modern research techniques and are educated in the application of basic knowledge to goal-directed research. This allows an intellectually productive interaction between academic teachers, investigators, and students with industrial research activities.

In 1987, Ciba-Geigy created an endowed chair for the Department of Biochemistry directed toward basic research in osteoarthritis and cartilage metabolism. Klaus was able to obtain two additional endowed chairs. All of these have the proviso that the income is applied to investigations and training in the field of Orthopaedic Sciences, especially osteoarthritis research.

Figure 4
The most important factor in the life of Klaus Kuettner has been the 11 years of his marriage to Erzsi, whom he met on a visit to the Center of Rheuma Pathology (WHO Center) in Mainz, Germany. She was the personal assistant to Professor Hans Georg Fassbender. The latter is a lifelong friend of Klaus, and a "father" of research in rheumatoid arthritis pathology.

More recently, Klaus's department entered into a unique relationship with the pharmaceutical company, Glaxo Wellcome of North Carolina. A contract was signed entitled "Osteoarthritic Changes in Human Cartilage: Research for the Development of New Medicines." The basic agreement focuses on methodology transfer combined with an exchange and specific interaction between research scientists at both locations, as well as to provide a program for academic research fellows to experience industrial research methods and, for some, to explore opportunities for careers in this arena. Significant funding and resultant career learning opportunities ensued. After the merger with Smith-Kline the program is continuing under new management.

At the conference, Klaus clearly enunciated his intent to continue to provide a leadership role in the SCOR project and to participate in the ebb and flow of the exciting research developments in the osteoarthritis and articular cartilage biology arena. Thus, we can look forward to new adventures and achievements in addition to those highlighted in the tables.

Thus, it is very fitting that these proceedings from a Conference honoring Klaus's contributions, celebrate not only his past achievements, but also anticipate those that remain to be accomplished. Some of these adventures and achievements are listed in Tables 1 and 2, and each listing could be greatly expanded upon. Leading the list in importance is his marriage to Erzsi (Fig. 4).

David Howell

Table 1 - Some of Klaus E. Kuettner's principal honors

1978	Chairman, 25th Gordon Research Conference on the Chemistry, Physiology and Structure of Bones and Teeth
1978	Kappa Delta Award for outstanding orthopedic research (AAOS and ORS)
1984	Chairman 1st Gordon Research Conference on Proteoglycans
1987	Recipient, International Carol Nachman Prize for advancement of clinical, therapeutic, and experimental research in rheumatology
1988	Recipient Pauwels Memorial Medal – German Society for Orthopedics and Traumatology for outstanding research in orthopedic sciences
2000	Co-recipient of Bristol-Myers Squibb/Zimmer Award for distinguished achievement in orthopedic research
1987–2002	Program director, NIH, Specialized Centers of Research (SCOR), "Osteoarthritis: A continuum (from Cartilage Metabolism to Early Detection and Treatment)". National Institute of Arthritis, Musculoskeletal, and Skin Diseases (funded for three consecutive 5-year periods).
1970	Personal NIH grant in Biochemistry of Bone and Connective Tissue (funded consecutively for 22 years)
2001	Honorary Degree in Medicine, University of Berne, Switzerland

Table 2 - Forums

1970s-90s	17 Midwest Connective Workshops host/co-chairman
1970s-90s	Participant 25 years Gordon Conference on Bones and Teeth, a session Chairman, selected each of 15 years
1989	Organizer/Chairman, International Conference on Articular Cartilage Biochemistry, Wiesbaden, Germany
1989	Co-Organizer – Bat Sheva Seminar-Methods Used in Research on Cartilaginous Tissues, Ginosar, Israel
1991	Organizer/Chairman, International Workshop on Articular Cartilage and Osteoarthritis, Wiesbaden, Germany
1994	Organizer/ Cochairman – NIH/American Academy of Orthopedic Surgeons/ National Arthritis Foundation Cosponsored Workshop "New Horizons in Osteoarthritis" and "WHO Satellite Meeting on Standardization of Methods for Assessment of Articular Cartilage Changes in Osteoarthritis of the Knee and Hip", Monterey, CA
1999	Organizing Committee International Symposium on Many Faces of Osteoarthritis, Granlibakken, Lake Tahoe

Acknowledgements

No conference such as "The Many Faces of Osteoarthritis" can be truly successful without the generous support of companies that have a genuine interest in the research area in question. We are most grateful to our corporate contributors listed below, both for their support and their participation. They have made it possible to invite outstanding investigators as speakers and to support talented younger investigators to participate. Special thanks go to Aventis Pharma Deutschland GMBH and Glaxo-Wellcome-SmithKline for additionally underwriting most of the expenses for publishing these Proceedings of the conference through the efforts of Ruth Raiss (Aventis Pharma Deutschland GMBH) and Steve Stimpson (Glaxo-Wellcome-SmithKline).

Amgen
Aventis Pharma Deutschland GmbH
Eli Lilly and Company
F. Hoffman-LaRoche Ltd.
Genetics Institute
Genzyme Corporation
Glaxo-Wellcome-SmithKline

Matrix Biology Institute
Pfizer, Inc.
Pharmacia Corporation
Procter and Gamble Pharmaceuticals
Roche Bioscience
Seikagaku Corporation
Stryker Biotech

When conferences run so smoothly as ours did, it is the result of the tireless efforts of the local organizers. In this case all the credit and many thanks go to Hari Reddi and his conference consultant, Lana Rich. Their work made the conference a lot of fun as can be seen in the photo montages of many of the participants.

With the exception of the session organized by Hari Reddi, each session of the conference was organized by a member of the organizing committee, the Scoriers. This team comprises the External Advisory Board for the SCOR program at Rush Medical College at Rush-Presbyterian-St. Luke's Medical Center in Chicago. We thank them for putting together a state-of-the-art program that features the very latest research efforts in the arthritic diseases.

Finally, it is no exaggeration to say that it would have been impossible for us to organize the conference program and to handle all the endless details, such as transcribing and editing the discussions at the conference and corresponding with the contributors and publisher after the conference, without the help of our talented assistants, Kathy Vukovich (for VH) and Verhonda Hearon Eggleston (for KK). We thank them wholeheartedly for their efforts on our behalf.

Vincent Hascall
Klaus Kuettner

The many phases of osteoarthritis

Introduction

Jody Buckwalter

Dept. of Orthopaedic Surgery, University of Iowa Hospital, 200 Hawkins Drive, Iowa City, IA 52242-1088, USA

Advances in laboratory and clinical research have made it clear that osteoarthritis is not the result of mechanical wear and tear of synovial joints over time. Instead, this disease is a complex biological and mechanical disorder that leads to joint degeneration that in turn causes the clinical syndrome of joint pain and loss of function we recognize as osteoarthritis. It is also increasingly clear that progress in the prevention and treatment of osteoarthritis will require in-depth study of the basic biology and metabolism of joint tissues, development and study of animal models of joint degeneration similar to the joint changes that cause osteoarthritis in humans, new approaches to investigation of the processes that lead to osteoarthritis in different human joints, critical examinations of the inheritance pattern of human osteosarthritis and studies of risk factors and prevention in humans. The papers in this section represent this important spectrum of investigation.

The work of McClinton, Fukui and Sandell on the interactions of Type 2A procollagen and H2 propeptide and bone morphogenetic proteins increases understanding of the role of the bone morphogenetic proteins in articular cartilage development, metabolism and maintenance. Further investigations of this type offer great promise in clarifying the basic biologic processes necessary for preventing and eventually developing new treatment approaches to osteoarthritis. Animal models of osteoarthritis are critical to the advancement of understanding of the pathogenesis of this disease and development of new treatments. The work of Wachsmith et al. on the dynamics of matrix loss in the spontaneous osteoarthritic mouse strain present important information concerning this form of spontaneous osteoarthritis and provide insight into the processes that lead to spontaneous joint degeneration. Warmath's paper presents an overview of approaches to gaining insights into the genetic basis of osteoarthritis and, in particular, the role of genes that increase the risk of osteoarthritis.

The report of Ingvarsson demonstrates the importance of a population wide study of the genetic component of osteoarthritis and provides an exciting basis for further investigation of the population of patients in Iceland. This population will give investigators the opportunity to clarify inheritance patterns and launch new investigations that will provide further insight into the genetic patterns of human osteoarthritis. It is clear that the development, pattern and progression of osteoarthritis varies considerably among joints within given individuals.

Cole et al. have contributed a number of important observations concerning the basic metabolic differences between knee and ankle articular cartilage that help explain the dramatic differences in the incidence and prevalence of knee and ankle osteoarthritis. Further pursuit of these ideas should help clarify the reasons for the differences in pattern, severity and progression of osteoarthritis among joints. Further development of basic laboratory investigations, animal models, studies of human inheritance patterns, differences among joints should further strengthen efforts to understand the mechanisms responsible for osteoarthritis.

The report of Hoakbart et al. takes a different important direction, it helps clarify current knowledge of risk factors for human osteoarthritis and possible preventive approaches, in particular, decreasing body mass for overweight adolescents and preventing limb injuries. Their work further suggests that individuals that have limb injuries at a young age should have appropriate treatment and education concerning future joint use. Further work of this kind may lead to strategies that focus on diet and nutritional supplementation. Taken together the studies in this section show the exciting potential progress of a multi-disciplinary wide spectrum of investigations in clarifying the pathogenesis of osteoarthritis and devising new and effective approaches to prevention and treatment.

Type IIA procollagen NH$_2$-propeptide functions as an antagonist of bone morphogenetic proteins

Audrey McAlinden, Naoshi Fukui and Linda J. Sandell

Department of Orthopedic Surgery, Washington University School of Medicine at Barnes-Jewish Hospital, St Louis, MO 63110, USA

Introduction

Knowledge of how growth factors are regulated in the extracellular matrix is crucial for understanding processes of tissue development and repair. A complex series of processes occurs in the extracellular matrix to control the activity and distribution of bone morphogenetic proteins (BMPs), which are important factors during cartilage and bone development. It is known that tissue regeneration is a recapitulation of processes in embryonic development and morphogenesis. Thus, in the case of cartilage repair during osteoarthritis, we can apply our knowledge of BMP function/regulation during embryogenesis to developing therapeutic strategies that will induce and even prolong natural tissue repair.

This chapter will introduce the reader to a novel concept for the role of a procollagen NH$_2$-propeptide in regulating BMPs in the extracellular matrix. Specifically, we will focus on the function of the developmentally-regulated NH$_2$-propeptide of type IIA procollagen and its ability to bind to and thereby regulate BMP concentrations and/or activity during skeletal development. We will also discuss ongoing research in our laboratory to investigate this functional role; the importance of type IIA procollagen during repair processes in osteoarthritis will also be addressed.

Function of bone morphogenetic proteins

BMPs are members of the transforming growth factor (TGF) superfamily and were originally identified as factors promoting the ectopic formation of cartilage and bone [1]. Major subdivisions within the superfamily include the TGF-βs, BMPs (excluding BMP-1), growth/differentiation factors (GDFs), inhibins, activins, Vg-related genes, nodal related genes, *Drososphila* genes (e.g., decaplentaplegic, dpp and *Drososphila* 60A) and glial-derived neurotropic factor [2]. BMPs contain a mature domain, which after cleavage, permits dimer formation *via* a cysteine-disulfide bridge. Protein assembly can produce homodimers, heterodimers and glycosylation variability which may influence the activity and effects of BMP.

BMPs in development

In addition to their role in promoting chondrogenesis and osteogenesis [2, 3], genetic evidence points to actions of BMPs in tissues other than cartilage and bone. Knockout of BMP-2 in mice displays malformation in development of the amnion, chorion and heart [4], and knockout of BMP-4 results in defects in mesoderm formation [5]. Mutations in BMP-5 in mice are responsible for short ear phenotype as well as aberrations in the ribs and vertebral processes [6] while BMP-7 null mice exhibit defects in eye, kidney and skeleton [7]. BMPs are necessary factors during embryogenesis of *Drosophila*, *Xenopus*, zebrafish and mammals [8], and they regulate cell-type specification during the organogenesis phase of embryogenesis [9]. These and many other studies have led to the understanding that BMPs are pleiotropic factors, and the precise concentration of active BMP is important for rendering a particular biological effect.

BMP signalling and regulation in the extracellular matrix

The Smad family of intracellular proteins plays a central role in the signal transduction of the TGF-β superfamily [10]. BMP-2 and -4, once bound to their cell receptor, result in phosphorylation of Smad 1. This phosphorylated protein will interact with Smad 4 during translocation to the nucleus where it activates target genes. A wide array of regulatory proteins exist in the extracellular matrix that can bind to the BMP molecule, thus preventing the growth factor from binding to its cell receptor. An additional level of complexity exists in the form of matrix proteinases which can disrupt the inactive BMP-binding protein complex and allow release of the BMP. The active BMP concentration in the extracellular matrix is controlled in part through the influence of BMP inhibitors. Several BMP binding proteins have been described including short gastrulation in *Drosophila* (sog; François et al., 1994 [11]) and its vertebrate homolog chordin [12], noggin [13, 14], follistatin [15], cerberus [16] and gremlin [17]. With the exception of follistatin, all of the binding proteins can inactivate BMP activity by preventing interaction with its cell receptor. Noggin and chordin have similar binding preferences and can antagonize BMP-2, -4 and –7, while follistatin strongly antagonizes activin activity.

BMP regulation by chordin and matrix metalloproteinases

There have been many reports describing the binding and regulation of BMPs by chordin. Studies of *Xenopus* development showed that over-expression of chordin *in vivo* dorsalized frog embryos by antagonizing the activity of BMPs [18]. Biochemical analyses by the same investigators showed that chordin interacts with BMP *via* the cysteine-rich, von Willebrand C (vWFC) type domains present in chordin (Fig. 1). Another level of complexity exists to release the BMP dimer from the binary complex and permit binding to its cell receptor. In *Xenopus*, over-expression of the astacin protease, Xolloid, resulted in ventralization of the embryos suggesting a BMP agonistic activity [19]. Xolloid cleaves chordin at two specific sites within the chordin molecule to release and hence activate the BMP dimer. Thus, not only does regulation of BMP activity depend on distribution of the various binding proteins present in the extracellular matrix, but also on the concentration and location of matrix metalloproteinases.

It is apparent that proteins containing these cysteine rich vWFC type domains play an important role in binding to and hence regulating other components of the matrix, whether during developmental or other processes. These domains are also found in other matrix proteins including thrombospondin-1, connective tissue growth factor and the amino propeptide of the fibrillar procollagens (types I, II, III and V; Fig. 1). Type II collagen is the main structural component of the extracellular matrix of articular cartilage and the following section will discuss the role of the developmentally-regulated type IIA procollagen molecule as a regulator of growth factor activity.

Type IIA procollagen as a BMP antagonist

In addition to the BMP binding proteins discussed previously, there is increasing evidence to suggest that the NH_2-propeptide of type IIA procollagen functions as a regulator of growth factor activity. This section will list conclusive findings from our laboratory and from other investigators, supporting our hypothesis that the type IIA fibrillar procollagen molecule has a unique functional role in the extracellular matrix during development of tissues including cartilage.

Type IIA procollagen synthesis is developmentally-regulated

The NH_2-propeptide of type II procollagen is subject to pre-mRNA alternative splicing resulting in two different molecules: type IIA procollagen containing exon 2, which encodes a cysteine-rich domain, and type IIB procollagen, devoid of exon 2 [20, 21]. Synthesis of these collagens is developmentally regulated in chondrogenesis: type IIA procollagen mRNA is synthesized by chondroprogenitor cells while type IIB procollagen mRNA is synthesized by mature chondrocytes (Fig. 2). In addition, type IIA procollagen mRNA has also been localized to non-cartilaginous epithelial and mesenchymal cells during development [21–23].

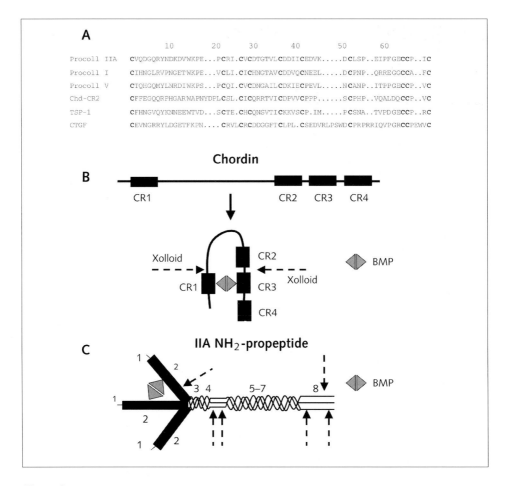

Figure 1
A: Sequence homology between the cysteine-rich (CR) domains in the amino propeptide of type IIA, I and V procollagens, chordin (Chd, CR2 domain), thrombospondin-1 (TSP-1) and connective tissue growth factor (CTGF). The aligned cysteine residues are in boldface.
B: Domain structure of chordin showing four CR domains. A conformational change in chordin has been hypothesized after binding to BMP such that interaction occurs via the CR1 and CR3 domains. This binary complex containing inactive BMP is susceptible to cleavage by the astacin protease, Xolloid (shown by dotted arrows) resulting in release of BMP for subsequent interaction with its cell receptor.
C: Structure of trimeric type IIA NH2-propeptide with the exon 2-encoded CR domains in black. Predicted interaction of the CR domains with a BMP molecule is shown. Dotted arrows represent potential matrix metalloproteinase (MMP) cleavage sites which may lead to release of the active BMP molecule. The MMPs found to cleave at these areas are listed in Figure 3D. Numbers represent the location of the exons encoding the IIA NH2-propeptide.

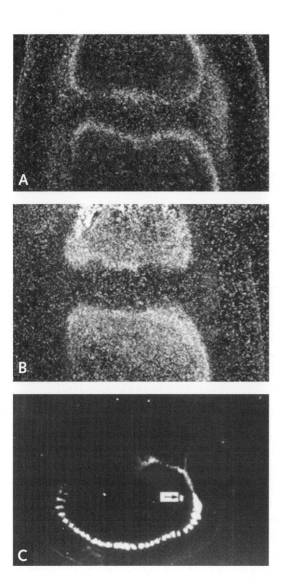

Figure 2
Localization of type IIA and type IIB procollagen mRNA during development.
A: In situ hybridization of type IIA procollagen mRNA in the proximal interphalangeal joint of 13 day (stage 39) chick embryo. Expression can be seen in the cells of the articular surface, perichondrium and capsule.
B: Type IIB procollagen mRNA expression in the same chick joint as shown in A. Note that type IIB procollagen mRNA expression is only found in the chondrocytes of the anlagen.
C: Expression of type IIA procollagen mRNA in an E11 mouse. Note expression in the somites and branchial arches (arrow). No type IIB expression is found at this stage of development.

The NH$_2$-propeptide of type IIA procollagen is present in the extracellular matrix during a specific developmental period

Antisera raised against the exon 2 encoded cysteine-rich domain of type IIA procollagen showed that type IIA procollagen protein is present in the extracellular matrix of pre-chondrogenic cartilage and epithelial tissues [24]. Type IIA procollagen was not detected in normal, mature cartilage tissue sections using the same antibody. Interestingly, Zhu et al. [25] identified free IIA NH$_2$-propeptide and type IIA pN-procollagen (the collagen triple helical domain attached to the IIA NH$_2$-propeptide, but lacking the COOH domain) in the extracellular matrix of developing fetal cartilage. In contrast, the COOH-propeptide was localized only inside of the cells. Recently, it has been shown in our laboratory that pre-chondrocytic cells in the inner annulus of the developing human intervertebral discs also secrete IIA NH$_2$-propeptide into the extracelluar matrix [26]. However, after a specific stage in development, cells of the inner annulus displayed altered collagen processing such that the IIA NH$_2$-propeptide was only localized inside of the cells and not in the matrix. Together, these immunolocalization studies suggest that the IIA NH$_2$-propeptide is present in the matrix during specific developmental stages and that it may play an important role in cartilage development during this time.

The cysteine-rich domain of type IIA procollagen is homologous to the BMP binding domains of chordin

The type IIA NH$_2$-propeptide is encoded by eight exons. The translated protein consists of a short globular domain followed by the cysteine-rich, exon 2 encoded domain, a Gly-X-Y triple-helical domain encoded by exons 3-7 and a short telopeptide domain encoded by exon 8. The spatial organization of the 10 cysteines within the exon 2 encoded domain is representative of a von Willebrand factor type C sequence and is homologous to the BMP binding domains of chordin, in particular the CR2 domain (Fig. 1).

Monomeric recombinant proteins containing the exon 2-encoded cysteine rich domain bind to BMP-2 and TGF-β *in vitro*

Immunoprecipitations and solid-phase binding assays carried out by Zhu et al. [25] showed that the type IIA procollagen cysteine-rich domain fused to GST bound to BMP-2 and TGF-β with a kD similar to that for an individual chordin CR domain binding to BMP [18]. From these results, we hypothesized that a correctly-folded, trimeric IIA NH$_2$-propeptide containing three cysteine-rich domains would be more efficient in binding BMP than the monomer. This led us to produce a recombinant, trimeric IIA NH$_2$-propeptide to study its growth factor binding potential compared to the monomers (see below and Fig. 3).

Over-expression of full-length *Xenopus* procollagen dorsalized frog embryos

Recent evidence suggests that full-length, trimeric *Xenopus* IIA procollagen has a similar role to that of chordin during embryonic development [18]. Interestingly, over-expression of *Xenopus* type IIB procollagen or a monomer fragment of type IIA procollagen in the same system had no dorsalizing effects. This shows that the dorsalizing (i.e., BMP binding) function resides in the exon 2-encoded domain and that function also depends on structural organization of the trimeric procollagen NH$_2$-propeptide containing three cysteine-rich domains. *Xenopus* procollagen IIA, like chordin, was localized in the dorsal mesoderm (notochord and somites) of frog embryos. The spatial distribution and timing of IIA procollagen expression is consistent with a role in dorsal development of the embryos, perhaps replacing the activity of chordin at later stages of *Xenopus* embryogenesis.

The NH$_2$-propeptide of type IIA procollagen is susceptible to cleavage by matrix metalloproteinases

Recent work from our laboratory [27] showed that human, trimeric type IIA NH$_2$-propeptide is susceptible to cleavage by various MMPs (Fig. 3). We synthesized a correctly-folded trimeric type IIA NH$_2$-propeptide [28] by fusing it to the neck domain and carbohydrate recognition domain of lung surfactant protein D [29, 30] and Fig. 3). The neck domain acts as a heterologous trimerization unit [31, 32], which is essential due to the absence of the procollagen COOH domain, the site of chain registration of the procollagen molecule. The resulting fusion protein, named IIA-SPD, was purified from conditioned media of stably-transfected Chinese hamster ovary cells. We have shown that this trimeric IIA NH$_2$-propeptide can bind to members of the TGF-β superfamily *in vitro* and that it is susceptible to cleavage by various MMPs (Fig. 3). These results suggest a similarity to the chordin/Xolloid scenario where the binding protein is cleaved by a protease, resulting in release of the growth factor. However, the site of IIA NH$_2$-propeptide cleavage is also an important issue in determining whether the growth factor will be released from the inactive complex or not. Figure 3 shows that MMP-7 (matrilysin) cleaves the IIA NH$_2$-propeptide within the cysteine-rich domain, which would result in production of monomeric procollagen α chains. However, MMP-9 (gelatinase B) can cleave within the telopeptide domain of the IIA NH$_2$-propeptide. In this case, the propeptide will remain in a trimeric configuration [28] and may possibly retain its BMP binding activity. We intend to explore these exciting findings further.

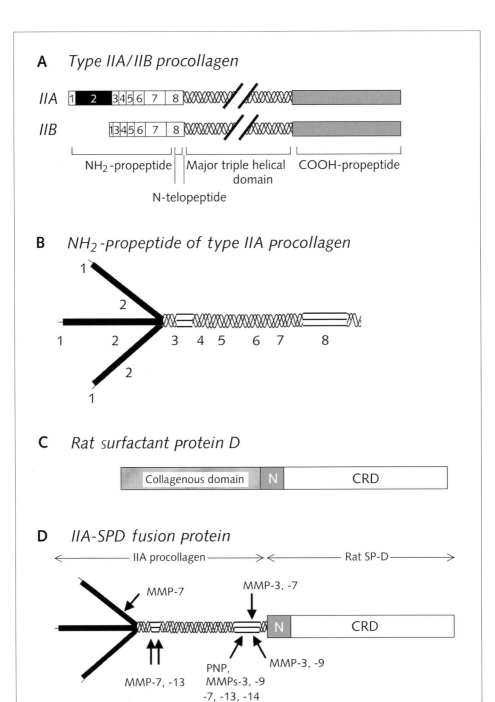

Studies of type IIA procollagen function: present and future

Structure/function studies

At present, research in our laboratory is specifically directed towards detailed analysis of growth factor interactions with the cysteine-rich domain of type IIA procollagen and determining its function in a biological system. For these studies, our IIA-SPD fusion protein containing the human, trimeric IIA NH_2-propeptide (Fig. 3) will be an invaluable tool. We are in the process of purifying the IIA NH_2-propeptide from the surfactant protein D domains to avoid the possibility of unwanted lectin activity in our binding and functional assays.

We intend to compare binding efficiencies of monomer *versus* trimer IIA NH_2-propeptide as well as the effect of various MMP cleavages on the growth factor binding potential. Mutational analysis will also be applied to examine which amino acids within the IIA procollagen cysteine-rich domain are important for growth factor binding. As for function, we are in the process of utilizing an *in vitro* biological assay that monitors the BMP-dependent elongation of embryonic lens epithelial cells [33]. Lens epithelial cell explants from embryonic day 6 chickens elongate in the presence of vitreous humor after just 5 h in culture. However, addition of the BMP binding protein, noggin, to vitreous humor resulted in inhibition of cell elongation. Preliminary experiments adding our recombinant IIA NH_2-propeptide to vitreous

Figure 3
A: Alternatively spliced forms of type II procollagen. Type IIA procollagen contains the alternatively-spliced exon 2 which encodes a 69 amino acid cysteine-rich (CR) domain within the NH2-propeptide. Type IIB procollagen lacks this CR domain.
B: Predicted protein structure of the IIA NH2-propeptide showing three CR domains, a minor collagen triple helix (encoded by exons 3–7) containing a short interruption, and the telopeptide domain (encoded by exon 8), which connects the propeptide to the major collagen triple helical domain.
C: Domain structure of rat lung surfactant protein D (SP-D) showing the collagenous domain, the trimerizing neck (N) domain and the carbohydrate recognition domain (CRD). cDNA encoding SP-D was used to synthesize a chimeric construct where the collagenous domain of SP-D was replaced by that encoding full length (exons 1–8) IIA NH2-propeptide.
D: IIA-SPD fusion protein containing a correctly-folded, trimeric IIA NH2-propeptide linked to the neck and CRD of SP-D. Chinese Hamster Ovary cells were stably-transfected with the chimeric construct to produce IIA-SPD, and the recombinant protein was efficiently purified by maltosyl-agarose chromatography due to the presence of the CRD. The NH2-propeptide of IIA-SPD was found to be susceptible to cleavage by various matrix metalloproteinases (MMPs) as shown by the arrows.

humor also showed inhibition in lens cell elongation, similar to noggin. Future studies will lead us to investigate the effect of trimeric IIA NH_2-propeptide *in vivo*. We intend to over-express the propeptide in mice, specifically in areas where we know type IIA procollagen is present during development such as cartilage and the eye.

IIA procollagen in disease and repair

We are also interested in the role of IIA procollagen during tissue repair. It is known that IIA procollagen is re-expressed in osteoarthritis [34] and thus may prove to be a useful candidate as a marker of degenerative disease. To date, little is known about the distribution and function of BMPs and BMP binding proteins in osteoarthritic tissue. If indeed there is a recapitulation of developmental processes to induce cartilage repair, then it is likely that IIA procollagen also plays a significant role. In terms of therapy, researchers are exploring ways to deliver BMPs to the repair site with the hope that the growth factor will remain at that site for long periods to induce an effect. It is possible that delivery of such BMPs in conjunction with a binding protein, such as IIA procollagen, will retain the growth factor in the matrix for a longer time. In addition, the presence of specific MMPs, which are known to be elevated in diseased cartilage, may function to cleave the IIA binding protein resulting in release of the active BMP. Thus, from development to disease, it is becoming apparent that type IIA procollagen is more to cartilage than just a pretty fiber!

References

1 Wang Z-Q, Grigoriadis AE, Wagner EF (1993) Stable murine chondrogenic cell lines derived from c-fos-induced cartilage tumors. *J Bone Min Res* 8: 839–847
2 Schmitt JM, Hwang K, Winn SR, Hollinger JO (1999) Bone morphogenetic proteins: an update on basic biology and clinical relevance. *J Orthopaed Res* 17 (2): 269–278
3 Reddi AH (1994) Cartilage morphogenesis: Role of bone and cartilage morphogenetic proteins, homeobox genes and extracellular matrix. *Matrix Biol* 14: 599–606
4 Zhang H, Bradley A (1994) Mice deficient for BMP2 are nonviable and have defects in amnion/chorion and cardiac development. *Development* 122: 2977–2986
5 Winnier G, Blessing M, Labosky PA, Hogan BL (1995) Bone morphogenetic protein-4 is required for mesoderm formation and patterning in the mouse. *Genes Dev* 9 (17): 2105–2116
6 Kingsley DM, Bland AE, Grubber JM, Marker PC, Russell LB, Copeland NG, Jenkins NA (1992) The mouse short ear skeletal morphogenesis locus is associated with defects in a bone morphogenetic member of the TGF beta superfamily. *Cell* 71 (3): 399–410
7 Dudley AT, Lyons KM, Robertson EJ (1995) A requirement for bone morphogenetic

protein-7 during development of the mammalian kidney and eye. *Genes Dev* 9 (22): 2795–2807

8   Harland R, Gerhart J (1997) Formation and function of Spemann's organizer. *Ann Rev Cell Dev Biol* 13: 611–667

9   Hogan BLM (1996) Bone morphogenic proteins: multifunctional regulators of vertebrate development. *Genes Dev* 10: 1580–1594

10 Heldin CH, Miyazono K, ten Dijke P (1997) TGF-beta signalling from cell membrane to nucleus through SMAD proteins. *Nature* 390 (6659): 465–471

11 Francois V, Solloway M, O'Neill JW, Emery J, Bier E (1994) Dorsal-ventral patterning of the *Drosophila* embryo depends on a putative negative growth factor encoded by the short gastrulation gene. *Genes Dev* 8: 2602–2616

12 Sasai Y, Lu B, Steinbeisser H, Geissert D, Gont LK, De Robertis EM (1994) *Xenopus* chordin: a novel dorsalizing factor activated by organizer-specific homeobox genes. *Cell* 79: 779–790

13 Smith WC, Harland RM (1992) Expression cloning of noggin, a new dorsalizing factor localized to the Spemann organizer in *Xenopus* embryos. *Cell* 70 (5): 829–840

14 Zimmerman LB, De Jesus-Escobar JM, Harland RM (1996) The Spemann organizer signal noggin binds and inactivates bone morphogenetic protein 4. *Cell* 86 (4): 599–606

15 Iemura S, Yamamoto TS, Takagi C, Uchiyama H, Natsume T, Shimasaki S, Sugino H, Ueno N (1998) Direct binding of follistatin to a complex of bone-morphogenetic protein and its receptor inhibits ventral and epidermal cell fates in early *Xenopus* embryo. *Proc Natl Acad Sci USA* 95 (16): 9337–9342

16 Bouwmeester T, Kim S, Sasai Y, Lu B, De Robertis EM (1996) Cerberus is a head-inducing secreted factor expressed in the anterior endoderm of Spemann's organizer. *Nature* 382 (6592): 595–601

17 Hsu DR, Economides AN, Wang X, Eimon PM, Harland RM (1998) The *Xenopus* dorsalizing factor Gremlin identifies a novel family of secreted proteins that antagonize BMP activities. *Mol Cell* 1(5): 673–683

18 Larrain J, Bachiller D, Lu B, Agius E, Piccolo S, de Robertis EM (2000) BMP-binding modules in chordin: a model for signalling regulation in the extracellular space. *Development* 127: 821–830

19 Piccolo S, Agius E, Lu B, Goodman S, Dale L, De Robertis EM (1997) Cleavage of Chordin by Xolloid metalloprotease suggests a role for proteolytic processing in the regulation of Spemann organizer activity. *Cell* 91 (3): 407–416

20 Ryan MC, Sandell LJ (1990) Differential expression of a cysteine-rich domain in the amino-terminal propeptide of type II (cartilage) procollagen by alternative splicing of mRNA. *J Biol Chem* 265 (18): 10334–10339

21 Sandell LJ, Nalin AM, Reife RA (1994) Alternative splice form of type II procollagen mRNA (IIA) is predominant in skeletal precursors and non-cartilaginous tissues during early mouse development. *Dev Dyn* 199 (2): 129–140

22 Sandell LJ, Morris N, Robbins JR, Goldring MR (1991) Alternatively spliced type II

procollagen mRNAs define distinct populations of cells during vertebral development: differential expression of the amino-propeptide. *J Cell Biol* 114: 1307–1319

23 Ng LJ, Tam PP, Cheah KS (1993) Preferential expression of alternatively spliced mRNAs encoding type II procollagen with a cysteine-rich amino-propeptide in differentiating cartilage and nonchondrogenic tissues during early mouse development. *Dev Biol (Orlando)* 159 (2): 403–417

24 Oganesian A, Zhu Y, Sandell LJ (1997) Type IIA procollagen amino-propeptide is localized in human embryonic tissues. *J Histo Cytochem* 45 (11): 1469–1480

25 Zhu Y, Oganesian A, Keene DR, Sandell LJ (1999) Type IIA procollagen containing the cysteine-rich amino propeptide is deposited in the extracellular matrix of prechondrogenic tissue and binds to TGF-beta1 and BMP-2. *J Cell Biol* 144 (5): 1069–1080

26 Zhu Y, McAlinden A, Sandell L (2001) Type IIA procollagen in development of the human intervertebral disc: regulated expression of the NH_2-propeptide by enzymic processing reveals a unique developmental pathway. *Dev Dyn* 220: 350–362

27 Fukui N, McAlinden A, Zhu Y, Crouch E, Broekelmann TJ, Mecham RP, Sandell L (2001) Processing of Type II procollagen amino propeptide by matrix metalloproteinases. *J Biol Chem* 277: 2193–2201

28 McAlinden A, Zhang P, Bann JG, Crouch E, Sandell L. Production of a functional amino propeptide of type IIA procollagen using a heterologous trimerization domain. *J Biol Chem*; in revision

29 Crouch E, Chang D, Rust K, Persson A, Heuser J (1994) Recombinant pulmonary surfactant protein D. Post-translational modification and molecular assembly. *J Biol Chem* 269 (22): 15808–15813

30 Crouch EC (1998) Structure, biologic properties, and expression of surfactant protein D (SP-D). *Biochim Biophys Acta* 1408 (2–3): 278–289

31 Hoppe HJ, Barlow PN, Reid KB (1994) A parallel three stranded alpha-helical bundle at the nucleation site of collagen triple-helix formation. *FEBS Letts* 344 (2–3): 191–195

32 Kishore U, Wang JY, Hoppe HJ, Reid KB (1996) The alpha-helical neck region of human lung surfactant protein D is essential for the binding of the carbohydrate recognition domains to lipopolysaccharides and phospholipids. *Biochem J* 318 (Pt 2): 505–511

33 Beebe DC, Feagans DE (1981) A tissue culture system for studying lens cell differentiation. *Vision Res* 21 (1): 113–118

34 Aigner T, Zhu Y, Chansky HH, Matsen FA, 3rd, Maloney WJ, Sandell LJ (1999) Reexpression of type IIA procollagen by adult articular chondrocytes in osteoarthritic cartilage. *Arthritis Rheum* 42 (7): 1443–50

Genetics and osteoarthritis

Matthew L. Warman

Departments of Genetics and Pediatrics and Center for Human Genetics, Case Western Reserve University and University Hospitals of Cleveland, Cleveland, OH, USA

Introduction

Most of the human genome has been sequenced [1, 2] and within the next few years it is anticipated that all human genes will be become known. This chapter discusses how this knowledge of the human genome might be used to identify genes that predispose to or affect the progression of osteoarthritis.

Genetic dissection of Mendelian diseases having joint failure as a component feature

Over 150 heritable disorders of connective tissue have been described [3], many of which have precocious failure of articular cartilage as a component feature. These diseases follow Mendel's laws of segregation: that is, dominant or recessive, sex-linked or autosomal. They tend to be highly penetrant, meaning that an individual with a disease-causing genotype invariably manifests the disease phenotype. Advances fueled by the Human Genome Project [4] in disease mapping and gene discovery have facilitated the study of these rare Mendelian diseases and led to the identification of novel genes that are essential to skeletal growth and homeostasis.

There are several advantages to studying diseases that are inherited in simple Mendelian patterns. One advantage is that the diseases result from genetic alterations that have large effects, reflecting the importance of the gene's protein product or the severity of a particular mutation on the protein's function. Another advantage is that a heritable genetic disease *a priori* results from a genetic alteration. Therefore, by beginning with a heritable phenotype it is certain that a "disease-causing" mutation exists and can be found. The study of genetic diseases affecting the skeleton is not restricted to humans, but can be applied to any organism in which inheritance of disease can be followed. Exciting advances in skeletal biology have come from genetic studies involving organisms as diverse as mice and zebrafish [5, 6]. However, *Homo sapiens* is likely to remain the best organism in which to understand human osteoarthritis. Human joint structure and stresses differ significantly from those of other species, and the onset of human osteoarthritis begins decades beyond the normal lifespans of most model organisms. The study of human heritable skeletal disease is also facilitated by the fact that humans seek medical attention. This contrasts to studies involving model organisms in which investigators must perform exhaustive screens to identify those few animals with heritable skeletal phenotypes. A final advantage of studying humans derives from the large size of the human population. With over 7 billion humans on Earth it is likely that similar phenotypes will result from independently arising mutations in different populations, such that a series of disease-causing mutations within a gene can be identified. Allelic series of mutations permit detailed genotype-phenotype correlation, which can yield insights about function that may not be gained from the study of a single mutation.

Genetic dissection of osteoarthritis, a complex genetic disease

It has been hypothesized that genes that cause Mendelian genetic skeletal diseases would also cause common skeletal diseases, such as osteoarthritis (Fig. 1). Rare diseases result from mutations having severe effects on gene function, while common diseases would be predicted to result from mutations having mild effects. Mutations in the gene encoding type II collagen (*COL2A1*) have been identified in a number of heritable skeletal diseases such as Stickler syndrome, spondyloepiphyseal dysplasia congenita, and precocious osteoarthritis with mild SED [7]. Studies have attempted to associate the *COL2A1* locus with common forms of osteoarthritis [8–10]. While the results from these studies are intriguing, they have not yet been independently confirmed. Consequently, it remains unclear whether any common form of OA will be attributed to mild variations in Mendelian genetic skeletal disease-causing genes.

Most osteoarthritis does not follow patterns of inheritance that would be attributable to single genes. Rather, common forms of osteoarthritis appear to result from interactions between multiple genes and between genes and environment (Fig. 2). Despite the absence of clear patterns of inheritance for common osteoarthritis, we infer that there is a genetic contribution because relatives of individuals with osteoarthritis are more likely to be similarly affected than age and gender-matched individuals from the general population [11, 12]. Since close relatives share genes in common, it is assumed that the clustering of osteoarthritis within families is due to the sharing of osteoarthritis predisposing genes. However, families share more than their genes in common. They can also share vocations, avocations, and other environmental factors. Consequently, clustering of osteoarthritis within families can only suggest, but not prove, a genetic contribution to osteoarthritis, even after controlling for known environmental factors.

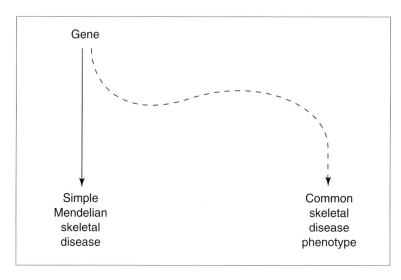

Figure 1
A direct connection (solid arrow) exists between gene mutations and simple Mendelian genetic skeletal diseases. These rare diseases follow simple modes of inheritance and are often highly penetrant, such that individuals with disease-causing genotypes invariably have disease phenotypes. Although it was hypothesized that other types of variation within Mendelian disease-causing genes would also cause common skeletal diseases (dashed arrow), at present there is little evidence supporting this hypothesis.

The failure to convincingly link known Mendelian disease-causing genes to common forms of osteoarthritis is disappointing; however, it is not surprising given the complexity of the osteoarthritis phenotype. Simple Mendelian diseases have clear phenotypes that permit them to be reliably identified and categorized. Complex diseases have phenotypes that are not easily recognized or categorized. Symptoms of osteoarthritis often differ significantly between individuals who have identical radiographic signs of osteoarthritis; conversely, radiographic signs of osteoarthritis can differ significantly between individuals who have identical symptoms. Equally perplexing is the spectrum of osteoarthritis that can exist within a family. Individuals with unilateral hip osteoarthritis, unilateral hip and knee osteoarthritis, and bilateral knee osteoarthritis can all be found within single families. Our current inability to reliably recognize osteoarthritis and to appropriately classify osteoarthritis phenotypes into correct pathogenic categories reduces our ability to find osteoarthritis-predisposing genes.

Current limitations in recognizing and categorizing osteoarthritis do not preclude finding disease-predisposing or disease-modifying genes; however, the number of individuals required for such genetic analyses could be orders of magnitude larg-

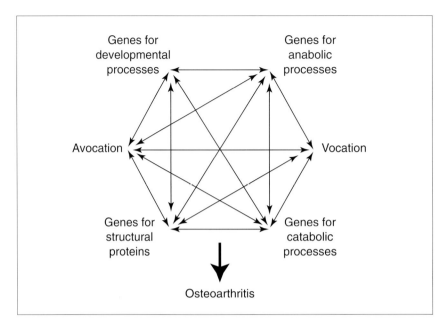

Figure 2
Common diseases often result from complex interactions between genes and environment. Genes, such as those encoding structural matrix proteins or proteins that participate in developmental, anabolic, or catabolic pathways, often interact when contributing to osteoarthritis onset and progression. Environmental factors can also modify the effects of genetic factors with respect to onset and progression of articular cartilage failure.

er than the number of individuals required to find genes which cause simple Mendelian genetic diseases. For example, the gene responsible for causing progressive pseudorheumatoid dysplasia was mapped by studying a single family [13]. In contrast, no definite osteoarthritis-predisposing loci could be detected after studying 500 affected sibling pairs who had undergone total joint replacements for osteoarthritis [14, 15]. This does not exclude the existence of osteoarthritis susceptibility genes, rather it indicates that a cohort size of 500 affected sibling pairs is probably too small to reliably detect these genetic risk factors [16]. Technologic advances during the past decade facilitated the identification of genes that cause simple Mendelian diseases and the upcoming decade promises advances that will facilitate the identification of genes that cause complex traits. Since technology will not limit our ability to find osteoarthritis genes, it is important that we not be limited by the clinical complexity of the disease. The ability to ascertain, recognize, categorize, and sample cohorts of patients with common forms of osteoarthritis will be essential to the osteoarthritis gene discovery process.

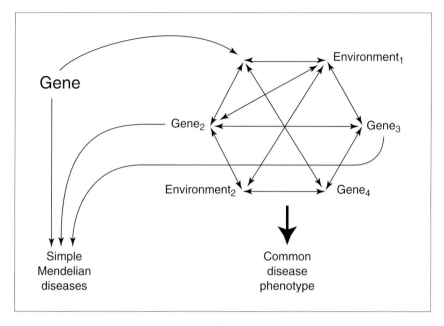

Figure 3
The study of Mendelian genetic disease can lead to the identification of novel genes and, perhaps more importantly, biologic pathways that are essential to human health. Other participants in these pathways, both genetic and environmental, may be better targets for therapeutic intervention. Genetic variation within other members of these pathways may be responsible for other simple Mendelian diseases in addition to being independent risk factors for common diseases.

Using genetics to access biologic pathways of joint homeostasis

Despite evidence for a genetic component to osteoarthritis, the extent to which normal genetic variation within the population contributes to osteoarthritis risk is unknown. Since age, gender, and environmental factors are strong contributors to osteoarthritis onset and progression [17], genetic risk factors may be minor and suboptimal targets for therapeutic intervention. Therefore, finding osteoarthritis-predisposing genes is only a stepping stone toward gaining a better understanding of the biologic pathways that participate in joint health.

Mendelian genetic diseases provide entry points with which to delineate biologic pathways involved in normal joint homeostasis. As knowledge regarding these pathways increases and other pathway participants are identified, new targets for osteoarthritis therapeutic intervention will emerge (Fig. 3). One means of identifying new pathway participants utilizes other developments fueled by the Human

Genome Project, including the elucidation of the human transcriptome and tools with which to analyze it. The transcriptome is the entirety of mRNA transcripts that are encoded and transcribed by cells [18, 19]. Silicon chip and glass slide microarray technologies now permit the simultaneous evaluation of thousands of mRNA species in a single experiment [20]. Such technology can be used to explore changes in gene expression between normal and diseased tissue, and between wild-type and mutant cells. The challenge will be to distinguish changes in gene expression which directly relate to the biologic pathway of interest from changes in gene expression which reflect "innocent bystanders" caught in biology gone awry. Nevertheless, the ability to assess changes in expression of genes that comprise the entirety of the human transcriptome is preferable to assessing changes in fewer numbers of genes, which could miss changes that are most important.

The complementarity of genetics to the study of skeletal biology

DNA sequence diversity in members of the same species contributes to phenotype diversity. As detailed above, phenotypes that can be recognized and classified can then be studied genetically to identify their causative genes. Conversely, genomes can be studied (e.g., interspecies comparisons) and genetically manipulated (e.g., transgenic mice) to identify how genes contribute to specific phenotypes. However, genetics must be viewed as one of many complementary approaches that are employed to understand fundamental biologic processes. Work on a protein known by a variety of names including lubricin, megakaryocyte stimulating factor, superficial zone protein, CACP protein, and now "officially" as PRG4, exemplifies the value of complementarity in biologic research.

Lubricin was purified from bovine synovial fluid based upon its biophysical lubricating properties [21]. Megakaryocyte stimulating factor was purified from human urine based upon its ability to stimulate megakaryocyte growth in bone marrow cultures [22]. Superficial zone protein was purified from bovine articular cartilage based upon its physical location in the superficial chondrocyte cell layer [23]. CACP protein was identified based upon genetic mapping and mutation detection in persons affected with the camptodactyly-arthropathy-coxa vara-pericarditis syndrome [24]. PRG4 was identified as a mRNA species exhibiting increased expression during ectopic bone formation in the tiptoe walking mouse [25]. Nucleotide and/or peptide sequencing ultimately led to the conclusion that lubricin, megakaryocyte stimulating factor precursor, superficial zone protein, CACP protein, and PRG4 were encoded by the same gene [26–28]. Yet, each approach (biophysical, functional, biochemical, and genetic) complemented the others by revealing novel features about the protein. The protein is essential for normal joint function since its genetic deficiency causes an autosomal recessive syndrome having hyperplasia and fibrosis of synovium and precocious joint failure [24, 29]. It is expressed by superficial zone chondrocytes and synovial cells [30] and has lubricating [31] and growth regulating activities [22]. The gene has now been knocked out in mice (Marcelino et al., unpublished). The mouse phenotype recapitulates what has been observed in human patients with disease, and should facilitate studies regarding the pathogenesis of the articular cartilage destruction.

Summary

Genetic approaches complement other approaches for gaining insights about the pathogenesis of osteoarthritis. Genetic mapping of simple Mendelian skeletal diseases can identify genes that participate in normal joint homeostatic pathways. Genetic studies in cohorts of patients having common forms of osteoarthritis will be more challenging, but may also identify genes that contribute to osteoarthritis predisposition and/or disease progression. Employing the entire human transcriptome to study changes in gene expression between normal and diseased tissues, or cells, can identify other participants in these complex pathways. Finally, naturally occurring mutations, or induced genetic variations, in other organisms can be used to model human disease phenotypes and further dissect the processes responsible for disease pathogenesis.

Acknowledgements
This work was supported by grants from the NIH (AR-45687), the Arthritis Foundation, and the Burroughs Wellcome Fund.

References

1. Lander ES, Linton LM, Birren B, Nusbaum C, Zody MC, Baldwin J, Devon K, Dewar K, Doyle M, FitzHugh W et al (2001) Initial sequencing and analysis of the human genome. *Nature* 409 (6822): 860–921
2. Venter JC, Adams MD, Myers EW, Li PW, Mural RJ, Sutton GG, Smith HO, Yandell M, Evans CA, Holt RA et al (2001) The sequence of the human genome. *Science* 291 (5507): 1304–1351
3. International working group on constitutional diseases of bone (1998) International nomenclature and classification of the osteochondrodysplasias (1997). *Am J Med Genet* 79 (5): 376–382
4. Collins FS, Patrinos A, Jordan E, Chakravarti A, Gesteland R, Walters L (1998) New goals for the U.S. Human Genome Project: 1998–2003. *Science* 282 (5389): 682–689
5. Fisher S, Halpern ME (1999) Patterning the zebrafish axial skeleton requires early chordin function. *Nat Genet* 23 (4): 442–446

6   Ho AM, Johnson MD, Kingsley DM (2000) Role of the mouse ank gene in control of tissue calcification and arthritis. *Science* 289 (5477): 265–270
7   Holderbaum D, Haqqi TM, Moskowitz RW (1999) Genetics and osteoarthritis: exposing the iceberg. *Arthritis Rheum* 42 (3): 397–405
8   Meulenbelt I, Bijkerk C, De Wildt SC, Miedema HS, Breedveld FC, Pols HA, Hofman A, Van Duijn CM, Slagboom PE (1999) Haplotype analysis of three polymorphisms of the COL2A1 gene and associations with generalised radiological osteoarthritis. *Ann Hum Genet* 63 (Pt 5): 393–400
9   Priestley L, Fergusson C, Ogilvie D, Wordsworth P, Smith R, Pattrick M, Doherty M, Sykes B (1991) A limited association of generalized osteoarthritis with alleles at the type II collagen locus: COL2A1. *Br J Rheumatol* 30 (4): 272–275
10 Uitterlinden AG, Burger H, van Duijn CM, Huang Q, Hofman A, Birkenhager JC, van Leeuwen JP, Pols HA (2000) Adjacent genes, for COL2A1 and the vitamin D receptor, are associated with separate features of radiographic osteoarthritis of the knee. *Arthritis Rheum* 43 (7): 1456–1464
11 Chitnavis J, Sinsheimer JS, Clipsham K, Loughlin J, Sykes B, Burge PD, Carr AJ (1997) Genetic influences in end-stage osteoarthritis. Sibling risks of hip and knee replacement for idiopathic osteoarthritis. *J Bone Joint Surg Br* 79 (4): 660–664
12 Lanyon P, Muir K, Doherty S, Doherty M (2000) Assessment of a genetic contribution to osteoarthritis of the hip: sibling study. *Bmj* 321 (7270): 1179–1183
13 el-Shanti H, Murray JC, Semina EV, Beutow KH, Scherpbier T, al-Alami J (1998) Assignment of gene responsible for progressive pseudorheumatoid dysplasia to chromosome 6 and examination of COL10A1 as candidate gene. *Eur J Hum Genet* 6 (3): 251–256
14 Chapman K, Mustafa Z, Irven C, Carr A J, Clipsham K, Smith A, Chitnavis J, Sinsheimer JS, Bloomfield VA, McCartney M (1999) Osteoarthritis-susceptibility locus on chromosome 11q, detected by linkage. *Am J Hum Genet* 65 (1): 167–174
15 Loughlin J, Mustafa Z, Irven C, Smith A, Carr AJ, Sykes B, Chapman K (1999) Stratification analysis of an osteoarthritis genome screen-suggestive linkage to chromosomes 4, 6, and 16 [letter]. *Am J Hum Genet* 65 (6): 1795–1798
16 Risch N (1990) Linkage strategies for genetically complex traits. II. The power of affected relative pairs. *Am J Hum Genet* 46 (2): 229–241
17 Felson DT, Zhang Y (1998) An update on the epidemiology of knee and hip osteoarthritis with a view to prevention. *Arthritis Rheum* 41 (8): 1343–1355
18 Camargo AA, Samaia HP, Dias-Neto E, Simao DF, Migotto IA, Briones MR, Costa FF, Nagai MA, Verjovski-Almeida S, Zago MA et al (2001) The contribution of 700,000 ORF sequence tags to the definition of the human transcriptome. *Proc Natl Acad Sci USA* 98 (21): 12103–12108
19 Hillier LD, Lennon G, Becker M, Bonaldo MF, Chiapelli B, Chissoe S, Dietrich N, DuBuque T, Favello A, Gish W et al (1996) Generation and analysis of 280,000 human expressed sequence tags. *Genome Res* 6 (9): 807–828
20 Schulze A, Downward J (2001) Navigating gene expression using microarrays – a technology review. *Nat Cell Biol* 3 (8): E190–195

21. Swann DA, Hendren RB, Radin EL, Sotman SL, Duda EA (1981) The lubricating activity of synovial fluid glycoproteins. *Arthritis Rheum* 24 (1): 22–30
22. Turner KJ, Fitz LJ, Temple P, Jacobs K, Larson D, Leary AC, Kelleher K, Gionnotti J, Calvetti J, Fitzgerald M et al (1991) Purification, biochemical characterization, and cloning of a novel megakaryocyte stimulating factor that has megakaryocyte stimulating activity. *Blood* 78: 279a
23. Schumacher BL, Block JA, Schmid TM, Aydelotte MB, Kuettner KE (1994) A novel proteoglycan synthesized and secreted by chondrocytes of the superficial zone of articular cartilage. *Arch Biochem Biophys* 311 (1): 144–152
24. Marcelino J, Carpten JD, Suwairi WM, Gutierrez OM, Schwartz S, Robbins C, Sood R, Makalowska I, Baxevanis A, Johnstone B et al (1999) CACP, encoding a secreted proteoglycan, is mutated in camptodactyly-arthropathy-coxa vara-pericarditis syndrome. *Nat Genet* 23 (3): 319–322
25. Ikegawa S, Sano M, Koshizuka Y, Nakamura Y (2000) Isolation, characterization and mapping of the mouse and human PRG4 (proteoglycan 4) genes. *Cytogenet Cell Genet* 90 (3–4): 291–297
26. Flannery CR, Hughes CE, Schumacher BL, Tudor D, Aydelotte MB, Kuettner KE, Caterson B (1999) Articular cartilage superficial zone protein (SZP) is homologous to megakaryocyte stimulating factor precursor and is a multifunctional proteoglycan with potential growth-promoting, cytoprotective, and lubricating properties in cartilage metabolism. *Biochem Biophys Res Commun* 254 (3): 535–541
27. Jay GD, Britt DE, Cha CJ (2000) Lubricin is a product of megakaryocyte stimulating factor gene expression by human synovial fibroblasts [see comments]. *J Rheumatol* 27 (3): 594–600
28. Merberg DM, Fitz LJ, Temple P, Giannotti J, Murtha P, Fitzgerald M, Scaltreto H, Kelleher K, Preissner K, Kriz R et al (1993) A comparison of vitronectin and megakaryocyte stimulating factor, in: KT Preissner, S, Rosenblatt, CW Kost, J Wegerhoff, DF Mosher (eds): *Biology of vitronectins and their receptors*, Elsevier Science Publishers BV, Amsterdam, 45–54
29. Bahabri SA, Suwairi WM, Laxer RM, Polinkovsky A, Dalaan AA, Warman ML (1998) The camptodactyly-arthropathy-coxa vara-pericarditis syndrome: clinical features and genetic mapping to human chromosome 1. *Arthritis Rheum* 41 (4): 730–735
30. Schumacher BL, Hughes CE, Kuettner KE, Caterson B, Aydelotte MB (1999) Immunodetection and partial cDNA sequence of the proteoglycan, superficial zone protein, synthesized by cells lining synovial joints. *J Orthop Res* 17 (1): 110–120
31. Swann DA, Silver FH, Slayter HS, Stafford W, Shore E (1985) The molecular structure and lubricating activity of lubricin isolated from bovine and human synovial fluids. *Biochem J* 225 (1): 195–201

Metabolic differences between knee and ankle

Ada Cole, Hans Häuselmann, Johannes Flechtenmacher, Klaus Huch, Holger Koepp, Wolfgang Eger, Matthias E. Aurich, Bernd Rolauffs, Arkady Margulis, Carol Muehleman, Allan Valdellon, Klaus E. Kuettner

Rush Medical College at Rush-Presbyterian-St. Luke's Medical Center, Chicago, IL 60612, USA

Our studies of human articular cartilages began approximately 13 years ago when collaboration was established between the Department of Biochemistry of Rush Medical College and the Regional Organ Bank of Illinois. As of April 2001, this collaboration has resulted in the acquisition of cartilages from more than 1900 different donors, including about 250 donors from whom we received both knees and ankles and 1656 donors from whom we received only ankles. These are primarily adult donors with a mean age of 51.5 years, 90% men, 99% Caucasian; these gender and race are a reflection of the population of donors received by the Regional Organ Bank of Illinois. The knee (tibiofemoral) and ankle (talocrural) joints are graded on a five-point scale of Collins [1] as modified by Muehleman et al. [2] and reported by Koepp et al. [3]. Of the ankles that have been graded, 53% are macroscopically normal and have received a Collins grade of 0, while only the 483 knee joints (35%) received a Collins grade of 0. For the knee and ankle metabolic, biochemical and biomechanical comparisons, we have used joints from the same limb (matched pairs) from donors with grade 0 in the ankle and grades 0 and 1 in the knee.

Our hypothesis is that the rarity of OA in the ankle joint may be due to the fact that its cartilage is more resistant than knee cartilage to progressive degeneration as a result of one or more of the following characteristics: 1) decreased responsiveness to catabolic factors, 2) increased metabolic potential for repair following damage (increased responsiveness to anabolic mediators), and 3) differences in biochemical composition, gene expression and/or biomechanical properties of the cartilage ECM as well as the underlying bone. Our data to date have supported all three components of this hypothesis.

For these studies, we have used primary chondrocytes and cartilage explants from matched pairs of normal knee and ankle cartilages to determine whether metabolic, biochemical and biomechanical differences exist between the two that might help explain the high incidence of osteoarthritis in knee *versus* the low incidence in ankle. We have been able to document qualitative differences between the cartilages from the two joints. These differences extend from the superficial zone to the sub-

Table 1. A summary of the comparisons between the knee and ankle joints and cartilages from donors

Joint	Degenerative changes	PG content	Dynamic stiffness	PG synthesis	IL-1, Fn-fs	OP-1
Knee	↑	↓	↓	↓	↑	↓
Ankle	↓	↑	↑	↑	↓	↑

chondral bone. In the superficial zone of ankle, but not knee, chondrocytes are organized into clusters suggesting proliferation. Compared to knee, the proteoglycan content is slightly higher in the ankle [4]; this higher content of proteoglycan may be, in part, responsible for the biomechanical properties of higher dynamic stiffness and lower hydraulic permeability of the ankle cartilage compared to the knee. In addition, proteoglycan synthesis is higher in ankle explants [5] suggesting a higher metabolic rate.

The responses to the catabolic mediators, interleukin-1 (IL-1) and fibronectin-fragments (fn-fs), are lower in the ankle measured as a decrease in proteoglycan synthesis [5] or proteoglycan loss from the cartilage [6]. Ankle chondrocytes have a higher IC_{50} for IL-1 than the knee chondrocytes when cultured either in explants [5] or in alginate beads [7] suggesting that these quantitative differences are programmed into knee and ankle chondrocytes just as they have been previously shown to be in chondrocytes from the superficial compared to deep zones [7]. The ankle chondrocytes respond faster with increased proteoglycan synthesis when anabolically stimulated with osteogenic protein-1 (OP-1) [5].

In addition, unlike the knee there was no detectable thickening in the subchondral bone of the ankle when the cartilage had degenerative changes [8]. Taken together these data suggest that there are a number of these subtle, quantitative differences that could help protect the ankle cartilage from degenerative changes that are diminished or absent in the knee.

Acknowledgements
This work was supported in part by NIH grants AR 39239 (M.A., C.M., A.M., K.E.K., A.A.C.), a grant from the Max Kade Foundation (M.A.). We wish to thank the Regional Organ Bank of Illinois and Dr. A. Valdellon and his staff, as well as the families of the cartilage donors.

References

1 Collins DH (1949) The pathology of articular and spinal diseases. London: Edward Arnold and Co., 76–79
2 Muehleman C, Barreither D, Huch K, Cole AA, Kuettner KE (1997) Prevalence of degenerative morphologic changes in the joints of the lower extremities. *Osteoarthritis Cartilage* 5: 23–37
3 Koepp H, Eger W, Muehleman C, Valdellon A, Buckwalter JA, Kuettner KE, Cole AA (1999) Prevalence of articular cartilage degeneration in the ankle and knee joints of human organ donors. *J Orthop Sci* 4: 407–412
4 Treppo S, Koepp H, Quan EQ, Cole AA, Kuettner KE, Grodzinsky AJ (2000) Comparison of biomechanical and biochemical properties of cartilage from human knee and ankle pairs. *J Orthop Res* 18: 739-748
5 Eger W, Schumacher BL, Margulis A, Aydelotte MD, Mollenhauer J, Kuettner KE, Cole AA (1999) Metabolic differences in the degradation progression between knee and ankle cartilage. *Trans Ortho Res Soc* 24: 446
6 Kang Y, Koepp H, Cole AA, Kuettner KE, Homandberg GA (1998) Cultured human ankle and knee cartilage differ in susceptibility to damage mediated by fibronectin fragments. *J Orthop Res* 16: 551-556
7 Häuselmann HJ, Mok SS, Flechtenmacher J, Gitelis SH, Kuettner KE (1993) Chondrocytes from human knee and ankle joints show differences in response to IL-1 and IL-1 receptor inhibitor. *Trans Ortho Res Soc* 18: 280
8 Sumner DR, Koepp H, Berzins A, Eger W, Muehleman C, Cole A, Kuettner KE (1998) Bone density of the human talus does not vary as a function of the cartilage damage score. *Trans Ortho Res Soc* 23: 722

Prevention of lower limb osteoarthritis: Data from the Johns Hopkins Precursors Study

Marc C. Hochberg

Division of Rheumatology and Clinical Immunology, Department of Medicine, University of Maryland School of Medicine, and Medical Service, Maryland Veterans Affairs Health Care System, Baltimore, MD, USA

Introduction

Osteoarthritis is the most common form of arthritis in the United States and other developed countries [1, 2]. The National Arthritis Data Workgroup, a committee convened by the Arthritis Foundation, Centers for Disease Control and Prevention, National Arthritis Advisory Board and National Institutes of Health, reviewed data on the prevalence of osteoarthritis by joint group from the Health Examination Survey, conducted from 1960–1962, and the First National Health and Nutrition Examination Survey (NHANES-I), conducted from 1971–1975 [1]. The case definition of osteoarthritis was based on radiographic changes in the hands and feet in the Health Examination Survey and on radiographic changes in knees and hips in NHANES-I. The prevalence of osteoarthritis among adults aged 25–74 years in the United States was 32.5, 22.2 and 3.8 per 100 for the hands, feet and knees, respectively. Prevalence increased with increasing age through the 65–74 years age group in both sexes; among persons aged 55–74 years of age, corresponding prevalence ratios were 70 percent for the hands, 40 percent for the feet, 10 percent for the knees, and 3 percent for the hips.

Selected risk factors for the development of lower limb osteoarthritis, identified from both case-control and prospective epidemiologic studies, are listed in Table 1, and details have been reviewed recently elsewhere [3, 4]. Most of these epidemiologic studies, such as the Baltimore Longitudinal Study on Aging, the Chingford Study, the Framingham Osteoarthritis Study, the Johnston County Osteoarthritis Project, the Rotterdam Study, and the Study of Osteoporotic Fractures, were conducted in populations of middle-age and older individuals, rather than in young adults [5]. As radiographic changes of osteoarthritis usually become evident in the fifth decade, studies of young adults may identify prevention opportunities that might be applied earlier in life. Reviewed herein are data from The Johns Hopkins Precursors Study, a longitudinal cohort study of young adult medical students that confirm and extend results from epidemiologic studies conducted in middle-aged

Table 1 - Risk factors for lower limb osteoarthritis

Older age
Sex
Women have a higher prevalence and incidence of knee, but not hip osteoarthritis
Race/ethnicity
African-Americans have higher prevalence of bilateral and more severe osteoarthritis
Asians have a lower prevalence of hip osteoarthritis than Caucasians
Genetic predisposition
Higher bone mineral density
Presence of radiographic hand osteoarthritis
Joint injury
Higher body mass index
Congenital and developmental disorders of bones and joints
Mild joint deformity
Valgus deformities of the knees in conjunction with overweight
Occupation-related repetitive joint usage

and older adults showing that overweight and joint injury are important, potentially modifiable, risk factors for the development of lower limb osteoarthritis [6–10].

Subjects and methods

Subjects

The Johns Hopkins Precursors Study was designed by the late Dr. Caroline Thomas to identify precursors of cardiovascular disease [11]. A total of 1337 students who matriculated at The Johns Hopkins University School of Medicine in the classes of 1948 through 1964 enrolled in the study. At study entry, participants underwent a complete medical history and physical examination, including measurement of height and weight. Following graduation, participants completed annual self-administered questionnaires to ascertain disease development and update information on risk factors. During each 5-year follow-up period, at least 85 percent of surviving participants completed one or more of these annual questionnaires.

Case identification

Incident cases of osteoarthritis were identified from these questionnaires. Before 1985, subjects were asked if they had developed "physical and musculoskeletal disorders." After 1985, participants were asked "Have you ever had arthritis?" and "If so, provide the type of arthritis, year of onset, and treatment received." An event of incident osteoarthritis was defined as the self-report of either osteoarthritis or degenerative arthritis. Two rheumatologists reviewed each physician report of osteoarthritis, assigned International Classification of Disease (ICD)-9 codes to designate the most specific site of osteoarthritis as reported by the participant, and determined the year of incidence by the reported year of osteoarthritis onset. For analysis of modifiable risk factors reviewed herein, cases coded as developing either knee osteoarthritis (ICD-9 code 715.96) or hip osteoarthritis (ICD-9 code 715.95) were included.

Risk factors

Body mass index was calculated as weight in kilograms divided by the square of height in meters. Physical activity during the month prior to the time of enrollment was categorized as "none, little, moderate, or much." Injury was defined as a report of trauma to the knee or hip joint or lower limbs, including fractures and internal derangement. Hip and knee injuries that occurred prior to enrollment were recorded on the baseline questionnaire; those occurring after graduation were assessed through the annual questionnaires. Injuries were also assigned ICD-9 codes.

Results

At the time of graduation from medical school, a total of 1216 men and 121 women with an average age of 26 years were enrolled in the cohort. Analyses were based on information from a maximum of 1321 participants, 1200 men and 121 women, who provided any follow-up data. All events reported through November 30, 1995, representing a median of 36 years of follow-up, were included. A total of 62 men and 7 women developed knee osteoarthritis, and 27 men and 5 women developed hip osteoarthritis; the cumulative incidence of knee and hip osteoarthritis by age 65 was 6.3 percent and 2.9 percent, respectively.

At baseline, 51 subjects reported a history of knee injury, and 13 reported a history of hip injury. Participants with a history of knee injury had a significantly greater incidence of knee osteoarthritis. The cumulative incidence of knee osteoarthritis by age 65 in those with a history of knee injury was 13.9 percent compared to 6.0 percent among those without a history of injury: relative risk (95% confidence interval) was 3.0 (1.4, 6.5), $p < 0.005$. After graduation, an additional 74 subjects reported a knee injury and 17 reported a hip injury, bringing the total number of subjects with knee or hip injuries to 125 and 30, respectively. The incidence of knee osteoarthritis was 7.5 per 1000 person-years among those with a knee

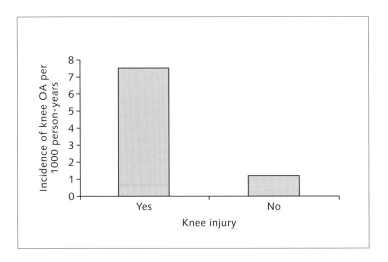

Figure 1
Incidence of knee OA by history of knee injury.

injury compared to 1.2 per 1000 person-years among those without a knee injury: relative risk 5.2 (3.1, 8.7) (Fig. 1). This increased risk remained significant even after adjustment for age, sex, body mass index and physical activity assessed at study entry: adjusted relative risk was 5.0 (2.8, 9.0). The incidence of hip osteoarthritis was 3.2 per 1000 person-years among those with a hip injury compared to 0.7 per 1000 person-years among those without a hip injury: relative risk was 3.5 (0.8, 14.7) (Fig. 2). There was evidence of a statistical interaction between hip injury and physical activity at entry. In those with little or more physical activity, the relative risk for hip osteoarthritis among those with a hip injury approached 10, while that among those with no physical activity at baseline was 5; these estimates were unstable, and the 95% confidence intervals were wide because of a few cases of hip osteoarthritis in this cohort.

Because there were so few cases of osteoarthritis that developed in women enrolled in this cohort, the analysis of the relationship of body mass index to osteoarthritis was conducted in men. Data on body mass index during medical school was available in 1132 men with a mean age of 23 years. Greater body mass index during medical school was associated with an increased risk of developing knee osteoarthritis; the cumulative incidence of knee osteoarthritis by age 65 for those in the highest tertile of body mass index was 12.8 percent compared to 4.0 percent among those in the lowest tertile of body mass index (Fig. 3). After adjustment for year of birth, physical activity in medical school and knee injury, those in the highest tertile of body mass index had a relative risk of developing knee osteoarthritis of 3.5 (1.8, 6.8) while those in the middle tertile had a relative risk of

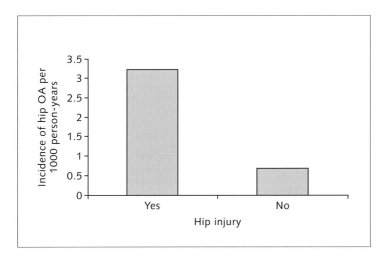

Figure 2
Incidence of hip OA by history of hip injury.

1.4 (0.7, 3.0). When body mass index was considered as a continuous variable, the relative risk for developing knee osteoarthritis was 1.7 (1.3, 2.1) for each standard deviation increase. In contrast, body mass index at study entry was not associated with risk of developing hip osteoarthritis.

Discussion

These data support a role for several primary prevention initiatives to decrease the incidence of knee osteoarthritis [12]. While these initiatives can be advocated for all young adults, they might be targeted to those at high risk. Examples of higher risk groups would include those with a parent or older sibling affected by osteoarthritis, or a person with a congenital or developmental disease of the hip, knee or lower limb.

First, being overweight during adolescence and young adulthood should be prevented through nutrition education programs reinforcing the need to follow a healthy diet. For those adolescents and young adults who are overweight, weight reduction should be recommended so they may achieve a normal weight, defined as a body mass index between 20 and 25 kg/m^2. It has been estimated that elimination of obesity would lead to the prevention of 25 to 50 percent of cases of osteoarthritis in adults [12].

Second, lower limb injuries should be prevented if possible. Again, educational programs focused on injury prevention and the use of safe sports equipment are

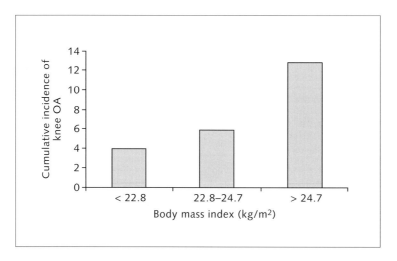

Figure 3
Incidence of knee OA by tertile of body mass index.

indicated. As for obesity, it has been estimated that elimination of major injuries would lead to the prevention of 25 percent of cases of osteoarthritis [12]. For those who sustain a major joint injury, physical therapy is indicated to stabilize the joint and maintain periarticular muscle strength. This is important as quadriceps muscle weakness has been suggested as a risk factor for knee osteoarthritis in middle-aged and older women [13]. Furthermore, the use of protective equipment and the avoidance of future excessive levels of physical activity is important, especially if the joint in unstable.

Beyond these recommendations for primary prevention, future studies may lead to additional strategies focused on dietary composition and nutritional factors, such as supplementation with vitamins C, D and E [14].

Acknowledgement

The analyses of data on the incidence of and risk factors for osteoarthritis in participants in The Johns Hopkins Precursors Study were conducted by Allan C. Gelber, M.D., M.P.H., Ph.D. in partial fulfillment of the requirements for his Ph.D. This work was supported by an Arthritis Foundation Postdoctoral Fellowship and a KO8 award from the National Institute of Arthritis, Musculoskeletal and Skin Diseases. The Johns Hopkins Precursors Study was funded in part by a grant from the National Institute on Aging; Michael J. Klag, M.D., M.P.H. was Director of the study. Both Dr. Klag and I served as mentors for Dr. Gelber on the above training awards.

References

1. Lawrence RC, Helmick CG, Arnett FC et al (1998) Estimates of the prevalence of arthritis and selected musculoskeletal diseases in the United States. *Arthritis Rheum* 41: 778–799
2. Scott JC, Lethbridge-Cejku M, Hochberg MC (1999) Epidemiology and economic consequences of osteoarthritis: the American viewpoint. In: JY Reginster, JP Pelletier, K Martel-Pelletier, Y Henroitin (eds): *Osteoarthritis: clinical and experimental aspects.* Springer-Verlag, Berlin, 20–38
3. Felson DT, conference chair (2000) Osteoarthritis: new insights. Part I: The disease and its risk factors. *Ann Intern Med* 133: 637–646
4. Hochberg MC (2001) Osteoarthritis. In: AJ Silman, MC Hochberg (eds): *Epidemiology of the rheumatic diseases*, 2nd ed., Oxford, Oxford University Press, 205–229
5. Hochberg MC, Vanchieri C. Longitudinal studies of osteoarthritis: state-of-the-science evaluation. Available at "http: //www.nih.gov/niams/news/oisg/oaepip.htm"
6. Gelber AC, Hochberg MC, Mead LA, Wang N-Y, Wigley FM, Klag MJ (1999) Body mass index in young men and the risk of subsequent knee and hip osteoarthritis. *Amer J Med* 107: 542–548
7. Gelber AC, Hochberg MC, Mead LA, Wang N-Y, Wigley FM, Klag MJ (2000) Joint injury in young adults and risk for subsequent knee and hip osteoarthritis. *Ann Intern Med* 133: 321–328
8. Gelber AC, Mead LA, Wigley FM, Hochberg MC, Levine DM, Wang N-Y, Klag MJ (1994) Gender differences in incidence of osteoarthritis: results from the Precursors Study. *Arthritis Rheum* 37 (9, Suppl): S238
9. Gelber A, Hochberg M, Mead L, Wigley F, Klag M (1997) Physical activity in young adulthood and risk of incident knee osteoarthritis: The Johns Hopkins Precursors Study. *Arthritis Rheum* 40 (9, Suppl): S118
10. Gelber A, Mead L, Klag M, Wigley F, Thomas J, Thomas DJ, Hochberg M (1998) Race and incidence of knee and hip osteoarthritis. *Arthritis Rheum* 41 (9, Suppl): S181
11. Thomas CB (1951) Observations on some possible precursors of essential hypertension and coronary artery disease. *Bull Johns Hopkins Hosp* 89: 419–441
12. Felson DT, Zhang Y (1998) An update on the epidemiology of knee and hip osteoarthritis with a view to prevention. *Arthritis Rheum* 41: 1343–1355
13. Hurley MV (1999) The role of muscle weakness in the pathogenesis of osteoarthritis. *Rheum Dis Clin NA* 25: 283–298
14. Sowers MF, Lachance L (1999) Vitamins and arthritis: the roles of vitamins A, C, D, and E. *Rheum Dis Clin NA* 25: 315–332

The inheritance of hip osteoarthritis in Iceland

Thorvaldur Ingvarsson

University Hospital, 600, Akureyri, Iceland

Introduction

The objectives of this study were

(1) to assess in a population-wide study the genetic component of hip OA in Iceland leading to total joint replacement (THR)
(2) to describe a large kinship with inherited hip OA and its associated susceptibility locus

Settlement in Iceland began AD 870 and was completed AD 930, at which time there were 10–20,000 inhabitants on the island. The Vikings brought slaves into the country, mostly from Ireland, during the same period. The population experienced several catastrophic decimations due to disease, such as bubonic plague, and famine in connection with volcanic eruptions. For example, some 30% of the Icelandic population perished in connection with eruptions around the year 1783. Until recently, much of the population lived in small isolated fishing and farming communities along the coast line. The Icelandic population, currently a total of some 275,000, thus fulfills several conditions for an enrichment, and identification, of inherited diseases.

Patients, materials and methods

Two population-based databases were combined to assess genetic contribution to THR for OA: (a) A national registry of THR – A computer-aided search identified hip arthroplasties from all the six orthopedic clinics in Iceland that performed this procedure 1972 – 1996. (b) A national genealogy database – Decode Genetics enters all available Icelandic genealogy records for the last eleven centuries into a computerized database. Each individual is given a personal identifier number (PN), connecting him/her to the corresponding PN of the father/mother. The genealogy database is encrypted by the Data Protection Commission of Iceland, before arriving to the laboratory of Decode Genetics (http://www.decode.is/ppt/protection/index.htm).
Genetic contribution to THR for OA was assessed by:

(1) *Identifying familial clusters of hip OA* – Data for the nationwide registry of 2713 patients with THR for OA were combined with the nationwide genealogy database and analyzed for familial clustering of THR. All family clusters were identified within a given meiotic distance. Recursive pedigree algorithms were developed to find all ancestors in the database related to each member of the input list within a given number of generations backwards. Using these identified groups, the cluster function then searches for ancestors who are common to any two or more members of the input lists.

(2) *Applying the minimum founder test (MFT)* – MFT assesses whether individuals on the patient list are more related to each other than the general population, indicating a relevant genetic component for hip OA. MFT thus compared the minimum number of founders (ancestors) accounting for a patient list with THR OA patients to the minimum number of founders for a set of matched control lists drawn from the national genealogical database. Each set of controls is of the same size as the patient list, and 1000 such separate control lists were constructed. The genealogy database allows a search back to a certain year to find the minimum number of founders, born in that year or later, required for the list of patients or controls. Founders are defined as the minimum set of individuals whose descendants include everybody on the list. Results are presented as a graph. The standard deviation for the 1000 control sets is calculated.

(3) *Calculating the average pairwise kinship coefficient (KC)* – KC was calculated to analyze the probability of identity by descent (IBD) of two randomly chosen chromosomal loci in any two related individuals. The average pairwise KC (a pair constitutes an individual with a hip replacement and a relative) was determined for individuals in the THR OA list and in 1000 matched control groups (generated as described for the MFT). For each pair, IBD was calculated, and the average for all pairs was calculated.

(4) *Estimating the relative risk (RR)* – RR for THR for OA was calculated.

(5) *Performing a genome-wide scan* – From one pedigree, four generations of a kinship with familial hip OA were identified. A genome-wide scan was performed using a framework set of microsatellite markers with average spacing of 4 cM.

All studies were approved by the Ethics Review Board at the Icelandic Health Ministry and the Data Protection Committee of Iceland.

Results

Familial clustering of THR for hip OA

At a meiotic distance of 5, about 3 out of 4 Icelandic patients with THR for OA were identified as belonging to a cluster (Tab. 1). At this meiotic distance, 471 clusters varying in size from 562 to 2 patients were identified.

Table 1 - Familial clustering of THR for hip OA in Iceland. Data for 2713 cases of THR for OA were examined in the national genealogy database for relationship, and the number and size of clusters were calculated at each given meiotic distance.

Meiotic distance	Number of clusters	Largest cluster size	Smallest cluster size	Total number of individuals in clusters
2	479	42	2	957
3	602	45	2	1196
4	649	54	2	1766
5	471	562	2	2010
6	250	1601	2	2202
7	156	1919	2	2241

Minimum founder test

To analyze the hereditary relationship for the OA THR patients, we assessed the MFT back to the year 1740 for all individuals in the THR registry operated on for OA, and compared the results to 1000 matched control sets. Results were plotted as the minimum number of founders against the earliest birth year of the founder (Fig. 1). There are 2713 original founders in the THR list. At the end of the 19th century, the curves begin to diverge, with the THR patients having fewer founders at all times as compared to the average for controls, back to the year 1740. The greatest difference is seen in the year 1860, where the calculated number of founders for the THR patient list diverges 12.3 control group standard deviations from the mean control list curve ($p < 0.00001$).

Data were also generated separately for the 1383 females and 1330 males. The greatest difference is again seen in the middle of the 19th century for both males and females. At the point of greatest difference THR males deviated 10.2 standard deviations from their matched control groups ($p < 0.00001$), while THR females showed a difference of 5.7 standard deviations from their controls ($p < 0.00001$) (Fig. 1).

Kinship coefficient

The kinship coefficient for both the THR OA patients and the 500 matched control lists were calculated in order to further analyze the hereditary relationship of the patients with hip replacements in Iceland. The controls had an average kinship coef-

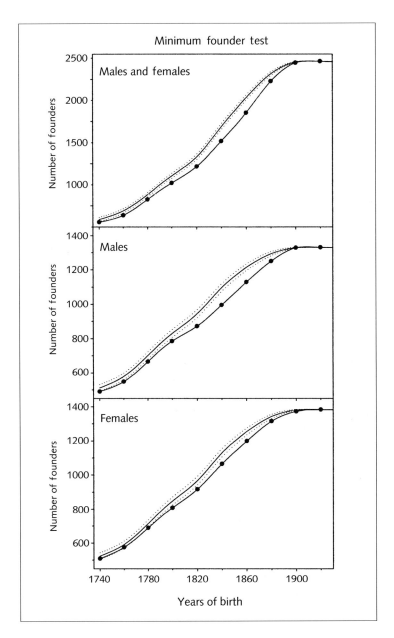

Figure 1
Minimum founder test for the complete 1967 to 1996 patient list of 2713 THR due to OA in Iceland. Mean data for 1000 same sized control lists are shown with two standard deviations (full and dotted lines, respectively), while THR patient data are denoted with line and symbols (— • —).

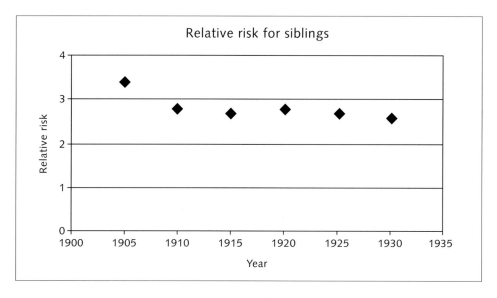

Figure 2
Siblings were about 2.6-fold likely to have the same operation, as compared to control. 95% CI 2.52-3.10 (p<0.000001).

ficient of 0.00019 (SD $5.12*10^{-6}$) while the THR OA patient list had an average kinship coefficient of 0.00027, or 14.5 control group standard deviations from the control average ($p < 0.00001$).

Relative risks

Based on the national THR data, the overall risk for hip replacement due to OA for the Icelandic population is 4% for individuals born around 1920.

Siblings of THR OA patients were about 2.6-fold more likely to have had the same operation, as compared to controls. Cousins and spouses of affected individuals showed no marked risk increase for receiving a hip replacement for OA as compared to controls ($p < 0.000001$) (Fig. 2).

Genome-wide scan

The genome-wide scan revealed a locus on chromosome 16p between 28 and 47cM from the telomere meeting the criteria for suggestive linkage (multipoint allele sharing lod score of 2.58, p-value of 1.6×10^{-4}). Two additional regions with a lod score above 1.5 were obtained.

Conclusions

This is the first population-based study using a national registry of THR for OA and an extensive genealogy database to examine the genetic contribution to a common form of OA. Previous studies have used only small or restricted samples.

The statistical methods applied in this study demonstrated that osteoarthritis patients are more related to each other than controls. Thus, we have shown that in the Icelandic population

- a large number of familial clusters exist of patients with THR for OA
- the number of founders that account for patients with THR for OA is less than for similar sized control groups in the average Icelandic population
- the degree of kinship among patients with THR for OA is greater than average in the Icelandic population
- the relative risk is 2.6-fold increased for siblings of patients with THR for OA to have undergone the same procedure
- Evidence for linkage in the described family suggests that a hip OA susceptibility gene exists on chromosome 16p

Taken together, our findings extend previous studies and support a significant genetic contribution to a common form of osteoarthritis. Environmental factors act on this genetic background. Our study further encourages the search for genes responsible for increased susceptibility to osteoarthritis of the hip in the Icelandic population.

References

1   Ingvarsson T, Stefánsson SE, Hallgrímsdóttir IB, Jónsson Jr H, Gulcher J, Jónsson H, Ragnarsson JI, Lohmander LS, Stefánsson K (2000) The inheritance of hip osteoarthritis in Iceland. *Arthritis Rheum* 43 (12): 2785–2792
2   Ingvarsson T, Stefánsson SE, Gulcher J, Jónsson HH, Frigge M, Lohmander LS, Stefánsson K (2001) A large Icelandic family with osteoarthritis of the hip associated with a susceptibility locus on chromosome 16p. *Arthritis Rheum* 44: 2548–2555

Dynamics of matrix loss in the spontaneous osteoarthritic mouse strain STR-1N

Lydia K. Wachsmuth[1], Beate Durchfeld-Meyer[2], Nadine I. Jahn[3], Nicole Verzijl[4], Uwe H. Dietz[5], Nicole Gerwin[5], Manfred Keil[2], Hans-Ludwig Schmidts[2] and Ruth X. Raiss[5]

[1]Institute of Medical Physics, University Erlangen, D-91054 Erlangen, Germany; [2]Drug Safety Evaluation, Aventis Pharma, D-65926 Frankfurt, Germany; [3]Department of Pharmacy, University Bonn, D-53113 Bonn, Germany; [4]TNO Prevention and Health, 2301 CE Leiden, The Netherlands; [5]Thrombotic Diseases/Degenerative Joint Diseases, Aventis Pharma, D-65926 Frankfurt, Germany

Animal models for osteoarthritis (OA) face the difficulty to display the key pathological events of a naturally slowly progressing disease in a protracted time frame. To serve as pharmacological models, in addition, these changes need to be meaningful and synchronous enough to allow predictions regarding the potential effects of new therapies. Therefore, high expectations have been placed in models where the OA-like pathology develops spontaneously, without any defined stimulus potentially limiting the true disease relevance of the tissue reactions thus induced.

Naturally developing spontaneous OA can be observed in a number of inbred strains of laboratory mice, one of which is the STR-1N mouse. We have bred this strain, originally obtained from the animal facility of the National Institutes of Health, Bethesda, MD, USA, for over 10 years inhouse (Aventis Pharma, Frankfurt, Germany), and have documented the time-course of knee joint pathomorphology semiquantitatively by a modified Mankin score [1]. The histological assessments at distinct timepoints have revealed a consistent pattern of disease progression reflecting many aspects of human OA. However, the exact localization, sequence, and hierarchy of molecular events underlying these changes are still poorly understood.

Therefore, additional diagnostic procedures, such as scanning electron microscopy (SEM) of the tibial plateau, immuno-histochemistry and in situ hybridization within the articular cartilage, as well as biochemical quantitation of urine-excreted degradation products of the cartilage loss and bone remodeling, have been included in this study. Of these, especially sampling and measuring collagen crosslinks excreted in urine provide a noninvasive and sensitive tool for longitudinal monitoring of individual animals in this model [2]. Several parameters depicting disease progression are then validated by correlation with semiquantitative histology, to which purpose a grading system discriminating degradative and repair processes in cartilage, bone, and synovial tissue is used.

Emerging from this synopsis is a dynamic picture: The onset of disease can already be histologically identified during adolescence in the articular cartilage as a loss of overall matrix staining with the proteoglycan-specific dye Safranin-O. In addition to the process of matrix degradation, an attempt of anabolic tissue remodeling can be observed: increased pericellular staining for proteoglycans is typical for stages of early OA. Type II collagen expression, documented by in situ hybridization, is turned on at certain areas and disease stages as well. Where proteoglycan loss is apparent, the cartilage oligomeric matrix protein (COMP), a promising biomarker for osteoarthritic pathology (see the chapter by T. Saxne et al., this volume), becomes more prominent (data not shown). The COMP localization pattern throughout articular cartilage, visualized by immuno-histochemistry, is complementary to Safranin-O staining and appears very similar to human OA. Distribution and intensity of staining do not change much over time.

Proteoglycan depleted areas are prone to progressive structural damage to the collagen network. Surface fibrillations are followed by deeper fissures, and subsequent cartilage loss occurs, especially in the medial compartment. Scanning electron microscopy reveals a spheric view upon the tibial plateau. Increasing areas with surface irregularities and erosions confirm the histological findings, but reach a maximum of surface disturbance well before the endpoint of cartilage deterioration is seen on the sectioned tissue (Fig. 1). Although skeletal growth and maturation overlap the disease-related pathomechanisms, an impressive bone remodeling can be distinguished in these mice – with increasing age, subchondral bone sclerosis and osteophyte formation become prominent features in the morphology of their knee joint. STR-1N mice exhibit a severe "late stage OA" at skeletal maturity.

Collagen crosslinks, lysylpyridinoline (LP), derived from bone, and hydroxylysylpyridinoline (HP), originating from bone, tendon, and cartilage, both display a consistent time-course when measured individually and weekly, and standardized to the respective urinary creatinine content. Initial, growth-derived high levels of collagen crosslink excretion decline to levels of normal matrix maintenance, but rise again to a second peak not observed in healthy animals. In the ratio HP/LP, reflect-

Figure 1
Histological stages of spontaneous osteoarthritic changes in the medial compartment of the knee joint of STR-1N mice, paired with age-matched scanning electron microscopical (SEM) views of the cartilagenous surface of the tibial plateaus. Safranin-O/fast green staining of sagittal joint sections depicts progressive loss of proteoglycans and tissue from articular cartilage, and intense subchondral remodeling. SEM of carefully dissected tibias in a standardized position allows the structural impairment to be followed in three dimensions. At the age of ~100 days the tibial surface damage exhibits its maximum, around 150 days the histopathology reaches the final disease score, when assessed by semiquantitative staging in groups of 10 mice per timepoint.

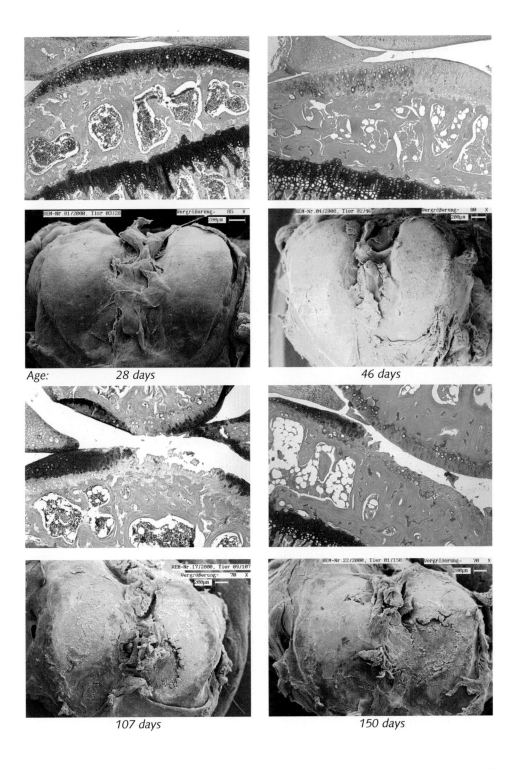

Age: 28 days 46 days

107 days 150 days

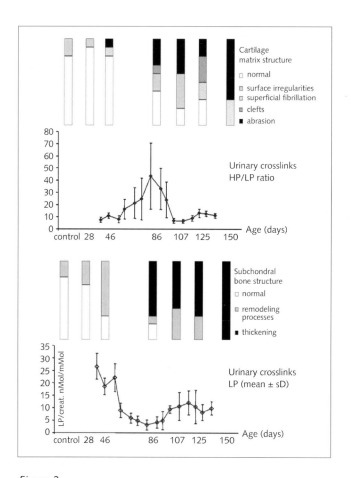

Figure 2
Correlation of the time-course of urinary collagen crosslink excretion with the occurrence of histopathological events in the respective joint tissues of supposed origin. The columns illustrate for articular cartilage and subchondral bone the proportional distribution of the structural changes within each group of 12 mice at their timepoints of sacrifice. "Control" represents a group of 12 nonarthritic (NMRI) mice at the age of 130 days. Adapted to the same time axis, the amount of excreted urinary crosslinks is plotted underneath. These crosslink data are mean values of 12 individuals assessed weekly over the whole period indicated. (They are not identical to the animals sacrificed for histology.) Values for lysylpyridinoline (LP) and hydroxylysylpyridinoline (HP) are standardized to the respective creatinine content of each sample. LP derives mainly from bone tissue and is therefore matched with the subchondral bone assessment. The HP/LP ratio is claimed to reflect mainly changes in cartilage, and is therefore placed below the histological scoring of the cartilage matrix. For both tissues the timepoint of the rise in biomarker concentration coincides well with the tissue dynamics as documented by histopathology. Changes in cartilage seem to preced those in bone.

ing mainly cartilage-derived crosslinks, this second peak (~80 days) is concommitant to the structural changes histologically seen in articular cartilage. It is prior to the peak of the bone-derived LP crosslinks (~115 days), which occurs when bone remodeling becomes obvious from microscopy (Fig. 2).

From these data we conclude that the matrix dynamics in the knee joint of STR-1N mice reflect both degradative as well as repair processes, and that the disease phenotype and progression share many features with the pathophysiology of human OA. Comprehensive histological scoring of defined joint areas is needed to document a complex disease and its development over time. The collagen breakdown products, HP and LP, detected in urine, match tissue degradation and remodeling documented by histology, SEM, and immuno-histochemistry. This strengthens the potential role for urinary crosslinks as surrogate parameters for the respective tissue changes. Time-courses of collagen crosslink levels as well as histological scoring of specific tissue alterations support cartilage degradation preceding subchondral bone remodeling. Proteoglycan and collagen type II neosynthesis occur in distinct areas, interpreted as attempts of repair. Similarity to human OA pathology also extends to typical disease markers, such as COMP, further increasing the validity of this mouse strain as an animal model of osteoarthritis.

References

1   Wachsmuth LK, Raiss RX, Berg-Scholl I, Keiffer R (1999) Histological characterization of disease progression and therapeutic intervention in the spontaneous osteoarthritic STR-1N mouse. *Transact Orthop Res Soc* 24: 461
2   Verzijl N, Wachsmuth L, TeKoppele JM, Raiss RX (2001) Urinary collagen cross-link excretion in STR/1N mice indicates cartilage destruction as an early event in the development of spontaneous osteoarthritis. *Transact Orthop Res Soc* 26: 271

Discussion

Type IIA procollagen NH$_2$-propeptide functions as an antagonist of bone morphogenetic proteins

Q: *Chandrasekhar Srinivasan*: There are now known to be quite a few BMP binding proteins in the extracellular matrix. Do they compete for binding with BMP? In other words, do they compete with each other?

A: *Linda Sandell*: Depending on the dissociation coefficient, there is no reason why the BMP binding proteins wouldn't compete for binding of the various BMPs. The binding domains of most of the BMP-binding proteins are not known. For type IIA procollagen and chordin, we do know that it is the cysteine-rich domain of approximately 70 amino acids, called von Willibrand Factor C. We don't know exactly where in the cysteine-rich domain the binding occurs. The activin-binding domain of follistatin has been recently shown, but none of the others have been determined as far as I know. CTGF and chordin can both bind to BMP. Whether they compete for the same binding sites, I don't know.

Q: *Warren Knudson*: Is there any organization or distribution that you might see in the extracellular matrix that might demonstrate a potential role for the type IIA procollagen N-propeptide?

A: *Linda Sandell*: It is localized in all chondrogenic tissues and remains in the tissue as protein well after the mRNA has switched to the type IIB procollagen form. It's also in the perichondrium where it may provide a reserve of BMP. In fact, it has been shown recently that MMP-9 is necessary to liberate anabolic factors from the perichondrium in fracture healing. We have shown that MMP-9 cleaves the N-propeptide and could function to release bound BMP.

Q: *Dick Heinegard*: Is the type IIA N-propeptide retained in the matrix?

A: *Linda Sandell*: That is an interesting question. We have immunohistochemistry results that indicate that the N-propeptide is in intervertebral disc without the fibrillar domain, so it may be retained in other places as well. We are currently trying to analyze matrix lysates by western blots.

Q: *Arthur Veis*: In the type IIA collagen N-propeptide, are there any interchain disulfide bonds?

A: *Linda Sandell*: In analogy with type I procollagen N-propeptide, there should not be interchain disulfide bonds, but we have not proven that in type IIA N-propeptide.

Genetics and osteoarthritis

Q: *Vince Hascall*: Is it time to begin to distinguish those rare genetic defects that lead to a condition we define as "osteoarthritis," as something other than osteoarthritis? For a while we talked about them as chondrodysplasias, for example. Clearly, as you've pointed out, they don't reflect the common problem of osteoarthritis.

A: *Matt Warman*: Clearly they are not, but I do think that at some level we can think of osteoarthritis as being a common endpoint to a variety of conditions that lead to joint failure. The conditions can be poor matrix, poor joint shape, poor lubrication, poor homeostasis, etc. However, the one benefit of finding these diseases that have an early onset is that perhaps one can use them to look at markers that down the road would be more generally applicable to the common forms of osteoarthritis.

Q: *Paul Dieppe*: The emphasis on the difficulty of the phenotype is very important, but I have a further worry with your approach to focus on people with joint replacement. I think you may be finding a gene for "shouting a lot." If you go for the extreme of people that have two joint replacements, you might be looking at genes for pain because, of course, we operate on pain and not on joint destruction. Or, you may just be looking at people who shout a lot and have the gene for being angry, because you've got to be angry to get a joint replacement.

A: *Matt Warman*: One control we did is to look at the rates of knee replacement in our hip proband siblings, and their rates of knee replacement are not higher than the general population, whereas hip replacement rates are higher. That leads me to think that shouting may be a component, but it's not a good component. The other thing is that we do many more knees than you do in England. The rate of knee replacement in the U.S., as best we can tell, does not have a strong genetic component. It's really explained by age and gender, whereas hip seems to have at least some genetic contribution.

Q: *Frank Luyten*: If we go to the clinic, we decide on our diagnosis based on a number of criteriA: One of the criteria that are important is the pattern of joint involvement. If you look in the PPD (progressive pseudorheumatoid dysplasia) population, for instance, the pattern is not common in OA, but is more common in rheumatic arthritis. That's why we call it pseudo-rheumoarthritis. How do we explain that genetically?

A: *Matt Warman*: I don't have a good suggestion as to why this is the case unless the etiologies for how the knee, hip, and finger joints form have different genetic contributions and variations, and that genes account for why those are involved. There is precedent for this. Certain forms of multiple epiphyseal dysplasias target the knees and spare the hips and the ankles. Others go right up to the hips and sort of spare the knees. Thus certain genes seem to be more important in certain joints. It's not that every joint has the exact same contribution. There is a study called the "go-go" study that is partly funded by Glaxo Smith Kline, in which they are trying to target nodal arthritis in combination with large joint arthritis. I don't know if it will succeed, but it is a really good approach, and it's based on good epidemiologic data that suggest that finger joints and large joints tend to go together more often than not.

Q: *Dick Heinegard*: In your knockout mouse, what is the pattern between joints and osteoarthritis? What about the ligaments in the joint – are they fibrotic also and could that be a factor?

A: *Matt Warman*: We have not had this mouse for very long, only 2.5 months. We tried to look at the cruciates, and they appear normal. We've tried to look at tendon sheaths, and to my eye, early on they looked normal in contrast to human patients who get the camptodactyly from scarring between the tendon and tendon sheath. But, as our mice have aged, they've also developed camptodactyly (flexion contracture of tendons), so we assume that for some reason, humans get it earlier than mice do.

Q: *Frank Wollheim*: Why don't they get inflammation?

A: *Matt Warman*: Clearly what leads to inflammation in inflammatory arthropathies is the presence of inflammatory cells that are releasing signals that are perhaps destroying proteins that inhibit intimal growth normally. So perhaps this protein is an inhibitor of intimal growth normally, and in inflammatory diseases enzymes degrade it, so it loses its inhibitory function. Another possibility is just as we get callouses when we don't have good grips and we have friction on our hands, perhaps what we are seeing is the equivalent of callous formation in a poorly lubricated joint. The way to know is to look at the mice earlier and later and see when these abnormalities are really occurring.

Q: *Joel Block*: When you are dealing at the interface between structure and clinical medicine, it gets into a pretty confusing area: When you look at asymptomatic osteophytes, there was high concordance. When you look at pain, there was no concordance. Nobody in the clinical world is going to call an asymptomatic osteophyte OA, and yet you are using pain as your endpoint – the surrogate for pain being total joint replacement. I'm wondering why you are looking for dramatic linkages of structure when really what you're looking at is genetic linkages of pain threshold?

A: *Matt Warman*: I'm not looking for any genetic linkages at the moment because what we wanted to try and figure out is what were the right phenotypes to go after. So, our studies are really to get a sense of the direction we need to go if we want to employ genetics. But, you are absolutely right in saying that in the large university case study; they looked at joint failure as being the endpoint, as opposed to looking at osteophyte measures or looking at rates of progression of OA, which might be more interesting. The reason I think they did joint failure is because I think it's much less expensive. The patients are coming to you, as opposed to having to go x-ray every patient and their siblings and do all of the quantitation. So my bet is that it is ten-fold less expensive to do it this way, but then again they might not get anything out of it.

Metabolic differences between knee and ankle

Q: *Vince Hascall*: There seems to be a higher metabolic status of the cells in the ankle cartilage than in the knee, and there is also a great deal of architectural difference in terms of the biomechanics of the two tissues. How would you assess whether the differences in terms of the development of OA relate either to the metabolic or to the organizational aspects of the cartilage?

A: *Ada Cole*: Right now we are speculating that this might be two-part. First of all, the higher metabolism in the ankle, if that follows through, seems to be surprising. We expected the ankle cells to be more quiescent, to have less proteoglycan turnover, but it seems to be the reverse. At one point we asked, could we make knee chondrocytes act more like ankle chondrocytes, and would this prevent degenerative changes? But, now we are not really sure if that is the proper question to ask because the ankle chondrocytes seem to be surprising us. We think also that higher metabolism of the proteoglycan may help with structural support to the ankle as well, so that higher content of GAG in the ankle may indeed help to protect it.

Q: *Vince Hascall*: What is the cell density of the two tissues? Is it similar in terms of cells per matrix?

A: *Ada Cole*: As far as we can tell right now, yes it is. There is no dramatic difference between them, and you certainly don't have a major decrease in cellularity that is different between them with age either.

Q: *Alice Maroudas*: Have you detected a difference between Grade 1 and Grade 0 in the case of the knee? Does that play any role in your relative rates of metabolism?

A: *Ada Cole*: Right now we can't tell the difference between the metabolic rate of the chondrocytes between a Grade 1 and a Grade 0, either in the knee or in the ankle. There are more differences in inter-individual rates than there are between the Grades 0 and 1.

Q: *Tom Andriacchi*: Two questions: If you look at the loading in the ankle vs. the knee during walking, the ankle is actually experiencing higher loads. Could the biochemical differences you are seeing be a response to that loading? Secondly, it was recently observed that a number of patients with ankle instability and, in particular flat foot deformity associated with a posterior tendon dysfunction, get OA at the ankle much more frequently. I'm wondering if instability, which is intrinsic to the knee and not so much at the ankle, may be a factor also in this mechanism?

A: *Ada Cole*: I would say yes to both questions. We would certainly think that there is something that is causing ankle chondrocytes to have a stiffer matrix. So, maybe it is the higher loading in the ankle that is causing that. The instability – absolutely – we think that we may be seeing degenerative changes with instability at the ankle as well. Although, most of the donors that we have, even with the higher grades, don't seem to have had fractures. There probably was damage, but such a small amount that they probably did not get treatment for it.

Prevention of lower limb osteoarthritis: data from The Johns Hopkins Precursors Study

Q: *Warren Knudson*: In those students who were overweight in the beginning of medical school, do you know anything about whether they lost weight later in life and did it matter?

A: *Marc Hochberg*: Yes, we know those data, and the answer is no, it didn't matter. This is in contrast to the data from the Framingham, where weight loss was associated with the odds of developing symptomatic osteoarthritis. I would say that in this cohort, we had very few individuals who lost weight over time, and that may limit our ability to show an association with weight loss in this group. There has been a lot of interest in this issue of weight loss for treatment of osteoarthritis, in addition to prevention; and we have shown as part of an exercise study that individuals who lose weight as part of an exercise program have a greater improvement in their symptoms in knee osteoarthritis than those who maintain their weight.

Q: *Paul Dieppe*: I would like to reinforce the importance of getting longitudinal data like yours rather than the previous cross-sectional data that we have been trying to analyze. The data on the knee injury I find particularly depressing; the idea that if you are injured at quite a young age you've got to sit and watch television for the rest of your life. It seems to run in the face of old data about meniscectomy that suggest that the younger you have the meniscectomy the less important it was. The belief was that if you are younger and you get injured or damaged, you've got more chance to adapt and to initiate effective repair. However, your data are suggesting that injury at a young age is very bad news.

A: *Marc Hochberg*: One explanation could be related to the type of activity engaged in following the injury. A lot of the meniscectomy data were done in athletes. Those individuals would be more prone to participate in reconditioning programs following their injury. A group of medical students, who subsequently become physicians, although they maintain some level of activity, they are not participating in competitive athletics, or even necessarily recreational athletics. So I think exercises that would stabilize the joint in terms of increase in quadriceps strength might be important in terms of preventing the development of OA.

Q: *Holger Koepp*: Do you have any information as to whether any of the women, particularly, have dysplasia of the hip or degenerative dysplasia of the hip?

A: *Marc Hochberg*: We did a prospective analysis where we looked at women who were free of radiographic osteoarthritis in 1986 to 1988 and who developed osteoarthritis over the course of a follow-up of 8 years. We did find that there was an association between having acetabular dysplasia, predominantly a shallow acetabulum on their x-rays. Those with a mean age of 70 that were free of radiographic OA had an increased risk of developing osteoarthritis on average 8 years later.

Q: *Phillip Osdoby*: You showed a dramatic increase in OA in women over 40; is there a comparable study in men? The reason I ask is that it is well known that from the age of 40 on there is a dramatic drop in growth hormone production. Is there any relationship or link between that observation and what you are observing?

A: *Marc Hochberg*: There are cross-sectional studies that also show an increase in radiographic osteoarthritis in men, although the dramatic increase is not as great.

Q: *Frank Luyten*: Maybe the numbers are not there, but I wonder if there is any relationship to the surgical procedures that these young patients underwent at a young age, and the development of OA: If patients with knee injuries end up in consultation with a rheumatologist, they get conservative treatment. If they end up with a consultation with an orthopedic surgeon, the chance of operating to improve the joint is very high. Do you have any data on that?

A: *Marc Hochberg*: We have data on the type of injury, and there are a lot of different types. So what we did was to cluster them for the purposes of the analysis. If we break it down by the type of injury, the numbers really get too small, and we don't have good data on the type of surgical procedures that were used to treat the injuries.

Q: *Thomas Andriacchi*: Your information raises the question about the causative factors, and particularly laxity associated with injury versus load associated with increased body mass index. I wonder, especially with respect to the increased incidents in women for increased OA, if laxity may be a contributing factor to this kind of OA?

A: *Marc Hochberg*: Personally, I think it does contribute. I think that at least some of the cross-sectional and progression data that Leena Sharma has published, that subtle deformity of the knee, those who have more of a deformity of the knee as opposed to straight knees, or valgus knees in patients who were overweight, are more likely to have progressive knee osteoarthritis. So I think that there is probably an interaction between deformity and instability with weight. Now we didn't find an interaction with injury and weight in these analyses, but we don't have any measures of instability in this cohort.

The inheritance of hip osteoarthritis in Iceland

Q: *Linda Sandell*: I've heard that there have actually been some genes pinpointed in the Icelandic database. Could you comment on that?

A: *Stefan Lohmander*: As far as I can tell you today, we have not pinpointed any genes in the course of the work that I presented. But, obviously the situation changes more or less from day to day. There are a number of very interesting loci in addition to the ones that I have shown here. My view on the genes that we will be looking for is that I am sure we are in for some surprises and some unknowns. I think that there are some strong candidates, but I also think that we need to be very open with what kind of relationships or genes we are going to find.

Q: *Matt Warman*: I have a comment that concerns the differences between studying a Mendelian disease that causes osteoarthritis, and multigenic or polygenic forms of osteoarthritis. With the Mendelian form, because of the size of the world population, mutations recur within a gene, and many different mutations within the gene can cause the same end phenotype. In the Icelandic population in the one family that you showed, there may be a single founder mutation that is associated with that phenotype, and you may not have the benefit of having families from around the world, given how many different causes for OA there may be, to confirm the finding. So when you do linkages and see a broad region of the genome, perhaps 20 million base pairs of DNA that contain the disease gene, what is the end game? How do you find out which of the 40 or 400 genes in that interval is actually the one that is disease-causing versus an innocent bystander that still was inherited from that common ancestor way back when?

A: *Stefan Lohmander*: That is clearly going to require a lot of work. For these particular families, what's being done today is what's being done everywhere else, trying to narrow down the region by applying a larger number of markers on the same family, and then obviously extending the families both through larger numbers of generations and adding other families to it. I think what you will be looking at is a couple of major genes, and then a larger number of minor genes where different combinations are at play in different patterns of osteoarthritis. In order to dissect what is going on, we are going to need a large number of families, we are going to need some very sophisticated biostatistics, and we are going to have to redefine our phenotype. I think at the end of this, what is going to happen is that we will be able to reverse engineer phenotypes from having identified genes and genetic laws, and these phenotypes may well be quite different from the ones we are working with clinically today.

Q: *Tibor Glant*: What happens with those Iceland people who have a joint replacement? I mean what happens with the cartilage? Do you collect that? Basically the micro array analyses could then be a possibility to match the linkage analyses, and to identify the up- and down-regulated genes of that particular disease sample.

A: *Stefan Lohmander*: We have big freezers and lots of material. But even with the resources for decoding this, it requires large resources, but it is ongoing.

Dynamics of matrix loss in the spontaneous osteoarthritic mouse strain STR-1N

Q: *Matthias Aurich*: Most of the histology you showed is from the tibial plateau. Have you looked at the other joint surfaces; especially the femoral condyles and the patella, since there are differences when you do the correlation with your markers?

A: *Ruth Raiss*: No, we have only looked at the tibial surface in that first study because we also saw the most consistent changes in the tibial compartment as compared to the femoral compartment. The first lesions actually occur on the medial compartment of the tibial plateau.

Q: *Michael Lark*: Have you had the opportunity to evaluate any compounds in this model and look at the effects on collagen markers?

A: *Ruth Raiss*: We are in the process of doing that, but we haven't yet completed that.

Q: *David Howell*: Have you looked or has it been shown in the literature whether this model has mineralization in the cartilage as part of the syndrome?

A: *Ruth Raiss*: We have looked into that because that is mainly the course with the STR ORT mice. We can't see as much mineralization in the STR-1N strain as is observed in that strain. Both strains originate from the same subset described in the 60s, but I think there is a split between the STR-1N and the STR ORT mice. We got our breeding group from the NIH more than 10 years ago, and bred them in-house. So we are quite sure that we also have a more protracted disease progression than originally observed in these mice.

Q: *Tibor Glant*: What was the ratio between the males and females?

A: *Ruth Raiss*: Females also developed the disease, but a little milder. We have included in the study males only. The males have a certain variation in onset and severity; but they have a 100% incidence, compared to other strains. We actually haven't looked into the females in more detail.

Q: *Tibor Glant*: Does the activity level affect progression?

A: *Ruth Raiss*: They seem to walk more painfully, so we also trained them on a rotating rod and looked at how they actually kept up their forced physical activity. We did two things; in the earlier times, we just recorded their spontaneous movement, and this goes down as the disease is actually progressing. But when you have them walking on a rotating rod, the time they stay on also gradually goes down. So functionally, yes, there is a correlation to these changes.

Q: *Frank Luyten*: Is this model now accepted by the regulatory agencies as a pre-clinical model for OA?

A: *Ruth Raiss*: We haven't applied for that yet. I think that the last regulations, which are still as a draft from the FDA, just list certain models as examples, but are open to other models as well.

Morphogenics, development and repair

Introduction

A. Hari Reddi

Center for Tissue Regeneration and Repair, Department of Orthopaedic Surgery, University of California, Davis, School of Medicine, Research Bldg. I, Room 2000, 4635 Second Avenue, Sacramento, CA 95817, USA

Introduction

The progressive degradation and ultimate damage to articular cartilage in osteoarthritis is well known. Regeneration and repair of articular cartilage in experimental animals is initiated when the subchondral bone is penetrated in full thickness defects, possibly due to factors emanating from subchondral bone matrix and cells from bone marrow. However, in partial thickness defects confined to articular cartilage alone there is no repair. Thus, although bone and articular cartilage are adjacent tissues there is a marked difference in their healing potential. Regeneration is a recapitulation of embryonic development and morphogenesis. This brief introduction to the session will present the thesis that endogenous morphogens in articular cartilage will play a role in regeneration, and the potential of the innate morphogenetic family is restrained by antagonists of morphogens such as noggin and resident proinflammatory cytokines such as IL-1 and IL-17.

Morphogens

Bone morphogenetic proteins (BMPs) are a family of morphogens and they induce new cartilage and bone development [1]. BMPs have chemotactic, mitogenic and differentiation-inducing properties. BMPs are dimeric proteins with a single interchain disulfide bond. The biological actions of BMPs are based on concentration-dependent thresholds. There are nearly 15 BMPs. Although they are called bone morphogenetic proteins they have actions beyond bone. Homologous recombination (gene knock-outs) revealed the role of BMP 2 in heart development. Gene knockout of BMP 7 results in impairment of eye and kidney morphogenesis.

Cartilage-derived morphogenetic proteins

Articular cartilage contains endogenous morphogens such as cartilage-derived morphogenetic protein (CDMP-1) and they have been isolated and cloned and found to be identical to growth/differentiation factor 5 [2, 3]. In human chondrodysplasias and type C Brachydactyly with severe limb shortening and impaired morphogenesis, mutations in CDMP-1 gene were identified [4, 5]. Thus, BMPs and CDMPs are critical for cartilage and joint morphogenesis.

Morphogens and articular cartilage

Recombinant BMP-4 stimulates chondrogenesis in chick limb bud mesenchymal cells [6]. In addition to initiation of chondrogenesis, BMP-4 is involved in the maintenance of chondrocyte phenotype and extracellular matrix homeostasis. In monolayer cultures of articular chondrocytes there is a progressive dedifferentiation of phenotype and loss of cartilage-specific type II collagen and proteoglycan aggrecan synthesis. On the other hand in articular cartilage explant cultures, as the extracellular matrix surrounds the chondrocytes the dedifferentiation is delayed or prevented. In serum-free chemically defined medium, recombinant BMP-4 and BMP-7 maintain proteoglycan synthesis [7, 8]. In addition BMPs inhibited the rate of degradation of pre-labelled proteoglycans, demonstrating that BMPs stimulate proteoglycan synthesis and inhibit proteoglycan degradation and, therefore, play a critical role in cartilage homeostasis.

BMP antagonists

During a systematic search for neural inducers, noggin a BMP antagonist was discovered. Noggin induces neural tissue by antagonizing the actions of the dominant morphogen BMP-4. Noggin binds to BMP-2 and BMP-4 with high affinity and blocks the interaction with BMP receptor [9]. Thus, the bioavailability of BMP for its cognate receptor is reduced by noggin. Chordin was discovered to be a BMP-4 binding protein in *Xenopus*. A homologous gene in *Drosophila* is short gastrulation (Sog) which binds to BMP homolog decapentaplegic (dpp) in *Drosophila*. DAN is another BMP antagonist isolated as a tumor suppressor gene [10]. The *Xenopus* genes Cereberus and gremlin are related to DAN [11]. BMPs also interact with extracellular matrix components such as type IV collagen and type II procollagen [12, 13].

Cytokines

In addition to antagonists of morphogens, proinflammatory cytokines such as IL-1 and IL-17 may modulate actions of morphogens. For example, IL-1 inhibits the matrix biosynthesis and cytoskeletal synthesis in articular chondrocytes [14]. IL-

17B and a novel IL-17 receptor like protein are expressed by articular chondrocytes and may play a role in articular cartilage homeostasis [15].

Regeneration of articular cartilage

Regeneration in general recapitulates embryonic development and morphogenesis. The three key ingredients for regenerative medicine and tissue engineering of articular cartilage are: morphogenetic signals (cues), responding stem cells (cells) and the extracellular matrix scaffolding (context). Thus, the integration of cues, cells and context is critical for articular cartilage structure and function. The biological actions of morphogens are constrained by antagonists of morphogens such as noggin, chordin and DAN and pro-inflammatory cytokines such as IL-1 and IL-17 family members. Morphogens, such as BMPs and CDMPs, act on progenitor cells to channel them into chondrogenic lineage. Transcription factors such as Sox-9 are critical in the control of cell lineages by morphogens. The extracellular matrix scaffolding ensures the maintenance of the articular chondrocyte phenotype. Since chondrocytes devoid of matrix dedifferentiate into fibroblast-like mesenchymal cells. BMPs and CDMP-1 can reprogram the dedifferentiated mesenchymal cell to a chondrogenic phenotype. In conclusion, the emerging information in morphogenesis, including BMPs, BMP antagonists, BMP receptors, signaling, stem cells and extracellular matrix, has set the stage for systematic work on articular cartilage regeneration in osteoarthritis and new approaches to tissue engineering of functional joints.

Acknowledgements
I thank Rita Rowlands for excellent help in the preparation of this manuscript. Our research is supported by grants from Shriners Hospitals, and the Lawrence Ellison Chair in Musculoskeletal Molecular Biology and NIH.

References

1 Reddi AH (1998) Role of morphogenetic proteins in skeletal tissue engineering and regeneration. *Nature Biotechnology* 16: 247–252
2 Chang SC, Hoang B, Thomas JT, Vukicevic S, Luyten FP, Ryba NJ, Kozak CA, Reddi AH, Moos M Jr (1994) Cartilage-derived morphogenetic proteins. New members of the transforming growth factor-beta superfamily predominantly expressed in long bones during human embryonic development. *J Biol Chem* 269: 28227–28234
3 Storm EE, Huynh TV, Copeland NG, Jenkins NA, Kingsley DM, Lee SJ (1994) Limb

alterations in brachypodism mice due to mutations in a new member of the TGF-beta-superfamily. *Nature* 368: 639–643

4   Thomas JT, Lin K, Nandedkar M, Camargo M, Cervenka J, Luyten FP (1996) A human chondrodysplasia due to a mutation in a TGF-beta superfamily member. *Nat Genet* 12: 315–317

5   Thomas JT, Kilpatrick MW, Lin K, Erlacher L, Lembessis P, Costa T, Tsipouras P, Luyten FP (1997) Disruption of human limb morphogenesis by a dominant negative mutation in CDMP1. *Nat Genet* 17: 58–64

6   Chen P, Carrington JL, Paralkar VM, Pierce GF, Reddi AH (1992) Chick limb bud mesodermal cell chondrogenesis: inhibition by isoforms of platelet-derived growth factor and reversal by recombinant bone morphogenetic protein. *Exp Cell Res* 200: 110–117

7   Lietman SA, Yanagishita M, Sampath TK, Reddi AH (1997) Stimulation of proteoglycan synthesis in explants of porcine articular cartilage by recombinant osteogenic protein-1 (bone morphogenetic protein-7). *J Bone Joint Surg Am* 79: 1132–1137

8   Luyten FP, Yu YM, Yanagishita M, Vukicevic S, Hammonds RG, Reddi AH (1992) Natural bovine osteogenin and recombinant human bone morphogenetic protein-2B are equipotent in the maintenance of proteoglycans in bovine articular cartilage explant cultures. *J Biol Chem* 267: 3691–3695

9   Zimmerman LB, De Jesús-Escobar JM, Harland RM (1996) The Spemann organizer signal noggin binds and inactivates bone morphogenetic protein 4. *Cell* 86: 599–606

10 Nakamura Y, Ozaki T, Nakagawara A, Sakiyama S (1997) A product of DAN, a novel candidate tumour suppressor gene, is secreted into culture medium and suppresses DNA synthesis. *Eur J Cancer* 33: 1986–1990

11 Reddi AH (2001) Interplay between bone morphogenetic proteins and cognate binding proteins in bone and cartilage development: noggin, chordin and DAN. *Arthritis Res* 3: 1–5

12 Paralkar VM, Weeks BS, Yu YM, Kleinman HK, Reddi AH (1992) Recombinant human bone morphogenetic protein 2B stimulates PC12 cell differentiation: potentiation and binding to type IV collagen. *J Cell Biol* 119: 1721–1728

13 Zhu Y, Oganesian A, Keene DR, Sandell LJ (1999) Type IIA procollagen containing the cysteine-rich amino propeptide is deposited in the extracellular matrix of prechondrogenic tissue and binds to TGF-beta1 and BMP-2. *J Cell Biol* 144: 1069–1080

14 Vinall RL, Lo SH, Reddi AH (2001) Regulation of articular chondrocyte phenotype by BMP 7, interleukin 1 and cellular context is dependent on the cytoskeleton. *Exp Cell Res* 272: 32–44

15 Haudenschild DR, Moseley TA, Rose LM, Reddi AH (2001) Soluble and transmembrane isoforms of novel interleukin-17 receptor-like protein by RNA splicing, and expression in prostate cancer. *J Biol Chem* 277: 4309–4316

Identification and characterization of human cell populations capable of forming stable hyaline cartilage *in vivo*

Frank P. Luyten, Cosimo De Bari and Francesco Dell'Accio

Laboratory for Skeletal Development and Joint Disorders, Department of Rheumatology, University Hospitals KU Leuven, Gasthuisberg, Herestraat 49, B-3000 Leuven, Belgium

Introduction

Articular cartilage of adult individuals has a limited capacity for repair. Although not well studied, it appears likely that in many cases joint surface defects (JSD) progress over the years, ultimately evolving into frank osteoarthritic joints. Developing parameters that will identify the individuals at risk, i.e., those individuals who will develop osteoarthritis after joint injury, is a challenge in modern medicine, and using new technologies such as genomics, proteomics and bioinformatics may lead in the coming decade to the identification of prognostic predictors. In addition, to prevent this joint deterioration, several therapeutic strategies have been developed, microfracture, mosaicplasty, and autologous chondrocyte transplantation (ACT) presently being the most popular. The microfracture technique is mostly reserved to small size lesions, and consists in perforating the bottom of the JSD to recruit skeletal precursors present in the underlying bone marrow. The mosaicplasty procedure consists of collecting several osteochondral plugs obtained from low weight bearing areas of the joint, and subsequently fitting them into the JSD, like in a mosaic. This transforms a relatively large defect into several small defects that can be repaired spontaneously by the surrounding tissue and by the invading bone marrow derived skeletal precursors/mesenchymal stem cells [1].

ACT in humans is becoming a more established technique for repair of articular cartilage defects. In the original description, autologous chondrocytes are obtained arthroscopically with a cartilage biopsy from a minor-weight bearing area of the joint and expanded *in vitro* in the presence of autologous serum. In a second open knee surgery procedure, expanded chondrocytes are injected into the JSD under a periosteal flap [2]. The results of a 9-year follow up are promising, indicating at least a durable symptomatic relief [3]. Nevertheless, to achieve better reproducibility and consistency of the clinical outcome, some aspects of this procedure need to be improved. The knee surgery itself is not without its challenges. The cartilage biopsy results in additional damage to the joint surface and could be potentially minimized by optimizing cell expansion. Cell expansion, however, presents two major restrictions. First, the expandability of chondrocytes, as with most somatic cells, is limited by the occurrence of cell senescence [4, 5]. More importantly, chondrocyte expansion in monolayer cultures is known to induce a progressive phenotypic derangement, often referred to as de-differentiation [6, 7]. There is the possibility that the loss of phenotypic traits *in vitro* could reflect a progressive loss of potential to form cartilage tissue *in vivo*, thereby jeopardizing the long-term outcome of ACT. The use of mesenchymal cells with chondrogenic potential, e.g., obtained from periosteum, synovial membrane, or bone marrow, could circumvent these problems. These cells are more easily available and can be expanded extensively with minimal senescence [8, 8a, 9]. To date, the use of mesenchymal cells is restricted by the yet insufficient knowledge of the long-term stability of the repair tissue and by their tendency to differentiate towards undesired cell lineages. Their multilineage potential may present a risk of heterotopic tissue formation [10]. In particular, a tendency towards mineralization and ossification has been demonstrated when periosteum or bone marrow-derived mesenchymal cells [11] or periosteal and perichondrial grafts [1, 12, 13] are used.

Autologous chondrocyte transplantation: characterization of appropriate, consistent and reproducible chondrocyte populations

ACT offers the advantage of using cells that are stably committed to the cartilage phenotype, naturally resistant to vascular invasion, mineralization, and ossification. However, histologically, the repair tissue after ACT appears as a hypercellular hyaline-like cartilage lacking characteristic columnar organization [3]. A partial loss of phenotypic stability of the expanded chondrocytes may account for large variability of the clinical results and the inadequate repair observed in some patients [3], with the formation of poorly differentiated and disorganized fibrocartilage. It is critical, therefore, to ensure the competence of expanded articular chondrocytes to form stable cartilage *in vivo*. The development of tools to predict the cartilage forming capacity of expanded articular chondrocytes, and thereby guaranteeing a predictable and consistent cell product, represents an important scientific and regulatory issue.

To date, the expression of cartilage matrix molecules such as collagen type II, and the anchorage-independent growth in low temperature-melting (LTM) agarose are the commonly accepted quality controls for the phenotypic stability of (expanded) chondrocytes [2, 14]. Anchorage independent growth and capacity to differentiate are likely to be required, but not necessarily sufficient for the formation of hyaline-like cartilage tissue *in vivo*, starting from a suspension of isolated chondrocytes. The expression of ECM molecules is a measure of chondrocytic differentiation, whereas the agarose assay rather detects the potential for differentiation. None of these parameters, however, has been linked directly to the capacity of isolated chondrocytes to generate cartilage tissue *in vivo*. Ideally, studies in large animals would represent the optimal system to validate chondrocyte expansion procedures for ACT. Unfortunately, due to significant anatomical and biological differences between human cartilage and the articular cartilage of other animals used (essentially rabbit, dog, sheep and goat), it seems that none of the animal models of ACT available today are close enough to the human to be considered truly representative [15]. With these limitations in mind, we set out to define parameters predictive of the cartilage forming activity of expanded human articular chondrocytes, developing and standardizing an *in vivo* model of hyaline-like cartilage formation.

Generation of a model to measure the capacity of adult human articular chondrocytes (AHAC) to form stable cartilage *in vivo*

Description of the assay

The assay consists of the intramuscular injection of AHAC suspensions in the posterior compartment of the thigh of nude mice [16] (see Fig. 1). The outcome is evaluated by the dissection of the muscular mass at the site of injection.

The intramuscular injection of freshly isolated AHAC results reproducibly in the retrieval of a distinct cartilage implant as early as 7 days after injection. The implant is stable for at least 12 weeks without signs of vascular invasion, dystrophic calcification, or bone formation. Calcium deposits are undetectable by alizarin red staining at all timepoints examined, up to 12 weeks. Dose response experiments show that as little as 0.5×10^6 cells are sufficient to generate retrievable implants, and that the wet weight of the implants correlates reasonably well with the number of the cells injected. The number of 5×10^6 was chosen as a standard for the assay because it yielded a reproducible retrieval in 100% (15/15) of injections when freshly isolated chondrocytes were injected.

Characterization of the implant

When comparing the histological and histochemical characteristics of the implants to those of adult articular cartilage, Safranin O staining is approximately as strong in the implants as in normal adult articular cartilage, indicating the presence of sulfated proteoglycans, but the implants are hypercellular and lack the columnar organization. Chondrocytes are embedded in typical lacunae. Masson's trichrome staining reveals no vascular invasion or bone formation. The detection of collagen type II protein by immunostaining further supports the cartilaginous nature of the implant. The ECM is homogeneous with all the stains performed. Masson's

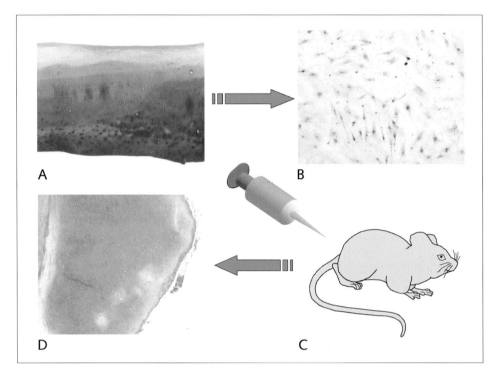

Figure 1
To evaluate the capacity of in vitro expanded chondrocytes to form cartilage in vivo, cells are released from adult human articular cartilage (A) by enzymatic digestion and expanded in vitro (B). Expanded chondrocytes are injected in suspension into the muscle of nude mice (C). The assay is considered positive when a cartilage implant (D) is retrieved in the site of injection after 14 days.

trichrome as well as hematoxylin and eosin preparations do not reveal the presence of dense collagen bundles as in fibrocartilage. The overall characteristics described above allow us to consider the cartilage implants of hyaline-like nature [16].

Specificity of the assay

To study the specificity of the assay, we injected human primary skin fibroblasts and other primary cells and cell lines known to have chondrogenic potential. The injection of primary human skin fibroblasts failed to produce any detectable implant. The injection of primary adult human periosteal cells resulted in fibrous tissue formation. The injection of cell lines with *in vitro* chondrogenic potential such as mouse teratocarcinoma cell line ATDC5 [17], or the rat calvaria cell lines RCJ [18], RCJ 3.1 C5.18 [19], and CFK2 [20] did not produce, in our conditions, any retrievable implants.

It is known that cartilage formation can be induced in similar systems by non viable materials containing chondrogenic factors such as transforming growth factor-β (TGFβ) and bone morphogenetic proteins [21–23] or non-chondrogenic cells producing those factors [24–26]. To address whether, in our system, living chondrocytes were required, we implanted in parallel and in duplicate, living chondrocytes and chondrocytes that had been devitalized by repeated freezing and thawing [27]. Only injection of living chondrocytes resulted in cartilage implants. To investigate whether cell proliferation was required, we X-irradiated chondrocytes with a single 50 Gy dose that blocks proliferation, but does not kill the cells [28, 29]. Under our conditions, chondrocytes that had been made unable to proliferate by X-irradiation could still generate a cartilage implant *in vivo*.

Proteoglycan synthesis and characterization.

High molecular weight sulfated proteoglycans constitute an important component of the cartilage matrix [30] and are critical for the biomechanical properties of articular cartilage, such as elasticity and resistance to load and impact [31]. Incorporation of [^{35}S] sulfate and size fractionation of macromolecules from cartilage displays a characteristic high molecular weight peak, which is mainly due to sulfate incorporation into proteoglycan aggrecan, and a second peak representing small proteoglycans comprising decorin, biglycan, fibromodulin, lumican, and others. In our experiments, we demonstrated the presence in the implants of high molecular weight sulfated proteoglycans comparable in hydrodynamic size to human articular cartilage from an age-matched control. Digestion with appropriate enzymes, such as chondroitinase AC and ABC, reveals a chemical composition of the proteoglycans largely comparable with the proteoglycans found in control human articular cartilage explants (unpublished data).

Contribution of the injected cells to the formation of the implant

In our *in vivo* system the human injected cells and the surrounding muscle are free to interact. Chondrogenesis in the mouse muscle could potentially be induced by the injected chondrocytes, e.g., by sustained secretion of growth factors and morphogens [23]. We explored the relative contribution of donor-derived (human) and host-derived (mouse) cells to the formation of the implant by the detection of human and mouse specific genomic repeats by in situ hybridization. Consecutive sections were hybridized separately with DNA probes specific for human-specific ALU and mouse-specific L1 genomic repeats. Images were aligned, converted to artificial colors, and superimposed. Although some chondrocytes infiltrated the muscle tissue, there was no significant contribution of mouse-derived cells to the implant.

Molecular markers that predict the *in vivo* cartilage forming ability of articular chondrocytes

It is well known that expansion of chondrocytes in monolayer culture results in the loss of their phenotype as measured by collagen type II expression and proteoglycan synthesis [7]. To investigate whether this phenotypic alteration was associated with a loss of the capacity to generate stable cartilage *in vivo*, we tested serially passaged human articular chondrocytes in our *in vivo* assay. Primary AHAC lost their *in vivo* cartilage forming potential after two to three passages *in vitro* (about four to six population doublings). In these samples chondrocytes generated an implant that stained poorly with alcian blue and safranin O. Chondrocytes from later passages failed to generate any retrievable implant.

To identify molecular markers that are predictive of cartilage forming capacity in our assay, we monitored, throughout *in vitro* expansion, the expression of molecules that are known to play a role in the formation and maintenance of the cartilage phenotype. The molecular analysis included the expression of matrix molecules and ligands, receptors and intracellular components of signaling pathways (TGFβ superfamily, fibroblast growth factor, parathyroid hormone/ parathyroid hormone-related peptide, and hedgehog) [32]. The analytical studies included differential display approaches as well as micro-array analysis. These studies resulted in the identification of a number of molecular markers, the expression of which tightly and reproducibly correlates with the capacity of expanded chondrocytes to form stable cartilage *in vivo*. For instance alpha1(II) collagen (COL2A1) and fibroblast growth factor receptor-3 (FGFR3) correlated positively with our *in vivo* assay. These markers were consistently downregulated when the cartilage forming potential of expanded chondrocytes was reduced and immature cartilage implants were retrieved. The expression of these markers became undetectable when chondrocytes were unable to organize a retrievable implant.

The expression of activin receptor-like kinase-1 (ALK-1) mRNA appeared to be a negative marker of chondrocyte stability, being steadily upregulated during expansion. ALK-1 has been reported to be expressed by endothelial cells [33].

Under our conditions, this set of molecular markers appeared to be predictive of the *in vivo* capacity to generate cartilage tissue reproducibly, independently of age [16].

Interestingly, the agarose assay appears to be insufficient to predict the potential of expanded chondrocytes to generate stable cartilage *in vivo*. Anchorage independent growth and rescue of collagen type II expression in anchorage independent culture [7] have been considered predictive of the phenotypic stability of chondrocytes [2]. To validate the agarose assay as a predictor of the cartilage forming potential in our assay, we challenged AHAC at passage 5 (about 10 population doublings) with our *in vivo* assay and with the agarose assay. Although passage 5 AHAC in LTM agarose were still able to form colonies and to recover to the original levels of a number of markers such as COL2A1 and FGFR3 mRNAs, no cartilaginous implant was retrieved in the injected mice in two independent experiments each performed in duplicate. Interestingly, the upregulation of a number of some negative markers such as ALK-1 in serially passaged chondrocytes could not be reverted by the agarose culture.

Human mesenchymal stem cells are incapable of generating stable hyaline cartilage in the *in vivo* mouse model

The use of mesenchymal cells with chondrogenic potential, e.g. obtained from periosteum or bone marrow, could be an alternative cell source and a valuable approach to repair joint surface defects (JSD). Because of their plasticity, one could anticipate a better incorporation of the newly formed tissue into the defect. However, there is clearly insufficient knowledge of the long-term stability of the repair tissue. In view of this, we challenged human MSCs of different origins, i.e., bone marrow, periostium and synovial membrane derived MSCs in the nude mouse assay as described above, both before as well as after induction into the chondrogenic pathway as described [8, 8a, 34]. Interestingly, in most cases we could not retrieve any significant cartilage implant. In a number of cases, especially when using periosteal derived mesenchymal stem cells, we could identify by histochemical staining and *in situ* hybridization the presence of human derived fibrocartilaginous tissue pieces infiltrating in the injected muscles. We could never detect any significant hyaline-like cartilage tissue as in the case of human articular chondrocytes (De Bari C, Dell'Accio F, Luyten FP, manuscript in preparation). These data indicate that human MSCs, grown under "standard" conditions, and induced to undergo chondrogenesis do not become "stable" chondrocytes as defined by our *in vivo* model, and therefore may not be sufficient to organize stable hyaline-like tissue *in vivo* in JSD. These data corroborate previous findings that full thickness articular cartilage defects in rabbits do result in an impressive spontaneous JSD repair early on, but subsequently display a complete loss of the cartilage repair tissue after 12 and 18 months, resulting in an even larger defect and a degenerative joint (Luyten et al, unpublished observations). These findings may indicate that precursor cells only contribute to the formation of cartilaginous scar tissue (callus cartilage?) and are incapable of forming stable articular cartilage.

Conclusions and future perspectives

Articular cartilage defects do not repair, cause clinical symptoms and signs, and probably lead in a significant number of patients to frank osteoarthritis of the involved joint. Although the data on the repair of articular cartilage in the knee joint using autologous chondrocytes are promising, there is an important variability in the outcome of the procedure. The development of standardized tools, such as the above-mentioned approaches to produce consistent and reproducible cell populations, will most probably improve the outcome of cartilage repair procedures. The link between molecular markers and biological behavior of cell populations is anticipated to provide a better understanding of the biology of cartilage in particular, and other tissues in general, and will lead to more successful tissue engineering approaches.

References

1   Minas T, Nehrer S (1997) Current concepts in the treatment of articular cartilage defects. *Orthopedics* 20: 525–538
2   Brittberg M, Lindahl A, Nilsson A, Ohlsson C, Isaksson O, Peterson L (1994) Treatment of deep cartilage defects in the knee with autologous chondrocyte transplantation. *N Engl J Med* 331: 889–895
3   Peterson L, Minas T, Brittberg M, Nilsson A, Sjogren-Jansson E, Lindahl A (2000) Two- to 9-year outcome after autologous chondrocyte transplantation of the knee. *Clin Orthop* 374: 212–234
4   Dominice J, Levasseur C, Larno S, Ronot X, Adolphe M (1986) Age-related changes in rabbit articular chondrocytes. *Mech Ageing Dev* 37: 231–240
5   Froger-Gaillard B, Charrier AM, Thenet S, Ronot X, Adolphe M (1989) Growth-promoting effects of acidic and basic fibroblast growth factor on rabbit articular chondrocytes aging in culture. *Exp Cell Res* 183: 388–398
6   von der Mark K, Gauss V, von der MH, Muller P (1977) Relationship between cell shape and type of collagen synthesised as chondrocytes lose their cartilage phenotype in culture. *Nature* 267: 531–532
7   Benya PD, Shaffer JD (1982) Dedifferentiated chondrocytes reexpress the differentiated collagen phenotype when cultured in agarose gels. *Cell* 30: 215–224
8   De Bari C, Dell'Accio F, Luyten FP (2001) Human periosteum-derived cells maintain phenotypic stability and chondrogenic potential throughout expansion regardless of donor age. *Arthritis Rheum* 44: 85–95
8a De Bari C, Dell'Accio F, Luyten FP (2001) Multipotent mesenchymal stem cells from adult human synovial membrane. *Arthritis Rheum* 44 (8): 1928–1942
9   Pittenger MF, Mackay AM, Beck SC, Jaiswal RK, Douglas R, Mosca JD, Moorman MA,

Simonetti DW, Craig S, Marshak DR (1999) Multilineage potential of adult human mesenchymal stem cells. *Science* 284: 143–147

10   Gilbert JE (1998) Current treatment options for the restoration of articular cartilage. *Am J Knee Surg* 11; 42–46

11   Wakitani S, Goto T, Pineda SJ, Young RG, Mansour JM, CaplanAI, Goldberg VM (1994) Mesenchymal cell-based repair of large, full-thickness defects of articular cartilage. *J Bone Joint Surg Am* 76: 579–592

12   Homminga GN, Bulstra SK, Bouwmeester PS, van der Linden AJ (1990) Perichondral grafting for cartilage lesions of the knee. *J Bone Joint Surg (Br)* 72: 1003–1007

13   Nehrer S, Spector M, Minas T (1999) Histologic analysis of tissue after failed cartilage repair procedures. *Clin Orthop*: 149–162

14   Shortkroff S, Barone L, Hsu HP, Wrenn C, Gagne T, Chi T, Breinan H, Minas T, Sledge CB, Tubo R, Spector M (1996) Healing of chondral and osteochondral defects in a canine model: the role of cultured chondrocytes in regeneration of articular cartilage. *Biomaterials* 17: 147–154

15   Hunziker EB (1999) Biologic repair of articular cartilage. Defect models in experimental animals and matrix requirements. *Clin Orthop*: S135–S146

16   Dell'Accio F, De Bari C, Luyten FP (2001) Molecular markers predictive of the capacity of expanded human articular chondrocytes to form stable cartilage *in vivo*. *Arthritis Rheum* 44: 1608–1619

17   Atsumi T, Miwa Y, Kimata K, IkawaY (1990) A chondrogenic cell line derived from a differentiating culture of AT805 teratocarcinoma cells. *Cell Differ Dev* 30: 109–116

18   Grigoriadis AE, Heersche JN, Aubin JE (1988) Differentiation of muscle, fat, cartilage, and bone from progenitor cells present in a bone-derived clonal cell population: effect of dexamethasone. *J Cell Biol* 106: 2139–2151

19   Lau WF, Tertinegg I, Heersche JN (1993) Effects of retinoic acid on cartilage differentiation in a chondrogenic cell line. *Teratology* 47: 555–563

20   Bernier SM, Desjardins J, Sullivan AK, Goltzman D (1990) Establishment of an osseous cell line from fetal rat calvaria using an immunocytolytic method of cell selection: characterization of the cell line and of derived clones. *J Cell Physiol* 145: 274–285

21   Urist MR (1965) Bone formation by autoinduction. *Science* 150: 893–899

22   Thyberg J, Moskalewski S (1979) Bone formation in cartilage produced by transplanted epiphyseal chondrocytes. *Cell Tissue Res* 204: 77–94

23   Chang SC, Hoang B, Thomas JT, Vukicevic S, Luyten FP, Ryba NJ, Kozak CA, Reddi AH, Moos MJ (1994) Cartilage-derived morphogenetic proteins. New members of the transforming growth factor-beta superfamily predominantly expressed in long bones during human embryonic development. *J Biol Chem* 269: 28227–28234

24   Boyan BD, Swain LD, Schwartz Z, Ramirez V, Carnes DLJ (1992) Epithelial cell lines that induce bone formation *in vivo* produce alkaline phosphatase-enriched matrix vesicles in culture. *Clin Orthop* 277: 266–276

25   Van Noorden CJ, Jonges GN, Vogels IM, Hoeben KA, Van Urk B, Everts V (1995)

Ectopic mineralized cartilage formation in human undifferentiated pancreatic adenocarcinoma explants grown in nude mice. *Calcif Tissue Int* 56: 145–153

26 Ostrowski K, Wlodarski K, Aden D (1975) Heterotopic chondrogenesis and osteogenesis induced by transformed cells: use of nude mice as a model system. *Somatic Cell Genet* 1: 391–395

27 Boskey AL, Doty SB, Stiner D, Binderman I (1996) Viable cells are a requirement for *in vitro* cartilage calcification. *Calcif Tissue Int* 58: 177–185

28 Matsumoto T, Iwasaki K, Sugihara H (1994) Effects of radiation on chondrocytes in culture. *Bone* 15: 97–100

29 Cornelissen M, Thierens H, de Ridder L (1993) Effects of ionizing radiation on the size distribution of proteoglycan aggregates synthesized by chondrocytes in agarose. *Scanning Microsc* 7: 1263–1267

30 Luyten FP, Yu YM, Yanagishita M, Vukicevic S, Hammonds RG, Reddi AH (1992) Natural bovine osteogenin and recombinant human bone morphogenetic protein-2B are equipotent in the maintenance of proteoglycans in bovine articular cartilage explant cultures. *J Biol Chem* 267: 3691–3695

31 Schmidt MB, Mow VC, Chun LE, Eyre DR (1990) Effects of proteoglycan extraction on the tensile behavior of articular cartilage. *J Orthop Res* 8: 353–363

32 Dell'Accio F, De Bari C, Luyten FP (1999) Molecular basis of joint development. *Jap J Rheumatol* 9: 17–29

33 Oh SP, Seki T, Goss KA, Imamura T, Yi Y, Donahoe PK, Li L, Miyazono K, Ten Dijke P, Kim S, Li E (2000) Activin receptor-like kinase 1 modulates transforming growth factor-beta 1 signaling in the regulation of angiogenesis. *Proc Natl Acad Sci USA* 97: 2626–2631

34 Luyten FP, Lories R, De Valck D, De Bari C, Dell'Accio F (2002) Bone morphogenetic proteins and the synovial joints. In: S Vukicevic, KT Sampath (eds): *Bone morphogenetic proteins*. Birkhäuser, Basel, 223–248

Osteogenic protein-1 promotes proteoglycan synthesis and inhibits cartilage degeneration mediated by fibronectin-fragments

Holger E. Koepp[1,3], Johannes Flechtenmacher[2,3], Klaus Huch[1,3], Eugene J.-M.A. Thonar[3,4], Gene A. Homandberg[3] and Klaus E. Kuettner[3,5]

[1]Department of Orthopedic Surgery, University of Ulm, D-89081 Ulm, Germany; [2]Orthopaedic Clinic, D-76133 Karlsruhe, Germany; [3]Department of Biochemistry, Rush University, Rush-Presbyterian-St. Luke's Medical Center, Chicago, IL 60612, USA; [4]Department of Internal Medicine, Rush University, Rush-Presbyterian-St. Luke's Medical Center, Chicago, IL 60612, USA; [5]Department of Orthopedics, Rush University, Rush-Presbyterian-St. Luke's Medical Center, Chicago, IL 60612, USA

Introduction

Osteoarthritis (OA) is characterized by the failure of anabolic processes to keep up with an acceleration in catabolic processes. The role that inflammation, trauma, instability and other factors play in the initiation and progression of OA are still poorly understood [1, 2]. A possible approach to slow down the degeneration of the articular surface might be to administer locally or systemically anabolic factors that promote matrix synthesis and repair [3]. Such factors include osteogenic proteins and bone morphogenic proteins (OPs/BMPs) that have been shown capable of inducing cartilage and bone formation *in vivo* [4, 5].

OP-1 and normal cartilage

Our group has demonstrated that OP-1/BMP-7 is most effective in stimulating proteoglycan (PG) synthesis by normal human articular chondrocytes of different ages, provided by our collaborators at the Regional Organ Bank of Illinois (ROBI). In this study, chondrocytes from grossly normal human cartilage were isolated, enclosed in alginate, and cultured for 3 days in the presence or absence of OP-1. While the addition of OP-1 did not have an effect on chondrocyte proliferation, it stimulated PG synthesis by chondrocytes from newborn cartilage more than threefold as compared to controls, while in adult cartilage, the increase was up to 160% [6].

OP-1 and degenerated cartilage

In degenerated cartilage, as well as in cartilage showing signs of end stage OA, both obtained through ROBI, the cells were much less responsive to OP-1-induced stimulation of PG synthesis. Much higher doses were needed and, even then, the increase in PG synthesis was much less pronounced. In two out of nine samples of damaged or osteoarthritic cartilage, OP-1 actually caused further depletion of PGs from the tissue.

OP-1 and IL-1β

OP-1 was also able to counteract the downregulation of PG synthesis caused by IL-1β at up to 10 pg/ml. This effect did not seem to be PG-E_2 related [7].

OP-1 and fibronectin fragments (Fn-fs)

It has been shown that addition of Fn-fs to cultured articular cartilage leads to cartilage degradation *via* release of matrix metalloproteinases and other catabolic cytokines as well as to a suppression of PG synthesis. Therefore, we used the fibronectin fragment model as a model for experimental OA [8, 9]. After damaging bovine cartilage with Fn-fs for 7 days, the addition of OP-1 restored the PG content back to normal by day 14 and reached above normal levels by day 21 (Fig. 1). When cultured simultaneously in the presence of Fn-fs and OP-1, the explants did not show any net PG depletion [10]. Thus, OP-1 was able to compensate for the extensive PG degradation caused by the Fn-fs. Since Fn-fs elevate cytokine release, this observation is consistent with the ability discussed above, for OP-1 to also compensate for cytokine mediated PG degradation.

Discussion

OP-1 is a potent stimulator of proteoglycan synthesis in non-osteoarthritic human cartilage. We were also able to demonstrate that OP-1 overcame the effect of IL-1β and also blocked the Fn-f mediated depletion of PG and is thus able to promote reparative responses. We conclude that OP-1 may not only be a supportive therapeutic agent in degenerative joint disease, but also may compensate for the effect of deteriorating agents. Future studies should determine if it is safe to administer OP-1 *in vivo*, how best to limit its effects to specific sites or tissues, what concentrations should be used, and what modes of delivery should be developed [11, 12].

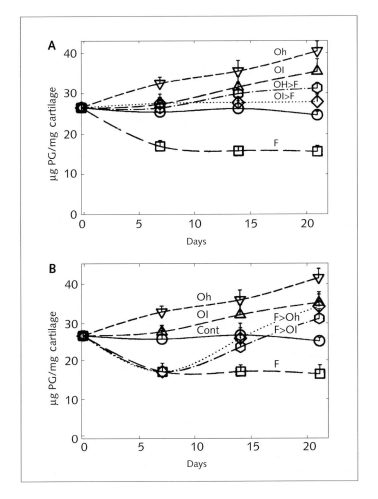

Figure 1
Effect of OP-1 on blocking Fn-f mediated decreases in cartilage PG content (A) and on promoting restoration of PG in Fn-f treated cartilage (B). For A, bovine articular cartilage cultures in 10% serum/DMEM were adjusted to 100 nM Fn-f (□) alone, or 50 ng/ml OP-1 alone (△) or to 200 ng/ml OP-1 alone (▽) or to 50 ng/ml OP-1 with 100 nM Fn-f (◇) or to 200 ng/ml OP-1 with 100 nM Fn-f (◇). Control with nothing added except 10% serum is shown (○). Media were changed every other day with fresh reagents added. The cartilage was continually incubated with these compounds. Cartilage was asseyed for total PG using the DMB assay. For B, cartilage cultures were prepared in 10% serum/DMEM and adjusted to 100 nM Fn-f for day 0–7 only (□) or 50 ng/ml OP-1 alone (△) or to 200 ng/ml OP-1 alone (▽) or to 50ng/ml OP-1 with 100 nM Fn-f (◇) or to 200 ng/ml OP-1 with 100nM Fn-f (◇). Control with nothing added except 10% serum is shown (○). Media were changed every other day with fresh Fn-f added. Cartilage was assayed for total M using the DMB assay. From [10].

References

1 Bayliss MT (1992) Metabolism of animal and human osteoarthritic cartilage. In: KE Kuettner, R Schleyerbach, JG Peyron, VC Hascall (eds): *Articular cartilage and osteoarthritis*. Raven Press, New York, 487–499

2 Luyten FP, Hascall VC, Nissley SP, Morales TI, Reddi AH (1988) Insulin-like growth factors maintain steady state metabolism of proteoglycans in bovine articular cartilage explants. *Arch Biochem Biophys* 267: 416–415

3 Lietman SA, Yanagishita M, Sampath TK, Reddi AH (1997) Stimulation of proteoglycan synthesis in explants of porcine articular cartilage by recombinant osteogenic protein-1 (bone morphogenetic protein-7). *J Bone Joint Surg* 79-A, 8: 1132–1137

4 Urist MR (1965) Bone: formation by autoinduction. *Science* 150: 893–899

5 Reddi AH, Cunningham NS (1993) Initiation and promotion of bone differentiation by bone morphogenic proteins. *J Bone Min Res* 8 (Suppl. 2): S499–S502

6 Flechtenmacher J, Huch K, Thonar EJ-MA, Mollenhauer JA, Davies SR, Schmid TM, Puhl W, Sampath TK, Aydelotte MB, Kuettner KE (1996) Recombinant human osteogenic protein 1 is a potent stimulator of the synthesis of cartilage proteoglycans and collagens by human articular chondrocytes. *Arthritis Rheum* 39: 1896–1904

7 Huch K, Wilbrink B, Flechtenmacher J, Koepp HE, Aydelotte MB, Sampath TK, Kuettner KE, Mollenhauer J, Thonar EJ-MA (1997) Effects of recombinat human osteogenic protein 1 on the production of proteoglycan, prostaglandin E_2, and interleukin-1 receptor antagonist by human articular chondrocytes cultured in the presence of interleukin-1β. *Arthritis Rheum* 40: 2157–2161

8 Bewsey KE, Wen C, Purple C, Homandberg GA (1996) Fibronectin fragments induce the expression of stromelysin-1 mRNA and protein in bovine chondrocytes in monolayer culture. *Biochim Biophys Acta* 1317 (1): 55–64

9 Homandberg GA, Hui F, Wen C, Purple C, Bewsey HK, Koepp H, Huch K, Harris A (1997) Fibronectin-fragment induced cartilage chondrolysis is associated with release of catabolic cytokines. *Biochem J* 321 (pt 3): 751–757

10 Koepp HE, Sampath TK, Kuettner KE, Homandberg GA (1999) Osteogenic protein-1 (OP-1) blocks cartilage damage caused by fibronectin fragments and promotes repair by enhancing proteoglycan synthesis. *Inflamm Res* 48: 199–204

11 Groeneveld EH, Burger EH (2000) Bone morphogenetic proteins in human bone regeneration. *Eur J Endocrinol* 142 (1): 9–12

12 Wozney JM, Rosen V (1998) Bone morphogenetic protein and bone morphogenetic protein gene family in bone formation and repair. *Clin Orthop* 346: 26–37

Osteogenic protein-1 and its receptors in human articular cartilage

Susan Chubinskaya[1,2], David C. Rueger[4], Richard A. Berger[3] and Klaus E. Kuettner[1,3]

Departments of [1]Biochemistry, [3]Orthopedic Surgery, and [2]Section of Rheumatology, Rush Medical College at Rush-Presbyterian-St. Luke's Medical Center, 1653 W. Congress Parkway, Chicago, IL 60612, USA, [4]Stryker Biotech, Hopkinton, MA 01748, USA

Cartilage regeneration and repair is one of the major obstacles to treat arthritic diseases. The members of the bone morphogenetic protein (BMP) family of the transforming growth factor-β (TGFβ) superfamily received a high degree of attention among factors potentially capable of inducing and promoting anabolic processes in articular cartilage [1, 2]. BMPs were first isolated from demineralized bone matrix and shown to induce ectopic bone formation when implanted subcutaneously [3, 4]. The BMP family currently consists of more than 47 related molecules identified in different species [5, 6]. Although BMPs were initially found in bone matrix, it has been shown that they are present in most tissues and organs. They are involved in numerous physiological and pathological processes: differentiation, embryogenesis, morphogenesis, skeletal patterning, development, tissue regeneration, organogenesis, apoptosis, etc.

Our research interests are focused on the seventh member of this family, osteogenic protein-1 (OP-1 or BMP-7). OP-1 as well as BMP-2 and 4 were purified from bovine bone matrix. Recombinant OP-1 was cloned from human placental and brain cDNA libraries [10]. OP-1 exhibits the highest homology to BMP-6 (88%) and BMP-5 (87%) and a lesser homology to BMP-2 (60%) and BMP-4 (58%) [11,12]. OP-1 expression was found in numerous embryonic and adult tissues and organs, however we were the first to identify endogenous OP-1 in human adult articular cartilage [13].

Kidney is the major source of OP-1 protein in body fluids. It is the organ in which the expression and role of OP-1 were studied most extensively [14–16]. OP-1 was also detected in muscle, heart, brain, placenta, bone, ligament, osteo- and chondrosarcomas. Importantly, recombinant OP-1 possesses unique anabolic activity in a variety of biological systems and tissues. OP-1 improves renal function through the inhibition of fibrosis and inflammation by down-regulation of NF-κB and ICAM-1 [17]; induces dendritic growth and promotes recovery of the motor function in patients after stroke; initiates the differentiation of non-committed neural progenitor cells into functional neurons [18]; improves the function of cardiovascular and gastrointestinal systems, and promotes cartilage and bone regeneration and repair.

For the last 8 years we studied extensively in our department the effect of exogenous OP-1 on the metabolism of cartilage extracellular matrix (ECM) molecules and found that human recombinant OP-1 induces the synthesis of major components of cartilage ECM: collagen type II, aggrecan, small proteoglycans, hyaluronan, its receptor CD44 and hyaluronan synthase-2 [7–9]. This effect was evident in explants of human and bovine newborn, fetal and adult cartilage or on isolated chondrocytes cultured in alginate beads. Also, OP-1 was able not only to resist the catabolic processes caused by different catabolic mediators (IL-1, fibronectin [Fn-f] and collagen fragments), but more importantly to overcome their deleterious effects [8, 19, 20]. Flechtenmacher et al. [7] showed under the same experimental system that OP-1 is a more potent stimulator of the synthesis of cartilage-specific collagens and proteoglycans (PGs) than are anabolic mediators such as TGFβ1, activin, or fetal bovine serum (FBS). Among these factors, OP-1 was the only one able to stimulate chondrocytes beyond the levels of synthesis obtained in the presence of 10% FBS. The other factors demonstrated stimulation beyond control levels only in serum-free medium. When the effect of OP-1 on Fn-f-challenged cartilage was compared to that of TGFβ, it was found that TGFβ was not able to block Fn-f mediated PG depletion, but by itself promoted a decrease in cartilage PG content [19]. Noteworthy, our previous [7] and current data (Fig. 1) indicate that OP-1 does not induce proliferation of human and bovine adult articular chondrocytes, while TGFβ and BMP-2 have been shown to induce chondrocyte proliferation [21]. Although BMP-2 and BMP-4 stimulate the synthesis of aggrecan and small PGs in rabbit, murine, bovine newborn and adult cartilage [21–23] and TGFβ could counteract the catabolic effect of IL-1 [24], to our knowledge there are no data on endogenous expression of any of these growth factors in human adult articular cartilage. Furthermore, new data from our laboratory (Fig. 2) indicate that not only normal but also chondrocytes from osteoarthritic cartilage (OA) show an anabolic response to recombinant OP-1. Taken together these findings prompted us to focus on the characterization of endogenous OP-1 in normal and diseased human adult articular cartilage.

We identified the endogenous expression of OP-1 and its signaling molecules, type II and type I (ALK-2, 3, and 6) BMP receptors, in adult human normal, degenerative and OA cartilage [13, 25]. Normal cartilages and cartilages with signs of degeneration were obtained within 24 h of death from organ donors with no documented history of joint disease through the collaboration between the Department of Biochemistry at Rush Medical College (Chicago, IL) and Regional Organ Bank of Illinois (Chicago, IL). OA cartilages were obtained from patients who underwent knee arthroplasty due to diagnosed OA through the collaboration with the Department of Orthopedic Surgery at Rush Medical College. Expression of OP-1 and its receptors in adult articular cartilage was confirmed in mRNA and protein levels. By

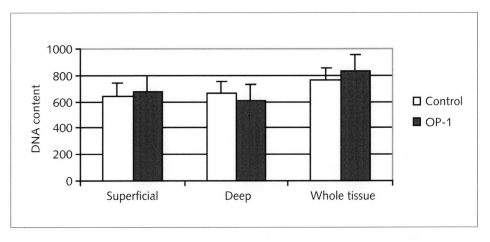

Figure 1
The absence of chondrocyte proliferation is shown by an insignificant change in DNA content between control and OP-1 treatment (p > 0.05). Chondrocytes (normal adult ankle) were isolated from the superficial or deep layers of cartilage or from the entire cartilage thickness and cultured for 4 days, 20,000 cells/bead, either in the presence of the serum-free medium (supplemented with ITS) or OP-1 (100 ng/ml).

in situ hybridization we localized the message for OP-1 and its receptors in normal (newborn and adult) and in OA cartilage. In normal adult cartilage, OP-1 message is primarily present in the superficial and upper middle layers of the tissue, while in degenerative and OA cartilage, OP-1 mRNA expression is detected throughout the entire tissue section. The message for BMP receptors is evident in chondrocytes from all layers of normal, degenerative and OA cartilage samples.

Very intriguing data on mRNA expression of OP-1 and its receptors were obtained by semi-quantitative RT-PCR. OP-1 expression was decreased with aging of normal cartilage as well as with the progression of cartilage degeneration. However, in OA tissue, OP-1 mRNA levels were higher than in the most degenerative donor cartilages ($p < 0.02$). Elevated mRNA expression of OP-1 in advanced OA cartilage could be due to a higher metabolic activity of OA chondrocytes, especially those that are encapsulated in the cell clusters (brood capsules) and, perhaps, undergo cell division. While mRNA expression of endogenous OP-1 was decreased with cartilage degeneration, mRNA expression of OP-1 receptors in most cases was increased. Type I receptors, ALK-2 and ALK-3, and type II receptor were 2–3-fold up-regulated in degenerative and OA cartilage as compared to normal ($p < 0.05$). However, surprisingly, expression of ALK-6 (type I receptor to which OP-1 shows the highest affinity) did not display any statistical changes between cartilage sam-

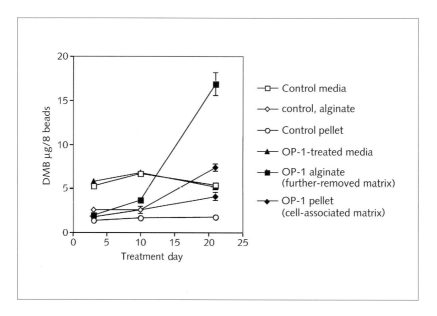

Figure 2
OP-1 induces the synthesis of PGs in OA chondrocytes primarily in the further-removed matrix (closed squares), some accumulation of PGs is seen in the cell-associated matrix (closed rhombs). Proteoglycan content detected by DMB assay in OA chondrocytes cultured in alginate beads for 21 days in serum-free culture medium supplemented with ITS or in the presence of OP-1 (100 ng/ml) added every other day of culture. Alginate, further-removed matrix; pellet, cell-associated matrix.

ples of different histomorphological appearance suggesting that a signaling through this particular receptor is not a primary anabolic event in the matrix synthesis in degenerative cartilage. Hence, in degenerative and OA cartilage, expression of endogenous anabolic factor OP-1 appears to be down-regulated, while the expression of most of its receptors seems to be up-regulated. Such reverse correlation might represent one of the mechanisms regulating the anabolic response of OA chondrocytes to the recombinant OP-1.

By studying endogenous OP-1 protein we found that in human articular cartilage OP-1 could be detected in two forms: pro-form (unprocessed, inactive) and mature (processed, active). These two forms are localized in an inverted manner in human normal and OA cartilage (Fig. 3). Mature, active form of OP-1 was primarily identified in the chondrocytes from the superficial layer, some mature OP-1 was also detected in the interterritorial matrix. Pro-OP-1 was predominantly observed intracellularly in the middle and deep layer chondrocytes (Fig. 3). No matrix localization of pro-OP-1 was found. In addition, some degradation products of the

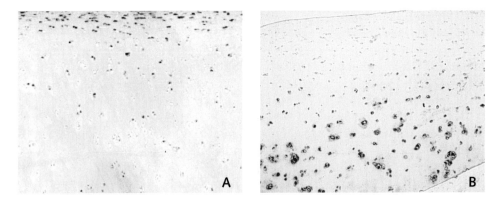

Figure 3
Immunohistochemistry on human normal adult articular cartilage obtained from the femoral condyle of organ donors with anti-OP-1 antibodies. A, tissue section stained with anti-mature OP-1 antibody; B, section stained with anti-pro-OP-1 antibody.

mature domain were extracted from OA and degenerated cartilage, and detected by Western Blotting. The presence of both pro- and mature OP-1 in human articular cartilage was also confirmed by Western blotting. In order to estimate quantitatively the concentration of endogenous OP-1 protein in cartilage and body fluids, we developed a novel sandwich ELISA method in our laboratory. This assay allows the detection of pg quantities of OP-1 in cartilage extracts, synovial fluid and conditioned media. By this method, we detected that the concentration of endogenous OP-1 protein, extractable by 1 M guanidine HCl buffer from human normal adult articular cartilage, is varied from pg to ng quantities per g of tissue dry weight. Comparing human donor cartilages of different ages and different histomorphological grades, an age-dependent and grade-dependent decrease in endogenous OP-1 message ($p<0.001$ and $p<0.02$) and protein ($p<0.02$) was found. This decrease was significant even with the aging of normal cartilages ($p<0.02$). Western Blot analysis of the same cartilage samples with anti-pro- and anti-mature OP-1 antibodies showed the reduction in the intensity of all immunoreactive bands specific for OP-1. However, with age and disease the major qualitative changes were detected in the bands that correspond to mature and partially processed forms of OP-1. Specifically, in aging normal cartilages, the intensity of the band representing mature (active) OP-1 was dramatically reduced; in extracts from degenerated and OA cartilage, this band was absent. Thus, in older and degenerative cartilages there is not only an overall decrease in the quantities of endogenous OP-1, but, critically, a lack of active OP-1, the form that through the binding to the receptors could induce the repair of cartilage ECM components. The findings of the endogenous expression of

OP-1 by articular chondrocytes suggest that cartilage has the potential to repair. However, the deficiency in the active form of OP-1 might indicate at least one of the reasons of why articular cartilage has the limited regenerative capacity. The unique role of this BMP in cartilage tissue protection and regeneration was supported by recent data [26] where over-expression of OP-1 in mice led to the increased synthesis of collagen type II and proteoglycans.

We also detected endogenous OP-1 mRNA and protein in tendon, ligament, synovium, and meniscus. OP-1 protein was identified in synovial fluid from the knee joint of normal donors and patients diagnosed with OA and rheumatoid arthritis in the same forms (pro- and mature) as in cartilage. The content of total OP-1 in synovial fluid is much higher than in cartilage extracts, and there are statistical differences in the concentration of OP-1 in synovial fluid between normal and diseased samples.

To understand the response of endogenous OP-1 to early degenerative changes, different models were used. *In vivo* treatment of rabbit joint cartilage with chymopapain leads to PG depletion, formation of chondrocyte clusters and cartilage fibrillation. The immunostaining of rabbit cartilage with antibodies to OP-1 and its receptors revealed an up-regulation of mature OP-1 protein as a temporary protective response to the degenerative changes caused by the chymopapain injection. No such changes were noticed in the pro-OP-1 and BMP receptors.

In a second *ex vivo* model, the response of endogenous OP-1 was tested to low (0.1 ng/ml) and high (1.0 ng/ml) doses of the catabolic mediator, interleukin-1β (IL-1β). IL-1β was selected due to a widely held view that IL-1, acting in an autocrine or paracrine fashion, contributes to the pathogenesis of OA, specifically during inflammatory episodes. IL-1 may disturb tissue homeostasis by suppressing the synthesis of the ECM molecules and thus the repair mechanisms of cartilage. Since exogenous OP-1 can overcome the deleterious effects of IL-1β [8], it was of interest to find the response of the endogenous repair factor OP-1 to the catabolic events, especially at the early stages of degeneration. As we found, low concentration of IL-1β (0.1 ng/ml) in a short-term culture (48 h) led to a statistically significant up-regulation (two to three-fold) of total OP-1 protein ($p < 0.02$) as compared to cultured and uncultured controls. On the level of message, only a trend of the increased expression (approximately two-fold) was observed in cartilage cultures treated with a low dose of IL-1β. In cartilages treated with a high dose of IL-1β (1 ng/ml; 48 h), OP-1 mRNA expression did not vary statistically between control and experimental groups. However on the protein level, there was a statistically significant decrease (1.5 fold) in total OP-1 under the influence of a high dose of IL-1β when compared to cultured and uncultured controls ($p < 0.001$). No differences in endogenous OP-1 mRNA and protein expression were found between uncultured and cultured controls, since in agreement with our recent findings, OP-1 maintains its near steady-state for at least 25 days of culture. Very low levels of OP-1 were detected in the media in the presence of both doses of IL-1β. Together, the results obtained from IL-1β culture experiments indicate an immediate anabolic response of endogenous OP-1 (both message and protein) only in the presence of low (0.1 ng/ml) doses of IL-1β. Short exposure and low concentrations of IL-1 turn off the synthesis of PGs but do not lead to the depletion of human cartilage ECM *via* induction of matrix metalloproteinases (MMP). High doses of IL-1β caused a decrease in endogenous OP-1 suggesting an elevation of catabolic events such as matrix degradation and up-regulation of MMPs under these conditions.

Collectively, the data generated from the experimental approaches discussed in this chapter indicate that adult articular chondrocytes have the pathways that might control or initiate their anabolic response to degenerative processes either induced or occurring naturally. However, under stronger destructive influences cartilage repair mechanisms may be overwhelmed. Further studies are needed to elucidate the role of endogenous OP-1 in adult articular cartilage and to identify factors that could induce or activate this anabolic factor thus protecting cartilage from degradation and promoting its repair.

Acknowledgments

We would like to thank Drs. Richard Loeser, Carol Muehleman and Stefan Soeder for very fruitful collaborations; Dr. Allan Valdellon from the Regional Organ Bank of Illinois and his staff for making human tissue available for our research; Dr. Arkady Margulis for procurement of human cartilage tissues and Mrs. Bhavna Kumar, Anu Hakimiyan, Katherine Heretis and Rita Mikhail for excellent technical assistance. We would also like to acknowledge Mrs. Charis Merrihew, Ph.D. candidate, for sharing some of the data from her Ph.D. Thesis. This work is supported by 2-AP-39239-11 NIH/NIAMS SCOR Grant and a Grant from Stryker Biotech KK-001.

References

1 Wozney JM, Rosen V, Celeste AJ, Mitsock LM, Whitters MJ, Kriz RW, Hewick RM, Wang EA (1988) Novel regulators of bone formation: molecular clones and activities. *Science* 242: 1528–1534

2 Hogan BLM (1996) Bone morphogenetic proteins: multifunctional regulators of vertebrate development. *Genes Dev* 10: 1580–1594

3 Urist MR, Mikulski A, Lietze A (1979) Solubilized and insolubilized bone morphogenetic protein. *Proc Natl Acad Sci USA* 76: 1828–1832

4 Sampath TK, Reddi AH (1981) Dissociative extraction and reconstitution of extracellular matrix components involved in local bone differentiation. *Proc Natl Acad Sci USA* 78: 7599–7603

5   Urist MR (1997) Bone morphogenetic protein: the molecularization of skeletal system development. *J Bone Miner Res* 12(3): 343–346
6   Reddi AH (1998) Role of morphogenetic proteins in skeletal tissue engineering and regeneration. *Nature Biotechnol* 16: 247–252
7   Flechtenmacher J, Huch K, Thonar EJ-MA, Mollenhauer JA, Davies SR, Schmid TM, Puhl W, Sampath TK, Aydelotte MB, Kuettner KE (1996) Recombinant human osteogenic protein 1 is a potent stimulator of the synthesis of cartilage proteoglycans and collagens by human articular chondrocytes. *Arthritis Rheum* 39: 1896–1904
8   Huch K, Wilbrink B, Flechtenmacher J, Koepp HE, Aydelotte MB, Sampath TK, Kuettner KE, Mollenhauer JA, Thonar EJ-MA (1997) Effects of recombinant human osteogenic protein 1 on the production of proteoglycan, prostaglandin E2, and interleukin-1 receptor antagonist by human articular chondrocytes cultured in the presence of interleukin-1β. *Arthritis Rheum* 40: 2157–2161
9   Nishida Y, Knudson CB, Eger W, Kuettner KE, Knudson W (2000) Osteogenic protein-1 stimulates cell-associated matrix assembly by normal human articular chondrocytes: upregulation of hyaluronan synthase, CD 44 and aggrecan. *Arthritis Rheum* 43: 206–214
10 Özkaynak E, Rueger DC, Drier EA, Corbett C, Ridge RJ, Sampath TK, Oppermann H (1990) OP-1 cDNA encodes an osteogenic protein in the TGF-β family. *EMBO J* 9: 2085–2093
11 Sampath TK, Rueger DC (1994) Structure, function, and orthopedic applications of osteogenic protein (OP-1). *Complications in Orthopedics*, Winter, 101–107
12 Cook SD, Rueger DC (1996) Osteogenic protein-1. Biology and applications. *Clin Orthop Rel Res* 324: 29–38
13 Chubinskaya S, Merrihew C, Cs-Szabo G, Mollenhauer J, McCartney J, Rueger DL, Kuettner KE (2000) Human articular chondrocytes express osteogenic protein-1. *J Histochem Cytochem* 48(2): 239–250
14 Reddi AH (2000) Bone morphogenetic proteins and skeletal development: the kidney-bone connection. *Pediatr Nephrol* 14: 598–601
15 Vukicevic S, Kopp JB, Luyten FP, Sampath TK (1996) Induction of nephrogenic mesenchyme by osteogenic protein 1 (bone morphogenetic protein 7) *Proc Natl Acad Sci USA* 9333 (17): 9021–9026
16 Godin RE, Takaesu NT, Robertson EJ, Dudley AT (1998) Regulation of BMP7 expression during kidney development. *Development* (United Kingdom) 125 (17): 3473–3482
17 Hruska KA (2000) BMP-7 prevents tubulointerstitial nephritis associated with chronic renal disease. International Conference on BMPs, June 7–11, Lake Tahoe, CA: 135
18 Kaplan PL (2000) BMPs – brain morphogenetic proteins – potential therapeutic utility in treating neurological damage and disease. International Conference on BMPs, June 7–11, Lake Tahoe, CA: 138
19 Koepp HE, Sampath KT, Kuettner KE, Homandberg GA (1999) Osteogenic protein-1 (OP-1) blocks cartilage damage caused by fibronectin fragments and promotes repair by enhancing proteoglycan synthesis. *Inflamm Res* 47: 1–6

20 Jennings L, Madsen L, Mollenhauer J (2001) The effect of collagen fragments on the extracellular matrix metabolism of bovine and human chondrocytes. *Connective Tissue Res* 42 (1): 71–86
21 Van Beuningen HM, Glansbeek HL, van der Kraan PM, van den Berg WB (1998) Differential effects of local application of BMP-2 or TGF-β on both articular cartilage composition and osteophyte formation. *Osteoarthritis and Cartilage* 6: 306–317
22 Glansbeek HL, van Beuningen HM, Vitters EL, van der Kraan PM, van den Berg WB (1998) Stimulation of articular repair in established arthritis by local administration of transforming growth factor b into murine knee joints. *Lab Invest* 78 (2): 133–142
23 Sailor LZ, Hewick RM, Morris EA (1996) Recombinant human bone morphogenetic protein-2 maintains the articular chondrocyte phenotype in long-term culture. *J Ortho Res* 14: 937–945
24 Van Beuningen HM, van der Kraan PM, Arntz OJ, van den Berg WB (1993) Protection from interleukin 1 induced destruction of articular cartilage by transforming growth factor β: studies in anatomically intact cartilage *in vitro* and *in vivo*. *Ann Rheum Dis* 52: 185–191
25 Chubinskaya S, Merrihew C, Mikhail R, ten Dijke P, Rueger D, Kuettner KE (2001) BMP receptors specific for OP-1 are identified in human normal, degenerative and OA cartilage. *Trans ORS* 47: 678
26 Hidaka C, Quitoriano M, Attia E, Hannafin J, Warren R, Crystal R (2000) Increased matrix synthesis and matrix gene expression in chondrocytes over expressing BMP-7. *Trans ORS* 46: 41

The transcription factors L-Sox5 and Sox6 are essential for cartilage formation

Véronique Lefebvre[1], Benoit de Crombrugghe and Richard R. Behringer

Department of Molecular Genetics, The University of Texas M.D. Anderson Cancer Center, Houston, TX 77030, USA; [1]Present address: Department of Biomedical Engineering, Lerner Research Institute, Cleveland Clinic Foundation, Cleveland, OH 44195, USA

Introduction

Cartilages form the primary skeleton of vertebrate embryos, ensure its rapid growth, and provide mandatory templates upon which definitive bone is progressively laid down. Chondrogenesis is effected by a single cell type, the chondrocyte, which is of mesenchymal origin. Chondrocytes sequentially fulfill the different functions of cartilages by undergoing multiple steps of differentiation. Our main interest over recent years has been to identify transcription factors that specifically control the chondrocyte differentiation pathway. We found that L-Sox5, Sox6, and Sox9 were able to bind and activate a cartilage-specific enhancer of collagen type 2 gene (*Col2a1*) and were specifically co-expressed with *Col2a1* during chondrogenesis *in vivo*. These data, supported by the notion that many Sox factors control cell fate determination in various lineages, strongly suggested that L-Sox5, Sox6, and Sox9 act as master chondrogenic transcription factors. This hypothesis was tested and confirmed by assessing the consequences of the null mutation of their genes on mouse embryo development. We review here these studies, with a special emphasis on L-Sox5 and Sox6. Akiyama and collaborators address the role of Sox9 in an accompanying review.

Identification of L-Sox5 and Sox6 as candidate chondrogenic transcription factors

To identify chondrocyte-specific transcriptional mechanisms, several groups, including ours, first delineated cartilage-specific regulatory sequences in the *Col2a1* gene [1–4]. The minimum sequence was a 48-bp enhancer located in the first intron [3]. No consensus binding site for transcription factors was recognized in this enhancer, but four sequences resembling a consensus binding site for HMG box proteins were distributed throughout the enhancer. The mutations of either one (the fourth one was not tested) totally abolished enhancer activity in cartilage of transgenic mouse embryos, indicating that these sites were all needed and likely acted in a cooperative manner [5].

In DNA-binding assays, proteins present only in chondrocytes were forming large complexes with the *Col2a1* enhancer [5]. Each of the four HMG box-like sites contributed to complex formation, a result consistent with functional cooperativity between the sites. Sox9, an Sry-related HMG box protein, became a strong candidate when it was demonstrated that its gene was expressed in chondrocytes and that SOX9 heterozygous mutations caused the severe human skeletal malformation syndrome called campomelic dysplasia [6, 7]. SOX9 indeed was able to bind and activate this enhancer and the *Col2a1* gene *in vitro* and *in vivo* [4, 8]. Sox9 was then shown to be required for chondrogenesis in the mouse by directing the differentiation of mesenchymal precursor cells into prechondrocytes [9].

In addition to Sox9, we had biochemical evidence that the large *Col2a1* protein/enhancer complexes also contained other HMG box proteins [5]. Complementary cDNAs for L-Sox5 and Sox6 were identified by Southwestern screening and DNA-binding supershift experiments using specific antibodies confirmed that L-Sox5 and Sox6 were the additional proteins binding to the *Col2a1* enhancer [10].

L-Sox5 was a new product of the Sox5 gene, longer than the previously reported Sox5 protein, which is made exclusively in testis [10, 11]. L-Sox5 and Sox6 share a high degree of overall identity with each other (67%). Their HMG box domains are more than 90% identical, but only 50% identical to that of Sox9. L-Sox5 and Sox6 are totally unrelated to Sox9 outside the HMG box domain. They harbor a coiled-coil domain that is also more than 90% identical between the two proteins. This domain mediates homodimerization and heterodimerization of the two proteins with each other. Sox9 does not have a dimerization domain. Dimers of L-Sox5 and Sox6 bind DNA to pairs of HMG box sites with a much higher affinity than to single sites. In contrast to Sox9, L-Sox5 and Sox6 have no transcriptional activation domain. They were, nevertheless, shown to cooperate with Sox9 to activate the *Col2a1* enhancer and also the *Col2a1* endogenous gene upon forced expression in transiently transfected cultured cells [10]. The molecular role of L-Sox5 and Sox6 in this transactivation is unknown. Whereas the DNA-binding domains of most other transcription factors contact the major groove of the DNA helix, the HMG box domain contacts the minor groove, and upon binding induces a strong DNA bend. These properties were shown for the LEF-1 HMG box protein to promote interaction between proteins bound to non-adjacent sites on either sides of the LEF-1 binding site and resulted in the formation of functional high-order transcriptional complexes [12]. It is possible that L-Sox5 and Sox6 play similar architectural roles in transcription (Fig. 1).

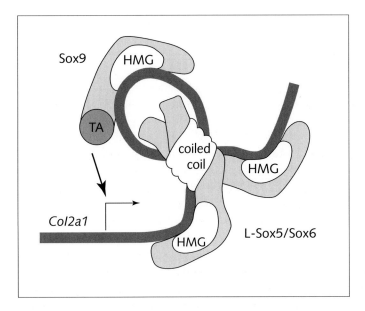

Figure 1
Model for the specific activation of Col2a1 in chondrocytes. A cartilage-specific enhancer located in the Col2a1 first intron binds L-Sox5, Sox6, and Sox9, whose genes specifically co-express with Col2a1 in all developing cartilages. The HMG box DNA-binding domain of the Sox proteins induces a 70–90 degree bend in the DNA helix. Sox9 harbors a potent transactivation domain (TA), whereas L-Sox5 and Sox6 do not. L-Sox5 and Sox6 homo- and heterodimerize through a highly conserved coiled-coil domain. They cooperate with Sox9 in the activation of Col2a1, possibly by contributing an architectural role in the organization of a functional protein/DNA transcriptional complex. Note that this model was constructed from indirect evidence provided by all presently available in vitro *and* in vivo *data, but additional molecular experiments* in vivo *are needed to directly prove or disprove it.*

RNA *in situ* hybridization experiments demonstrated that *Sox5* and *Sox6* were both expressed at high levels in every developing cartilage of mouse embryos [10]. Their expression coincided with that of *Sox9* and *Col2a1*, being turned on in pre-cartilaginous condensations and turned off in prehypertrophic chondrocytes. Altogether, these data were thus consistent with the hypothesis that L-Sox5 and Sox6 play redundant roles in activating *Col2a1* and possibly other genes of the chondrocyte differentiation program, in cooperativity with, but with different roles than Sox9. Sry and a number of other Sox proteins are known for their critical roles in determining cell fate in various lineages [13]. This notion thus added strength to our hypothesis.

L-Sox5 and Sox6 are required for chondroblast differentiation

To determine the roles of L-Sox5 and Sox6 in chondrogenesis *in vivo*, we inactivated their genes in the mouse using conventional techniques of DNA homologous recombination [14]. *Sox5*-null as well as *Sox6*-null mice were born with an overall fairly normal endochondral skeleton. They displayed only a few, mild skeletal abnormalities. $Sox5^{-/-}$ mice had a cleft palate, short ribs, and a short chondrocranium. $Sox6^{-/-}$ mice had a short and bent sternum. In contrast, $Sox5^{-/-}$; $Sox6^{-/-}$ embryos developed a generalized and very severe chondrodysplasia. From embryonic day 13.5 (E13.5), they featured a small and rounded head, and very short limbs, trunk and tail. As a consequence of their underdeveloped skeleton, the tongue was sticking out of a short snout and the abdomen was prominent. They died at E16.5 (3 days before birth), most likely from compression of internal organs within their underdeveloped skeleton. Their phenotype was thus much more severe than the sum of those of single-null mice, proving that *Sox5* and *Sox6* genetically interact in chondrogenesis.

$Sox5^{-/-}$; $Sox6^{-/-}$ embryos were analyzed in depth to define the precise roles of L-Sox5 and Sox6 in chondrogenesis. During the first steps of chondrogenesis, mesenchymal precursor cells are recruited to proper chondrogenic sites and differentiate into prechondrocytes. They form precartilaginous condensations, growth arrest, and activate the first cartilage markers, including *Col2a1* and *Sox9*. Histology and RNA *in situ* hybridization experiments demonstrated that these steps occurred normally in $Sox5^{-/-}$; $Sox6^{-/-}$ embryos. L-Sox5 and Sox6 were thus not needed for the early patterning of cartilages and, in contrast to Sox9 [9], were not needed for prechondrocyte differentiation.

Next, prechondrocytes convert into cells that we like to call chondroblasts instead of chondrocytes or proliferating chondrocytes, as they are often referred to. The "blast" suffix designates in many lineages (osteoblast, myoblast, adipoblast, etc.) cells that form *de novo* tissue by proliferating rapidly and actively expressing tissue-specific genes. Chondroblasts are characterized by these two properties. Their high rate of proliferation contrasts with the growth-arrest status of earlier-step prechondrocytes and later-step prehypertrophic and hypertrophic chondrocytes. At the same time that they proliferate, chondroblasts express a panel of extracellular matrix genes that allow them to build the typical, abundant and dense cartilage matrix. $Sox5^{-/-}$; $Sox6^{-/-}$ embryos revealed that L-Sox5 and Sox6 are crucial determinants of chondroblast differentiation. By E14.5, wild-type embryos have formed a fairly complete cartilage skeleton, and this skeleton intensely stains with alcian blue, a specific dye for aggrecan deposited in the cartilage matrix (Fig. 2A). This skeleton appeared rudimentary in $Sox5^{-/-}$; $Sox6^{-/-}$ embryos and most impressively, hardly stained with alcian blue. On histology sections performed from E12.5 to E16.5, mutant prechondrocytes were seen to mature very slowly to a stage at which cells surrounded themselves with a thin layer of extracellular matrix and resumed proliferation. The matrix, however, poorly retained alcian blue (Fig. 2B). The cartilage matrix deficiency was explained by inability of the $Sox5^{-/-}$; $Sox6^{-/-}$ cells to express extracellular matrix genes at appropriate levels (Fig. 2C). The genes for collagen type 2, aggrecan, and link protein were expressed, but at low levels, as seen in prechondrocytes, never at high levels, as typically seen in chondroblasts. Moreover, the cells failed to significantly express chondroblast-specific matrix markers, such as the genes for matrilin-1 and Comp. The poor development of cartilages was explained in part by this matrix deficiency and in part by the inability of $Sox5^{-/-}$; $Sox6^{-/-}$ cells to resume proliferation in time. L-Sox5 and Sox6 were thus required for chondroblast differentiation.

L-Sox5 and Sox6 are essential to maintain chondroblast pools in developing cartilages

The fate of each chondroblast in developing cartilages is tightly controlled, both temporally and spatially [15]. First, cells in the core region of cartilages switch off expression of *Sox5*, *Sox6*, and *Sox9* and activate new genetic programs to undergo successively prehypertrophic and hypertrophic chondrocyte differentiation and ultimately apoptosis. Prehypertrophic and hypertrophic cells send messages, including Indian hedgehog (Ihh), to the perichondrium to induce the recruitment and differentiation of bone-forming cells. An intramembranous bone collar forms, which, following chondrocyte apoptosis, sends cells into the cartilage to form the primary ossification center. Chondroblasts adjacent to this first ossification center engage in multiple rounds of unidirectional proliferation, forming parallel columns of flattened cells. One layer at a time, cells in these columns then proceed to the late stages of chondrocyte differentiation. The cartilage growth plates that they so form contribute most of the skeleton longitudinal growth. Their maintenance critically depends on the persistence of an epiphyseal pool of chondroblasts, on the rate at which this pool provides new cell layers, and the rate at which chondrocytes further mature and bone cells invade the cartilage.

The *Sox5*; *Sox6* double mutation dramatically impaired growth plate and endochondral bone formation. Despite their inability to undergo chondroblast differentiation, $Sox5^{-/-}$; $Sox6^{-/-}$ cells activated typical markers of prehypertrophic and hypertrophic chondrocytes. However, while genes such as *Ihh*, *Cbfa1*, and *Vegf* were expressed at normal levels, *Col10a1* expression was barely detectable. Moreover, expression of *Col2a1* and *Sox9* was maintained, instead of turned off. The cells growth-arrested, but underwent only slight morphological hypertrophy (Fig. 2D). This further maturation of mutant cells started, as expected, in the core region of cartilages as soon as the cells had accumulated a thin layer of extracellular matrix, and it resulted in the induction of bone collar formation. However, it very rapidly spread to the entire cartilage, preventing growth plate formation and leaving no epiphyseal pool of chondroblasts. By E16.5, when mutant fetuses died, mutant cartilages had formed no primary ossification center, in contrast to most wild-type cartilages, but bone collars were thick and starting to invade the cartilages. L-Sox5 and Sox6 thus play an essential role not only in promoting chondroblast differentiation, but also in preventing premature chondrocyte maturation in epiphyseal cartilages.

Discussion

Our studies have demonstrated that L-Sox5 and Sox6 have essential, mostly redundant roles in the chondrocyte differentiation pathway (Fig. 3). They act as transcriptional enhancers to stimulate chondroblasts in assuming their major differentiation functions, that is, expression of cartilage extracellular matrix genes and cell proliferation.

We initially identified them through their ability to bind to a cartilage-specific enhancer of the *Col2a1* gene. Based on these data and the significant downregulation of *Col2a1* in double mutant embryos, it is postulated that L-Sox5 and Sox6 directly target *Col2a1* in chondrocytes *in vivo*. However, additional experiments are needed to test this hypothesis and also to determine the molecular mechanisms whereby these factors, which have no transactivation domain, contribute to gene activation cooperatively with Sox9. It also remains to be determined whether L-

---

Figure 2
Severe cartilage formation defects in Sox5$^{-/-}$; Sox6$^{-/-}$ embryos. (A) Alcian blue staining of E14.5 control (left) and Sox5$^{-/-}$; Sox6$^{-/-}$ (right) embryos. Cartilages were rudimentary and barely stained in the mutant embryo. (B) Staining with alcian blue and nuclear fast red of histological sections through a tibial growth plate of an E16.5 control embryo (left) and through the proximal half of the tibia of a Sox5$^{-/-}$; Sox6$^{-/-}$ littermate (right). In the control, chondroblasts proliferating in columns are surrounded with abundant extracellular matrix staining intensely with alcian blue (toward the top). They mature one layer at a time into prehypertrophic and hypertrophic chondrocytes (toward the bottom). In the mutant, cells in the epiphyseal ends (toward the top) are still prechondrocytic and prechondrocytes in the core region (toward the bottom) have slowly matured to accumulate a thin layer of extracellular matrix staining weakly with alcian blue. (C) In situ hybridization of cross-sections through a thoracic vertebra of control (left) and Sox5$^{-/-}$; Sox6$^{-/-}$ E15.5 littermates (right). Col2a1 expression is severely downregulated and Comp expression undetectable in the mutant vertebra. (D) Alcian blue staining of a longitudinal section through the humerus of control (left) and Sox5$^{-/-}$; Sox6$^{-/-}$ E16.5 littermates (right). The mutant humerus is rudimentary, with cartilage poorly staining with alcian blue. No growth plates are distinguishable, but the cartilage is being invaded by a thick bone collar.

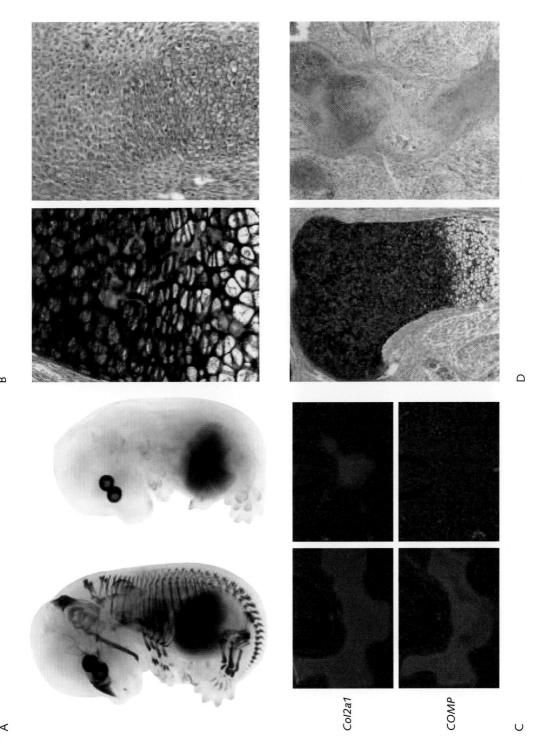

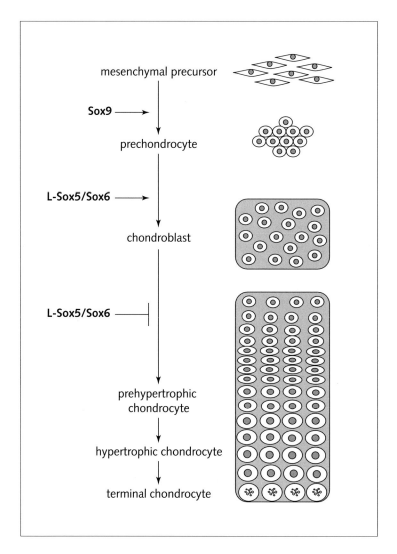

Figure 3
L-Sox5 and Sox6 act in redundancy in the chondrocyte differentiation pathway to promote chondroblast differentiation. See the discussion for details.

Sox5 and Sox6 also directly contribute to the activation of all other cartilage matrix genes, which were downregulated or inactive in mutant embryos, such as those for aggrecan, link protein, matrilin-1 and Comp. The effect of the double mutation of Sox5 and Sox6 on cell proliferation indicates that L-Sox5 and Sox6 may control cell cycle genes in chondroblasts. Their influence on such genes may be either direct or indirect, possibly mediated through their effect on extracellular matrix accumulation.

L-Sox5 and Sox6 were not needed for prechondrocyte differentiation. Sox9 thus controls this early step independently of L-Sox5 and Sox6. L-Sox5 and Sox6 were also not needed for prehypertrophic and hypertrophic chondrocyte differentiation, a result consistent with the inactivation of their genes during prehypertrophy. Although mutant cells activated genes of these late differentiation stages, their maturation was abnormal; they failed to form cartilage growth plate and become morphologically hypertrophic. Similar chondrocyte differentiation defects were described in mice lacking either one of the most abundant cartilage matrix components (collagen type 2, aggrecan, and link protein), thus strongly suggesting that the severe matrix deficiency of *Sox5*; *Sox6* double-null cartilages contributed in a large part for these defects. L-Sox5 and Sox6 were needed to prevent precocious chondrocyte maturation and thereby maintain chondroblast pools in cartilage epiphyses. PTHrP, a cytokine with an essential role in the control of chondrocyte maturation, was expressed at normal levels in mutant cartilages. It is thus possible that L-Sox5 and Sox6 work downstream of PTHrP signaling, as Sox9 does. Additional experiments are needed to test this hypothesis and also answer many other questions that are raised by the discovery of these two major players in endochondral skeleton development.

Acknowledgments

We thank Guang Zhou, Krishnendu Mukhopadhyay, Weimin Bi, Ping Li, Patrick Smits, Zhaoping Zhang, Jian Ming Deng, and Jennifer Mandel for valuable contributions to the work performed in the authors' laboratories and reviewed here. This work was supported by NIH grants AR42919 (to B.d.C. and R.R.B) and AR46249 (to V.L.) and an Arthritis Foundation Investigator Award (to V.L.).

References

1 Krebsbach PH, Nakata K, Bernier SM, Hatano O, Miyashita T, Rhodes CS, Yamada Y (1996) Identification of a minimum enhancer sequence for the type II collagen gene reveals several core sequence motifs in common with the link protein gene. *J Biol Chem* 271: 4298–4303

2 Zhou G, Garofalo S, Mukhopadhyay K, Lefebvre V, Smith CN, Eberspaecher H, de Crombrugghe B (1995) A 182 bp fragment of the mouse pro alpha 1(II) collagen gene is sufficient to direct chondrocyte expression in transgenic mice. *J Cell Sci* 108: 3677–3684

3 Lefebvre V, Zhou G, Mukhopadhyay K, Smith CN, Zhang Z, Eberspaecher H, Zhou X, Sinha S, Maity SN, de Crombrugghe B (1995) An 18-base-pair sequence in the mouse

pro-alpha1(II) collagen gene is sufficient for expression in cartilage and binds nuclear proteins that are selectively expressed in chondrocytes. *Mol Cell Biol* 16: 4512–4523

4   Bell DM, Leung KK, Wheatley SC, Ng LJ, Zhou S, Ling KW, Sham MH, Koopman P, Tam PP, Cheah KS (1997) SOX9 directly regulates the type-II collagen gene. 16: 174–178

5   Zhou G, Lefebvre V, Zhang Z, Eberspaecher H, de Crombrugghe B (1998) Three high mobility group-like sequences within a 48-base pair enhancer of the *Col2a1* gene are required for cartilage-specific expression *in vivo. J Biol Chem* 273: 14989–14997

6   Foster JW, Dominguez-Steglich MA, Guioli S, Kowk G, Weller PA, Stevanovic M, Weissenbach J, Mansour S, Young ID, Goodfellow PN et al (1994) Campomelic dysplasia and autosomal sex reversal caused by mutations in an SRY-related gene. *Nature* 372: 525–530

7   Wagner T, Wirth J, Meyer J, Zabel B, Held M, Zimmer J, Pasantes J, Bricarelli FD, Keutel J, Hustert E et al (1994) Autosomal sex reversal and campomelic dysplasia are caused by mutations in and around the SRY-related gene SOX9. *Cell* 79: 1111–1120

8   Lefebvre V, Huang W, Harley VR, Goodfellow PN, de Crombrugghe B (1997) SOX9 is a potent activator of the chondrocyte-specific enhancer of the pro-alpha1(II) collagen gene. *Mol Cell Biol* 17: 2336–2346

9   Bi W, Deng JM, Zhang Z, Behringer RR, de Crombrugghe B (1999) Sox9 is required for cartilage formation. *Nat Genet* 22: 85–89

10 Lefebvre V, Li P, de Crombrugghe B (1998) A new long form of Sox5 (L-Sox5), Sox6 and Sox9 are coexpressed in chondrogenesis and cooperatively activate the type II collagen gene. *EMBO J* 17: 5718–5733

11 Connor F, Cary PD, Read CM, Preston NS, Driscoll PC, Denny P, Crane-Robinson C, Ashworth A (1994) DNA binding and bending properties of the post-meiotically expressed Sry-related protein Sox-5. *Nucleic Acids Res* 22: 3339–3346

12 Giese K, Kingsley C, Kirshner JR, Grosschedl R. (1995) Assembly and function of a TCR alpha enhancer complex is dependent on LEF-1-induced DNA bending and multiple protein-protein interactions. *Genes Dev* 9: 995–1008

13 Wegner M (1999) From head to toes: the multiple facets of Sox proteins. *Nucleic Acids Res* 27: 1409–1420

14 Smits P, Li P, Mandel J, Zhang Z, Deng JM, Behringer RR, de Crombrugghe B, Lefebvre V (2001) The transcription factors L-Sox5 and Sox6 are essential for cartilage formation. *Dev Cell* 1: 277–290

15 Karsenty G (2001) Chondrogenesis just ain't what it used to be. *J Clin Invest* 107: 405–407

Amelogenin peptides have unique milieu-dependent roles in morphogenic path determination

Arthur Veis[1], Kevin Tompkins[1] and Michel Goldberg[2]

[1]Northwestern University, Chicago, IL, USA; [2]University of Paris V, Montrouge, France

In early studies of the bone morphogenetic proteins from various mineralized tissues, Urist [1, 2] reported that the proteins from dentin matrix had a higher bone inductive activity in implants than did bone matrix, and had different inductive effects than demineralized enamel matrix. We set out to determine the source of the activity in dentin, expecting to find a high concentration of BMP. Our strategy was to identify the active fraction from dentin extracts using an *in vitro* assay related to the presumed first step in ectopic implant bone induction, chondrogenesis. Thus, we developed an *in vitro* system in which embryonic muscle fibroblasts were cultured in monolayers, under conditions not conducive to the formation of chondrogenic nodules (as compared with suspension cultures), and then exposed to the various extract fractions [3]. "Activity" was defined as the enhanced incorporation of ^{35}S-SO$_4$ into cetyl pyridinium chloride-precipitable proteoglycan. A fraction with activity was finally isolated and shown by amino acid composition studies to be unrelated to the BMP family of proteins. Specifically, the active peptides had compositions devoid of Cys residues, a hallmark of the BMP-TGFβ family of proteins. Moreover, the active peptides had molecular masses in the range from 6,000 to 10,000, smaller than any of the BMPs [4]. After extensive fractionation and purification, an active peptide was finally isolated and subjected to amino terminal and internal sequencing following trypsin hydrolysis. To our amazement, the peptide sequence proved to be related to amelogenins, proteins produced by ameloblasts, specialized epithelial cells not thought to be present in dentin, which is produced by mesenchyme-derived odontoblasts [5].

The amelogenins are molecules postulated to have a structural role in the formation of the pre-mineralization enamel matrix, and an important function in enamel mineralization. They are proteolytically degraded during the process of mineralization, such that mature enamel has < 2% amelogenin content. Thus, the enamel and amelogenin experts thought that our data could not be correct with regard to either their biological activity or their dentin source. Nevertheless, we proceeded to examine a rat incisor cDNA library prepared from the pulp-dentin complex, the library from which all of the dentin-specific proteins had first been cloned [6]. Using

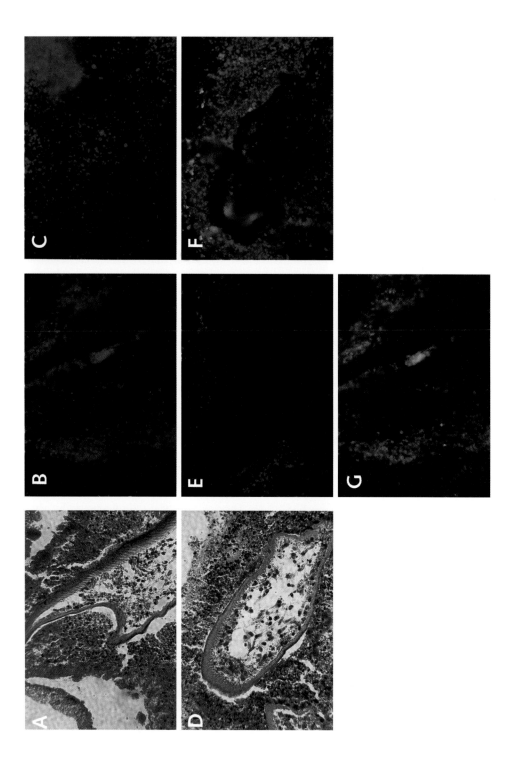

the appropriate PCR primers, it was demonstrated that the cDNA library contained two related cDNAs that were obviously specific splice products of the amelogenin gene. The formula masses of the two peptides, from signal sequence cleavage site to the stop codon, were 6,697 and 8,135 Da, in the correct range, and corresponded to exons 2,3,5,6d,7 and 2,3,4,5,6d,7, respectively, with amino acid compositions:

$$\underset{2\ 3}{\text{MP}}|\underset{}{\text{LPPHPGSPGYINLSYE}}|\underset{3\ 4}{\text{KSHSQAINTDRTAL}}|\underset{4\ 5}{\text{VLTPKWYQSMI}}$$
$$\underset{5\ 6d}{\text{RQP}}|\underset{6d\ 7}{\text{PLSPILPELPLEAWPATDKTKREEV}}|\text{D}$$

The difference is exon 4 (larger font), a charged hydrophilic domain of 14 amino acids. Both recombinant peptides were found to have activity equivalent to recombinant BMP2 in the *in vitro* sulfate incorporation assay, and activity in implants. However, the two peptides yielded somewhat different results, quantitatively and qualitatively [7]. For example, the 8 kDa peptide, designated [A + 4] upregulates SOX 9, whereas [A – 4] upregulates Cbfa1 in embryonic muscle fibroblasts under identical fibroblast culture systems.

In considering the possible role for such peptides in the tooth, we hypothesized that they might have some role in tooth morphogenesis. Lower first molars were collected from CD-1 mice at various embryonic stages and postnatal days, and grown on semisolid, serum-free medium supplemented with ascorbic and retinoic acids and transferrin, or in 20% FBS as a control, according to procedures of Bègue-Kirn et al. [8]. After an initial serum-free conditioning period of 24 h, r[A + 4] and r[A – 4] were added at 10 ng/ml. The tooth germs were cultured for 6 days, then fixed and paraffin embedded by standard procedures. The tissue blocks were serially sectioned and stained with hematoxylin-eosin, or antibodies.

In the supplemented-serum free medium the enamel organ does not synthesize proteins as efficiently as the dental papilla tissue [9] and thus the effects of the peptides are principally the result of direct action on the mesenchyme. Figure 1 shows serum-free-medium-cultured day 1 post-natal mouse molars, ± the peptides, then

---

Figure 1
Analysis of the results of tooth germ culture in the presence of the specific recombinant amelogenin peptides at 10 ng/ml. Culture was for 6 days, at 0% FBS in RPMI-1640, 180 µg/ml ascorbic acid, 2 mM L-glutamine, 1% penicillin/streptomycin, 1.5×10^{-7} M retinoic acid (All-trans, Sigma) and 0.5% agar [8]. Left column, H & E stain; Center column, stain for CAP, Texas Red, Col1, green; Right column, stain for DMP2, green. Upper row, r[A – 4]; Center row, r[A + 4]. Bottom, panel G, superposition of CAP and Col1 stains. Nuclei stained with DAPI (blue).

analyzed for the distribution of DMP2, a dentin matrix protein, CAP, cementum attachment protein, and Col1. Note that the enamel organ is essentially absent under the conditions used. It is evident that [A+4] induces the production of DMP2, whereas [A-4] upregulates CAP. Thus, the two small amelogenin splice product peptides have distinctly different effects on the path of tooth mesenchyme development. We have shown similar differences in implants in muscle, and most recently, in implants into teeth with created dentin defects [10] in which [A+4] induced formation of an intact dentin bridge, whereas [A-4] induced diffuse mineralization throughout the radicular pulp, results reminiscent of Urist's [1] initial study of the difference between crude enamel matrix protein implants and dentin matrix implants.

Thus, the specific amelogenin peptides both have cell signaling effects, but the presence of the exon 4 sequence makes an important difference in the specific effects. Further, since carrier supported implants in ectopic sites can lead to bone formation, it is evident that the milieu has a major role in determination of the morphogenic pathway of the affected cell population. It remains to determine the details of the interactions of [A±4] with their cell receptors and the pathways by which they interact.

References

1. Urist MR (1971) Bone histogenesis and morphogenesis in implants of demineralized enamel and dentin. *J Oral Surg* 29: 88–102
2. Conover MA, Urist MR (1979) Transmembrane bone morphogenesis by implants of dentin matrix. *J Dent Res* 58: 1911
3. Koskinen KP, Kanwar YS, Sires B, Veis A (1985) An electron microscopic demonstration of induction of chondrogenesis in neo-natal rat muscle outgrowth cells in monolayer cultures. *Connective Tiss Res* 14: 141–158
4. Amar S, Sires B, Sabsay B, Clohisy J, Veis A (1991) The isolation and partial characterization of a rat incisor dentin matrix polypeptide with *in vitro* chondrogenic activity. *J Biol Chem* 266: 8609–8618
5. Nebgen DR, Inoue H, Sabsay B, Wei K, Ho C-S, Veis A (1999) Identification of the chondrogenic inducing activity from bovine dentin (bCIA) as a low molecular mass amelogenin polypeptide. *J Dent Res* 78: 1484–1494
6. George A, Sabsay B, Simonian PAL, Veis A (1993) Characterization of a novel dentin matrix acidic protein. Implications for biomineralization. *J Biol Chem* 268: 12624–12630
7. Veis A, Tompkins K, Alvares K, Wei K, Wang L, Wang X-S, Brownell A, Jengh S, Healy KE (2000) Specific amelogenin gene splice products have signaling effects on cells in culture and in implants *in vivo*. *J Biol Chem* 275: 41263–41272
8. Bègue-Kirn C, Smith AJ, Ruch JV, Wozney JM, Purchio A, Hartman D, Lesot H (1992)

Effect of dentin proteins, transforming growth factor β1 (TGFβ1), and bone morphogenetic protein 2 (BMP2) on the differentiation of odontoblasts *in vitro*. *Int J Dev Biol* 36: 491–503

9   Zeichner-David M, Vo H, Tan H. Diekwisch T, Berman B, Thiemann F, Alcocer MD, Hsu P, Wang T, Eyna J et al. (1997) Timing of the expression of enamel gene products during mouse tooth development. *Int J Develop Biol* 41(1): 27–38

10 Goldberg M, Six N, Decup F, Bourd K, Palmier K, Salih E, Veis A, Lasfargues J-J (2002) Minéralisation de la pulpe dentaire: apports de l'ingénierie tissulaire aux thérapeutiques de demain en odontoligie. *Pathol Biol* 50: 194–203

Discussion

Identification and characterization of human cell populations capable of forming stable hyaline cartilage *in vivo*

Q: *Manas Majumdar*: During chondrogenesis you see collagen type II and type X come up. How would you consider type X as a marker for chondrocytes?

A: *Frank Luyten*: I think it's a late maturation marker.

Q: *Manas Majumdar*: Late maturation, in what way? Are the cells hypertrophic or is it going through articular chondrocytes?

A: *Frank Luyten*: It's pretty hypertrophic, but you also see that phenotype in articular cartilage.

Q: *Manas Majumdar*: When you say stem cells, have you done cloning studies with these cells? Do all clones have the same effect?

A: *Frank Luyten*: We cloned individual cells, and assayed the individual clones for the activity. The weakest was the myogenic activity. *In vivo* data with myogenesis and muscle repair experiments were impressive. However *in vitro*, we couldn't get them to work too well on individual clones. So, this tells you, in my opinion, that the *in vitro* myogenesis culture systems are not as sensitive as the *in vivo* systems.

Q: *Joel Block*: What is the prevalence in synovial tissues of these small type potential stem cells relative to bone marrow or periosteum?

A: *Frank Luyten*: We don't really know. From two small synovial biopsies, we isolate about 10,000 cells. Out of these 10,000 cells, maybe a few hundred attach and multiply. So, how many of these cells are truly present in the synovial membrane *in vivo*, we don't really know.

Q: *Joel Block*: Does it change with age?

A: *Frank Luyten*: It doesn't really change too much with age. The number of cells that are attaching may be changing, but if you go to the expansion procedure they take off.

Q: *Klaus Kuettner*: These pieces of cartilage that you form *in vitro* for transplantation have about 10% of the volume as cells, but in adult tissue, about 3–4% of the volume is cells. When you transplant your material into the defects, how do you expect that the percentage of cells will come down to the value in the original tissues?

A: *Frank Luyten*: In the nude mice model, we can go down to about half a million cells and still retrieve a very reasonable cartilage implant. How many cells do you really need and what else do you need in order to have a well-organized tissue is a big question.

Q: *Klaus Kuettner*: I meant the ratio of cells to matrix which is totally different *ex vivo* than *in vivo*. How can we get rid of about 80% of the cells?

A: *Frank Luyten*: I don't know. However, we coated the cells with a BMP-like gene and implanted these. If a BMP-like growth factor is present, the cell to matrix ratio may be closer to the *in vivo* situation. In these cases we had about the matrix to cell ratio that was needed. So, a single morphogen might answer that question. In vivo, the periosteal flap contains cells which are an excellent source of growth factors such as GDF5 and TGF(beta). A good periosteal flap implant may add an amount of growth factors that is sufficient to have a result on the cell to matrix issue, which is reasonable.

Q: *Jürgen Mollenhauer*: I don't like the term "stem cells," because you are talking about mesenchymal precursors. My question is, what is the percentage of cells that you have transformed into chondrocytes at the point of implantation? Most people think that you are transforming all these progenitors. What's the percentage?

A: *Frank Luyten*: Whatever you implant, when you take your serial sections, then a majority of the cells become chondrocytes.

A: *Juergen Mollenhauer*: If you harvest these cells, you don't have 100% chondrocytes. You should be more realistic. Even in micromass cultures, you only have about 40% transformation to cartilage cells.

Osteogenic protein-1 and its receptor in human articular cartilage

Q: *Hari Reddi*: Have you had the opportunity to compare knee cartilages with ankle cartilages, and did you find any patterns about endogenous OP-1 content?

A: *Susan Chubinskaya*: Yes, we have compared knee and ankle cartilage. In work in progress we see higher levels of OP-1 expression in the knee cartilage.

Q: *Hari Reddi*: You have put all your eggs in one basket – that is OP-1 (also known as BMP-7). It is well known that there are other BMPs, including BMP-2 and -4 in articular cartilage, and CDMP-1 (also known as GDF-5). What thoughts do you have about the relative importance of other endogenous BMPs?

A: *Susan Chubinskaya*: To our knowledge, all BMPs that you mentioned (excluding CDMPs) haven't yet been shown to be in the human adult articular cartilage. Also, only BMP-7 does not induce cell proliferation. That is the rationale for us to focus on OP-1. We have not compared expression of the other BMPs with OP-1.

Q: *Manas Majumdar*: OP-1 has to be processed. What is known about this?

A: *Susan Chubinskaya*: For BMP-4, TGFβ, and BMP-2, we know that furin type convertases are involved in their processing. To my knowledge, it hasn't been shown for BMP-7, but I wouldn't be surprised if it would be the same mechanism. It has been shown by others that furin is expressed in cartilage, but I'm not sure whether it has been shown to have differences in distribution in the tissues. Maybe it is related to that. It very well could be that the processing of BMPs is a critical and important step in articular cartilage protection.

Q: *Manas Mujumdar*: You showed that the superficial zone has more mature protein than the deep zone. Do you see more receptors in the superficial zone in relation to the deeper zone?

A: *Susan Chubinskaya*: We do not see such correlation between zonal receptor distribution and mature OP-1 localization. I think that in order to answer this question we have to separate superficial and deep zones. Then maybe we would be able to track receptors at that level. It doesn't seem as striking as for OP-1.

Q: *Hari Reddi asks Linda Sandell*: Have you found BMP-7 in articular cartilage? Or BMP-2 or -4?

A: *Linda Sandell*: We've been looking more at inhibitors. I agree with you, Hari, that by PCR you can find BMPs any place, certainly in almost any cartilage. On the other hand, it is very hard to immunolocalize BMPs. I think that Susan's work showing BMP-7 in the pro and mature forms in cartilage is really the best example of seeing BMPS in cartilage. That is not to say that others are not there. I think almost everybody has this issue with BMPs about whether it's masked, and why we don't really see BMPs outside the cell when we know that they are there. They are just very hard to immunostain.

The transcription factors L-Sox5 and Sox6 are essential for cartilage formation

Q: *Phil Osdoby*: Do you have any information on what happens to Sox expression in the PTHrP knockout animals? Is there a relationship or change in expression?

A: *Véronique Lefebvre*: We haven't looked at that and I don't think anyone with the mice have looked at that. The PTHrP mice don't have a cartilage matrix deficiency, so I think Sox5 and Sox6 must be expressed at pretty normal levels. In Sox5 and Sox6 mice, PTHrP is present at a normal level, but unable to maintain the chondroblasts. PTHrP might act to maintain Sox5 and Sox6 expression.

Q: *Chandrasekhar Srinivasan*: How do you distinguish a chondroblast and a chondrocyte?

A: *Véronique Lefebvre*: I would call a chondrocyte a permanent cell in articular cartilage because it doesn't proliferate much. A chondroblast in a developing cartilage actively proliferates and makes new tissue whereas a chondrocyte just maintains the tissue.

Q: Chandrasekhar Srinivisan: How would you define an articular cartilage cell that proliferates in an arthritic cartilage?

A: *Véronique Lefebvre*: If it resumes proliferation and starts making again higher amounts of cartilage matrix, it likely goes back to a chondroblast phenotype.

Q: *Vince Hascall*: Your knockout for Sox5 seemed to have pretty normal skeletal development, and the knockout for Sox6 seemed to have pretty good development also. Are they compensating by getting together as homodimers?

A: *Véronique Lefebvre*: Yes, our *in vitro* data showed that they can homodimerize as well as heterodimerize. The proteins are so similar that it is not surprising that they act independently of each other. They don't need to heterodimerize.

Q: *Vince Hascall*: Yet one of them was lethal shortly after birth. Does that mean that one of these is operating very importantly in another tissue?

A: *Véronique Lefebvre*: Yes, actually they are both lethal at birth. Sox5 mice have a cleft palate, can't breathe, and die within 1 h after birth. Most Sox6 mice die at birth also as they are unable to breathe. We don't know yet why they die.

Q: *Vince Hascall*: Were there other tissues, such as in the heart that is using these factors also in development? Are these factors operating in tissues other than skeleton that is important for the animal?

A: *Véronique Lefebvre*: For Sox5 we haven't seen abnormalities in any other tissues. We didn't look for other abnormalities in the case of Sox6. Another group has obtained Sox6 mutants by mutagenesis and have shown that some of these mice die at birth and others die in the second week. They have myopathy apparently and die from heart failure. Sox6 is also expressed in other tissues, but we haven't looked for other phenotypes yet.

Q: *Robin Poole*: I was very intrigued about the atypical expression of hypertrophy in the absence of type II collagen. It is interesting to note that in natural development, the hypertrophy occurs when the type II collagen is degraded extensively. We have discovered that if degradation of type II collagen is arrested, you suppress hypertrophy. In osteoarthritis where you have an extensive degradation of type II, that's exactly where you see expression of type X. So the suggestion is that hypertrophic differentiation of chondrocytes is regulated by the presence of type II collagen. Have you attempted to take chondrocytes from the double knockout and plate them in a type II environment, as compared to a type I environment and determine whether or not type II would then suppress and control differentiation of the chondrocyte to the hypertrophic phenotype?

A: *Véronique Lefebvre*: Yes, that is a good idea, but we haven't done any of this type of experiment.

Q: *Ted Oegema*: Did you look at tracheal cartilage? Is it possible that the Sox6 mouse do not have quite enough matrix to keep the trachea open?

A: *Véronique Lefebvre*: It's possible that there is a problem like this in the Sox6 mutant. We have made sections through the neck to look at the trachea, and we couldn't see anything abnormal in the mice, so maybe it's a muscle related problem as to why they can't breathe.

Amelogenin peptides have unique milieu-dependent roles in morphogenic path determination

Q: Linda Sandell: You no longer think that these amelogenin peptides are cleavage products, but rather a differential expression of certain forms?

A: Arthur Veis: We never did think that they were degradation products. Normal degradation of enamel proteins occurs as mineral begins to form, and they range all the way from 25 to 29 thousand down to about 10 thousand. These active peptides I described are 6,500 and 8,900, and they have both the N-terminal domain and the C-terminal domain. The degradation products that you see in normal enamel degradation have one or the other, never both. Thus, the bioactive peptides are specific pre-mRNA alternative splice products.

Q: Linda Sandell: So, then, what's the expression pattern?

A: Arthur Veis: That's probably the next thing we are going to look at.

Q: Linda Sandell: You're going to have to stay home more, Art, and do the scientific stuff.

A: Arthur Veis: Yes. I meant to welcome Klaus to what is going to be known as the "Class of Professors Emeritus" that are involved in his fields. He's got to prepare himself for "nasty" comments like that. Although it's very hard to give up all of these power positions that one has, it actually isn't so bad because you can get back into the lab and get some work done.

Q: Hari Reddi: Are there specific transcription factors for enamel as well as dentin?

A: Arthur Veis: It's never been shown. Everybody has been looking at the factors which are known, and it was only by accident that we came upon the amelogenin peptides – or perhaps persistence and stubbornness. Many factors such as BMP-2, -4, -7, and -3 cycle up and down at different developmental stages during normal tooth development. There are also position specifying genes, and other transcription factors that are involved at different times. None of these factors acts by itself during development, there is a continuum of signaling events. With regard to the amelogenin peptides, we do know that there are cell receptors that bind to both peptides. We haven't identified them yet, but that's also on our program of things to be done.

Matrix molecules in cartilage and other tissues

Introduction

Dick Heinegård

Lund University, Department of Cell & Molecular Biology, BMC, plan C12, SE-22184 Lund, Sweden

Connective tissues contain an abundant extracellular matrix that essentially determines the tissue properties. The few cells have key roles in building up the matrix and maintaining its structural integrity and function. This is accomplished by sensing the properties of the matrix *via* cell surface receptors as well as reacting to mechanical strain.

The matrix macromolecular organization is primarily assembled outside the cells and is regulated by the coordinated production of the appropriate building blocks as well as *via* synthesis of a number of factors regulating the assembly process.

The major elements of the matrix are collagens forming various fibrillar networks, extremely polyanionic proteoglycans, and a variety of so-called non-collagenous proteins with roles in matrix assembly as well as in crosslinking the various fibrillar networks.

There are two sets of fibrillar networks with collagen as the major constituent. One contains as the major constituent the typical fibril forming collagen type I in most tissues, while type II is the one prominent in cartilage. These fibrils, that have a key role in providing tensile strength to the tissue, contain additional collagens in small amounts, e.g., collagen V with collagen I, and collagen XI with collagen II. These have putative roles in regulating fiber assembly. Additional proteins including COMP, decorin, fibromodulin and lumican have all been shown to be able to regulate collagen fibrillogenesis. All these proteins appear to influence the rate of fibril formation as well as the dimensions of the completed fibril. Another feature is that many of these proteins occur bound to the collagen fibril surface in the tissue. They thereby modify its structure and its ability to interact with other matrix constituents.

An additional protein that has been demonstrated to occur bound at the collagen fibril surface in the tissue is collagen type IX. Also, thrombospondins other than COMP have been shown to bind to collagen. More recently several members of the matrilin family of proteins have been demonstrated to bind to collagens. These oligomeric proteins also bind to other matrix constituents and thereby have the potential to crossbridge between various organizational elements in the tissue.

Another filamentous structure in the tissue is represented by beaded filaments of collagen type VI, particularly prominent in weight bearing tissues. Also collagen VI interacts with other molecular constituents that offer the potential for crossbridging to other matrix elements.

Another structural element in the tissue is the large hyaluronan-binding proteoglycans, particularly aggrecan and versican. These have large numbers of anionic glycosaminoglycan chains that contribute fixed negatively charged groups to the tissue, which are essential for its resilience.

The presentations of this session deal with the key structural elements in the tissue represented by the collagens, by aggrecan, and by representative matrix proteins with distinct functional properties. They all demonstrate important aspects of the current standing and focus in extracellular matrix research where functional aspects have a central place.

Electron microscope studies of collagen fibril formation in cornea, skin and tendon: Implications for collagen fibril assembly and structure in other tissues

Karl E. Kadler and David F. Holmes

Wellcome Trust Centre for Cell-Matrix Research, School of Biological Sciences, University of Manchester, Stopford Building 2.205, Oxford Road, Manchester M13 9PT, UK

Introduction

Collagen is one of the most widespread structural proteins in animals, and more than 23 genetically-distinct types of collagen are found in man (for review see [1]). Collagens comprise three polypeptide chains in which glycine (the smallest amino acid) occurs at every third residue position. The repeating Gly-X-Y motif (in which X and Y can be any amino acid and is often proline and hydroxyproline amino acids) is required for three polypeptide chains to assemble into a triple helix. The most abundant collagens are the fibril-forming types I, II, III, V and XI, which contain three polypeptide chains, each containing ~1000 residues, wound into an uninterrupted triple helix of ~295 nm in length (for review see [2]). These collagens occur in the extracellular matrix as D-periodic fibrils (where $D = \sim 67$ nm, the axial periodicity), which are indeterminate in length [3], and have a near-uniform diameter in the range 12–500 nm depending on tissue and stage of development (see Fig. 1). The fibrils are heterotypic and contain more than one genetic type of collagen. For example, collagen fibrils in cartilage comprise type II collagen and minor quantities of type XI collagen and type IX collagen. The type IX collagen is an example of a fibril-associated collagen with interrupted triple helices (FACIT). Fibrils in other tissues contain type I collagen with minor amounts of type III and V collagen. The fibrils are stabilized by interchain covalent crosslinks, which require oxidative deamination of specific lysyl and hydroxylysyl residues by lysyl oxidase(s) (for review see [4]). The fibrils have binding sites on their surfaces for small leucine rich proteoglycans (SLRPs) [5].

A fundamental feature of fibrillar collagens is that they are synthesized as precursor procollagens containing N- and C-propeptides at each end of the triple helical domain (Fig. 2). The C-propeptides have at least two functions. Firstly,

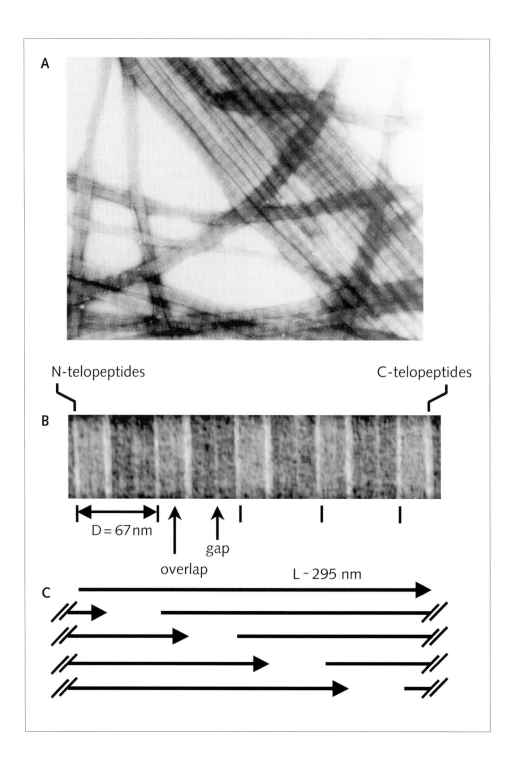

they are important in polypeptide chain recognition and in selection of type-specific proα-chains to form particular collagen molecules. For example, type I collagen comprises two proα1(I) chains and one proα2(I) chains, and the C-propeptides of these chains select these particular chains and exclude other proa chains, for example proα1(III) chains of type III procollagen. Site directed mutagenesis studies have identified amino acid sequences in the C-propeptides that specify polypeptide chain partners [6]. Another role of the C-propeptides is to keep procollagen soluble, particularly during intracellular transport. Consequently, removal of the C-propeptides is necessary and sufficient to initiate the assembly of collagen fibrils [7]. This is achieved physiologically by procollagen C-proteinases, but trypsin and pepsin will remove the propeptides *in vitro*. The procollagen C-proteinases in man include bone morphogenetic protein (BMP)-1 [8], mammalian tolloid (mTLD) [9], and mammalian tolloid-like (mTLL)-1 [10], which are members of the astacin family of metalloproteinases. The function of the N-propeptides is less clear. Their removal is not essential for fibril formation, and they have been implicated in regulation of fibril diameter [11], shape [12] and in binding of BMP molecules [13]. The N-propeptides are removed by at least two members of the ADAMTS family of metalloproteinases, namely ADAMTS2 [14, 15] and ADAMTS3 [16].

The tips of fibrils are sites of molecule addition and diameter regulation

Studies of collagen fibrillogenesis *in vitro* by cleavage of type I pCcollagen (a normal intermediate of type I procollagen containing the C-propeptides, but not the N-propeptides) with procollagen C-proteinase [17], showed that the fibrils are D-periodic (D = 67 nm), the fibrils grow from symmetrical pointed tips [18], the tips are paraboloidal in shape [19], and the fibrils are tactoidal, i.e., they did not exhibit a uniform diameter [20]. These studies showed that the information required for collagen self-assembly is contained within the collagen molecules, and presumably from the unique sequence of amino acids in each of the three chains that comprise the collagen molecule. Moreover, the studies showed that the fibrils grow from

Figure 1
Axial structure of type I collagen fibrils. (A) Transmission electron micrograph of reconstituted type I collagen fibrils. The fibrils were negatively stained with uranyl acetate to show the axial periodicity of the fibril and uniformity of fibril diameter. (B) The detailed axial stain pattern corresponding to the length of a collagen molecule. (C) Schematic diagram of the axial arrangement of collagen molecules with amino termini at the lefthand side of the schematic and the carboxyl termini at the righthand side. The schematic is axially aligned with the stain pattern in B.

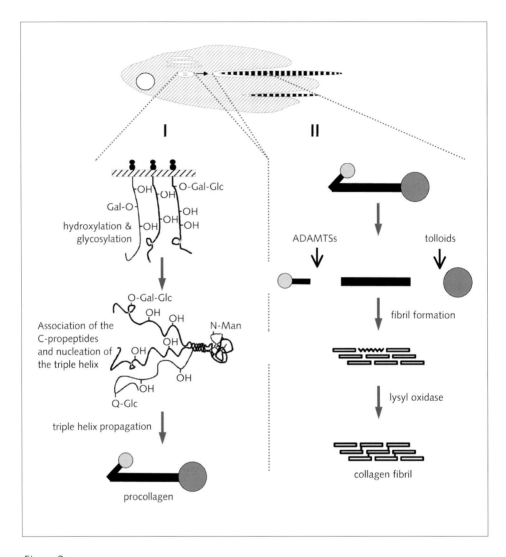

Figure 2
Schematic representation of some of the intracellular and extracellular events in collagen biosynthesis. The object at the top of the diagram represents a cell (diagonal stripes) synthesizing collagen fibrils (banded). (I) events occurring in the endoplasmic reticulum including: co-translational modification of proα chains, chain association at C-propeptides, and assembly of a triple helical procollagen molecule. (II) conversion of procollagen to collagen by cleavage of the N-propeptides (light grey circle and black rectangle) by ADAMTS proteinases and cleavage of the C-propeptides (dark grey circle) by tolloid-like proteinases. The resultant collagen molecules assemble into collagen fibrils and are subsequently stabilized by lysyl oxidase-derived crosslinks.

pointed tips. This raised the possibility that collagen fibrils might grow from pointed tips *in vivo*.

Examination of the tips of vertebrate collagen fibrils *in vivo* by transmission electron microscopy of embedded and sectioned tissue does not readily provide the quantitative data needed to determine the fine structure of the fibril tips, and the tips are difficult to identify with this approach. An alternative approach has been to examine individual fibrils by quantitative EM methods after they have been released from tissue using Dounce homogenization and/or ultrasound. Quantitative darkfield scanning transmission electron microscopy (QSTEM) has shown that the tips of tendon fibrils are paraboloidal in shape [21]. Therefore, the fibrils are like those formed *in vitro* by cleavage of pCcollagen with C-proteinase. However, a major difference between tendon fibrils and those formed *in vitro* is that tendon fibrils exhibit an abrupt limitation in diameter at the fibril tips. The abrupt limitation produces a shaft of uniform diameter. Furthermore, the fact that the abrupt limitation in diameter occurs at the fibril tips demonstrates that diameter limitation is tip mediated, and is an elegant mechanism for keeping the diameter of fibrils uniform at distances hundreds of microns away from the cell. Analysis of the staining pattern of fibrils released from tendon showed that pointed ends were either C-ends, i.e., they contained collagen molecules with carboxyl termini nearest the tip, or N-ends, i.e., they contained collagen molecules with amino termini nearest the tip [22] (see Fig. 3).

Collagen fibrillogenesis in vertebrate skin and tendon involves tip-to-tip fusion of collagen early fibrils: implications for other tissues

The earliest-formed collagen fibrils in chick tendon are ~1 μm in length [23]. These "collagen early fibrils" (CEFs) are shorter than the "fibril segments" described previously [24–26]. CEFs have two pointed tips, exhibit the 67 nm *D*-periodicity, and, therefore, exhibit features characteristic of collagen fibrils. However, they exhibit two features that make them different from mature fibrils, separate from the fact that they are only ~1 μm in length. Firstly, they are spindle shaped, i.e., they lack a region of constant diameter. Secondly, and perhaps most importantly, the diameter mid-way along their length is the same value as the diameter of mature fibrils. Thus, CEFs are disproportionately wide (e.g., ~50 nm) for their length (e.g., ~1000 nm), when compared to mature fibrils. These observations suggest that the diameter of mature fibrils is determined at the stage of CEF formation. The formation of CEFs, which might involve collagen, procollagen, pCcollagen or pNcollagen molecules, is most likely to be a decisive stage of fibrillogenesis.

Transmission electron microscopy shows that CEFs can be unipolar (with all collagen molecules oriented in the same direction) or N, N-bipolar (with two amino terminal tips and a switch region where the orientation of collagen molecules revers-

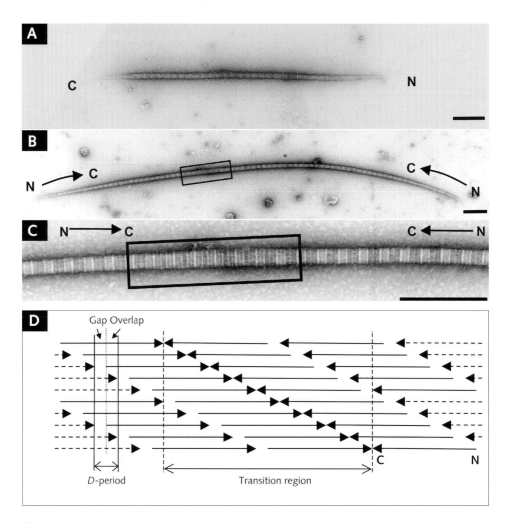

Figure 3
Collagen early fibrils (CEFs) from chick embryo leg tendon. Electron microscope images of the two basic forms are shown after negative staining. (A) a unipolar fibril where all collagen molecules point in the same direction. (B) an N-N bipolar fibril, which contains an axial zone of molecular polarity reversal (box) where anti-parallel packing occurs. Outside this transition region the molecules are in parallel register with their N-termini pointing towards the fibril tips. (C) enlargement of the boxed region in B showing the stain pattern throughout the molecular switch region. (D) a possible axial arrangement of collagen molecules in the transition region of an N, N bipolar fibril. The collagen molecule is represented by an arrow in which the arrow head is the C-terminus of the molecule. The arrangement is consistent with the observed axial extent and stain pattern of this region. Unipolar and N-N bipolar fibrils have a similar abundance in developing vertebrate tendon. Scale bar = 300 nm.

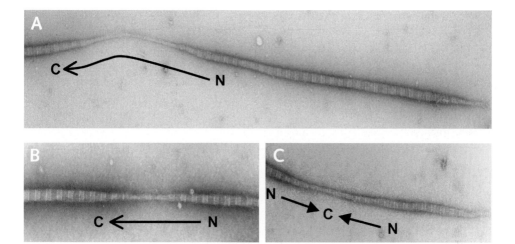

Figure 4
Examples of tip-to-tip fusion in fibrils from developing chick tendon. Collagen fibrils were released intact by mild mechanical (Dounce) homogenisation. Most fusion events involve N to C tip fusion (A and B), but occasionally antiparallel contact is observed with the fusion of two C-tips (C). Fusion between N-tips does not seem to be a normal event for type I collagen fibrils.

es 180°). Moreover, CEFs can fuse by tip-to-tip association to generate longer fibrils. Fusion involves the C-end of a unipolar fibril (Fig. 4). Tip-to-tip fusion occurs by overlap of the distal 4 D-periods of each fibril, which generates a thinned region in the fused fibril. A subsequent, and as yet poorly understood, process of remodeling restores the diameter of the fibril at the site of fusion.

The inability of two N, N-bipolar fibrils to fuse by tip-to-tip association suggests that the N-tips lack recognition sequences that are needed to stabilize the two tips during fusion. *In vitro* studies support the notion that two N-tips cannot fuse. N, N-bipolar fibrils formed by cleavage of pCcollagen with procollagen C-proteinase at 34° and 37°C contained only one molecular switch region, and were not seen to associate at their N-ends. Conversely, fibrils formed by cleavage at 29°C, which have a smoothly tapering N-end and a blunt C-end, were sometimes associated at their C-ends to form "star-like" aggregates [18].

It remains to be seen if collagen fibrillogenesis in tissues other than tendon proceeds by the assembly and fusion of CEFs. Tendon contains predominately type I collagen, and it might have been presumed that type I collagen fibrils in other tissues (e.g., skin, ligament, cartilage and bone) might assemble by the same route. Preliminary work in our laboratory shows that skin contains unipolar and bipolar fibrils as well as early fibrils. Furthermore, cultured skin fibroblasts deposit an extracellular matrix containing unipolar and bipolar fibrils, as well as short collagen fibrils that resemble CEFs. However, the fibrils in cornea appear to be long and predominately N, N-bipolar. This suggests that end-to-end fusion of fibrils is not an obligatory assembly mechanism for the formation of long fibrils and that accretion of collagen molecules onto the ends of the fibrils might be the predominant mode of fibrillogenesis in some tissues. The type II collagen-containing fibrils in cartilage and vitreous humor are very long. Further studies are needed to determine if these fibrils are formed by end-to-end fusion of CEFs or by continuous accretion of collagen molecules.

Collagen fibril structure in three dimensions

Studies of the high-resolution structure of collagen fibrils have used x-ray fiber diffraction of rat tail tendon (e.g. [27] and references therein) and lamprey notochord [28] in which the fibrils are composed predominately of one genetic type of collagen. This has yielded information about the triple helical structure of the collagen molecules. However, there has been controversy about the packing of the molecules in the fibril. In particular, about whether the fibril has a microfibrillar substructure or a crystalline packing of collagen molecules in a sheet structure. A five-stranded microfibrillar model was proposed by Smith [32] and developed by Piez and Trus [33, 34] on the basis of sequence regularities and fibre diffraction data. Analysis of synchrotron X-ray diffraction from rat tail tendon has yielded the best agreement with a microfibril substructure [35].

A recent study used automated electron tomography (AET) to study the three-dimensional organization of molecules in corneal collagen fibrils [36]. The reconstructions showed that the collagen molecules in the 36 nm-diameter collagen fibrils are organized into microfibrils (~ 4 nm diameter) that are tilted by ~15° to the fibril long axis in a right-handed helix (Fig. 5). Analysis of the lateral structure showed that the microfibrils exhibited regions of order and disorder relative to the fibril D-period, and were best ordered at the N- and C-telopeptides and the d-band of the gap zone (Fig. 6). The AET 3D reconstructions also showed macromolecules binding to the fibril surface at axial sites that correspond precisely to where the microfibrils are most ordered.

The discovery of the microfibrillar substructure to collagen fibrils raises new questions about the assembly and structure of collagen fibrils, including: (1) do all collagen fibrils, e.g., cartilage fibrils containing type II collagen, comprise microfibrils? (2) does the occurrence of microfibrils explain the paraboloidal shape of fibril tips? (3) what is the structure of a microfibril? (4) are the recognition sequences needed for interfibrillar fusion exposed at the C-terminal ends of unipolar microfibrils? Finding answers to these questions will require high-resolution structural studies of collagen fibrils and microfibrils using a combination of electron

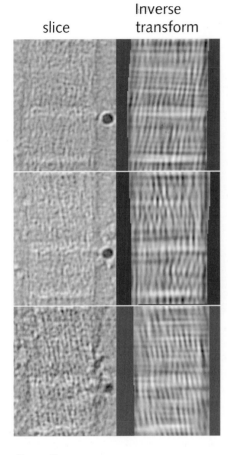

Figure 5
Three-dimensional reconstruction of isolated collagen fibrils using automated electron tomography (AET). A typical sample of longitudinal virtual slices through the reconstruction is shown (lefthand panels) from the upper, central and lower levels of a negatively stained fibril from bovine cornea. The corresponding Fourier-filtered images (righthand panels) show an enhancement of the filamentous substructure. The microfibrils have a lateral spacing of about 4 nm and are tilted at about 15° to the fibril long axis. The data support a constant tilt rather than a constant pitch model for these collagen type I/V heterotypic fibrils.

microscope and x-ray diffraction methods. The studies will be relevant to understanding the molecular basis of tissue assembly in health and disease, particularly those diseases caused by mutations in the genes encoding extracellular matrix macromolecules.

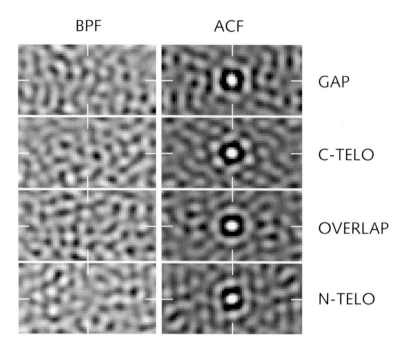

Figure 6
Transverse structure of isolated collagen fibrils. A sample of band pass filtered (BPF) images of virtual transverse slices at different locations within a single D-period are shown with their autocorrelation functions (ACFs). This analysis indicates three axial locations of consistently higher lateral packing order corresponding to the telopeptides (marked C-TELO and N-TELO respectively) and the d-band within the gap zone (marked GAP).

Acknowledgement

The work in our laboratory is generously supported by grants from The Wellcome Trust.

References

1   Kielty CM, Grant ME (2002) The collagen family: structure, assembly and organisation in the extracellular matrix. In: B Steinmann, PM Royce (eds): *Connective tissue and its heritable diseases*, 2nd ed. PM Wiley Liss, New York
2   Kadler KE, Holmes DF, Trotter J, Chapman JA (1996) Collagen fibril formation. *Biochem J* 316: 1–11
3   Parry DAD, Craig, AS (1984) Growth and development of collagen fibrils in connective

tissue. In: A Ruggeri, PM Motta (eds): *Ultastructure of the connective tissue matrix*. Martinus Nijhoff Publishers, 34–64

4   Smith-Mungo LI, Kagan HM (1998) Lysyl oxidase: properties, regulation and multiple functions in biology. *Matrix Biol* 16: 387–398

5   Iozzo RI (1999) The biology of the small leucine rich proteoglycans. *J Biol Chem* 274: 18843–18846

6   Lees JF, Tasab M, Bulleid NJ (1997) Identification of the molecular recognition sequence which determines the type-specific assembly of procollagen. *EMBO J* 16: 908–916

7   Kadler KE, Hojima Y, Prockop D J (1987) Assembly of collagen fibrils de novo by enzymic cleavage of the type I pCcollagen by procollagen C-proteinase. Assay of critical concentration demonstrates that the process is an example of classical entropy-driven self assembly. *J Biol Chem* 262: 15696–15701

8   Kessler E, Takahara K, Biniaminov L, Brusel M, Greenspan DS (1996) Bone morphogenetic protein-1: The type I procollagen C-proteinase. *Science* 271: 360–362

9   Li S-W, Sieron AL, Fertala A, Hojima Y, Arnold WV, Prockop DJ (1996) The C-proteinase that processes procollagens to fibrillar collagens is identical to the protein previously identified as bone morphogenetic protein-1. *Proc Natl Acad Sci USA* 93: 5127–5130

10 Scott IC, Blitz IL, Pappano WN, Imamura Y, Clark TG, Steiglitz BM, Thomas CL, Maas SA, Takahara K, Cho KW, Greenspan DS (1999) Mammalian BMP-1/tolloid-related metalloproteinases, including novel family member mammalian tolloid-like 2, have differential enzymatic activities and distributions of expression relevant to patterning and skeletogenesis. *Developmental Biol* 213: 282–300

11 Chapman JA (1989) The regulation of size and form in the assembly of collagen fibrils *in vivo*. *Biopolymers* 28: 1367–1382 (addition: 28: 2201–2205)

12 Holmes DF, Watson RB, Steinmann B, Kadler KE (1993) Ehlers Danlos syndrome type VIIB. Morphology of type I collagen fibrils is determined by the conformation of the N-propeptide. *J Biol Chem* 268: 15758–15765

13 Zhu Y, Oganesian A, Keene DR, Sandell LJ (1999) Type IIA procollagen containing the cysteine-rich amino propeptide is deposited in the extracellular matrix of prechondrogenic tissue and binds to TGF-beta 1 and BMP-2. *J Cell Biol* 144: 1069–1080

14 Coige A, Li S-W, Sieron A, Nusgens BV, Prockop DJ, Lapiere CM (1997) cDNA cloning and expression of bovine procollagen (N-proteinase: a new member of the superfamily of zinc-metalloproteinases with binding sites for cells and other matrix components. *Proc Natl Acad Sci USA* 94: 2374–2379

15 Colige A, Beschin A, Samyn B, Goebels Y, Beeumen JV, Nusgens BV, Lapiere CM (1995) Characterization and partial amino acid sequencing of a 107-kDa procollagen I N-proteinase purified by affinity chromatography on immobilized type XIV collagen. *J Biol Chem* 270: 16724–16730

16 Fernandes RJ, Hirohata S, Engle JM, Colige A, Cohn DH, Eyre DR, Apte SS (2000) Procollagen II amino propeptide processing by ADAMTS-3. Insights on dermatosparaxis. *J Biol Chem* 276: 31502–31509

17 Hojima Y, van der Rest M, Prockop DJ (1985) Type I procollagen carboxyl terminal proteinase from chick embryo tendons – purification and characterisation. *J Biol Chem* 260: 5996–6003

18 Kadler KE, Hojima Y, Prockop DJ (1990) Collagen fibrils *in vitro* grow from pointed tips in the C- to N-terminal direction. *Biochem J* 268: 339–343

19 Holmes DF, Chapman JA, Prockop DJ, Kadler KE (1992) Growing tips of type I collagen fibrils formed *in vitro* are near-paraboloidal in shape, implying a reciprocal relationship between accretion and diameter. *Proc Natl Acad Sci USA* 89: 9855–9859

20 Holmes DF, Watson RB, Chapman JA, Kadler KE (1996) Enzymic control of collagen fibril shape. *J Mol Biol* 261: 93–97

21 Holmes DF, Graham HK, Kadler KE (1998) Collagen fibrils forming in developing tendon show an early and abrupt limitation in diameter at the growing tips unobserved in cell-free systems. *J Mol Biol* 283: 1049–1058

22 Holmes DF, Lowe MP, Chapman JA (1994) Vertebrate (chick) collagen fibrils formed *in vivo* can exhibit a reversal in molecular polarity. *J Mol Biol* 235: 80–83

23 Graham HK, Holmes DF, Watson RB, Kadler KE (2000) Identification of collagen fibril fusion during vertebrate tendon morphogenesis. The process relies on molecular recognition sequences in unipolar fibrils and is regulated by collagen-proteoglycan interaction. *J Mol Biol* 295: 891–902

24 Birk DE, Nurminskaya MV, Zycband EI (1995) Collagen fibrillogenesis *in-situ* – fibril segments undergo postdepositional modifications resulting in linear and lateral growth during matrix development. *Developmental Dynamics* 202:229–243

25 Birk DE, Hahn RA, Linsenmayer CY, Zycband EI (1996) Characterization of collagen fibril segments from chicken embryo cornea, dermis and tendon. *Matrix Biology* 15: 111–118

26 Birk DE, Zycband EI, Woodruff S, Winkelmann DA, Trelstad RL (1997) Collagen fibrillogenesis *in situ*: Fibril segments become long fibrils as the developing tendon matures. *Developmental Dynamics* 208: 291–298

27 Wess TJ, Hammersley AP, Wess L, Miller A (1998) A consensus model for molecular packing of type I collagen. *J Struct Biol* 122: 92–100

28 Eikenberry EF, Childs B, Sheren SB, Parry DA, Craig AS, Brodsky B (1984) Crystalline fibril structure of type II collagen in lamprey notochord sheath. *J Mol Biol* 176: 261–277

29 Hulmes DJS, Miller A (1979) Quasi-hexagonal molecular packing in collagen fibrils. *Nature* 282: 878–880

30 Fraser RDB, MacRae TP, Miller A (1987) Molecular packing in type I collagen fibrils. *J Mol Biol* 193: 115–125

31 Miller A, Tocchetti D (1981) Calculated x-ray diffraction pattern from a quasi-hexagonal model for the molecular arrangement in collagen. *Int J Biol Macromol* 3: 9–18

32 Smith JW (1968) Molecular packing in native collagen. *Nature* 219: 157–158

33 Piez KA, Trus BL (1978) Sequence regularities and packing of collagen molecules. *J Mol Biol* 122: 419–432

34 Piez KA, Trus BL (1981) A new model for packing of type I collagen molecules in the native fibril. *Biosci Rep* 1: 801–810
35 Wess TJ, Hammersley AP, Wess L, Miller A (1998) Molecular packing of type I collagen in tendon. *J Mol Biol* 275: 255–267
36 Holmes DF, Gilpin CJ, Baldock C, Ziese U, Koster AJ, Kadler KE (2001) Corneal collagen fibril structure in three dimensions: structural insights into fibril assembly, mechanical properties, and tissue organisation. *Proc Natl Acad Sci USA* 98: 7307–7312

Hyaluronics and aggrecanics

Tim Hardingham

Wellcome Trust Centre for Cell-Matrix Research, School of Biological Sciences, University of Manchester, Manchester M13 9PT, UK

Introduction

Hyaluronan (HA) is a high molecular weight (10^5–10^7 Da) unbranched glycosaminoglycan, composed of repeating disaccharides of D-N-Acetylglucosamine and D-Glucuronic acid. It is a widely distributed component of the extracellular matrix of vertebrate tissues [1]. It also acts as a scaffold for the binding of selected matrix molecules including aggrecan and other members of the hyalectan family [2, 3]. HA forms viscoelastic solutions and there is much interest and speculation on the properties that contribute to its pronounced non-Newtonian behaviour.

In physiological solutions at neutral pH, HA acts as a stiffened random coil in solution, due to hydrogen bonding between adjacent saccharides and mutual electrostatic repulsion between carboxyl groups [4–11]. It has also been proposed that there is association between hyaluronan chains, which has been visualised in EM preparations and interpreted as anti-parallel double helices, bundles and ropes [12, 13] and NMR spectra have also been interpreted to suggest that chain-chain association occurs in solution [14]. However, the presence and stability of such structures in solution remains unclear, and how they contribute to the solution properties of hyaluronan has yet to be determined. We have investigated the solution properties of HA and compared them with aggrecan, which has some similar physical attributes, but contains a more compact branched structure.

Hyaluronan solution properties

In this study, evidence for inter-chain interactions in HA solutions were investigated with fluoresceinamine-labelled HA [15] by confocal fluorescence recovery after photobleaching (confocal-FRAP). This is a powerful method for determining concentrated solution properties in the absence of flow and shear forces and with no concentration gradients [16, 17]. Confocal-FRAP is an equilibrium method and provides measurements of lateral self diffusion, which thus reveal the extent of intermolecular interactions, and this can be followed to high concentration when interactions are maximal. It also provides a method to analyse the network formed by a polymer in solution by determining its effect on the diffusion within the network of fluorescently labelled tracer molecules of known size [10, 18].

Hyaluronan properties in concentrated solutions are determined by its structure and interaction with the surrounding water and ions. Amongst the interactions that contribute to its properties may be included:

1) electrostatic interaction of the regularly placed carboxyl groups;
2) hydrogen bonding between adjacent saccharides;
3) domain overlap and polymer entanglement;
4) chain-chain association through mechanisms such as interaction of hydrophobic patches.

An important aspect and, indeed, a major advantage of using confocal-FRAP, is because it permits analysis at concentrations of hyaluronan up to and far exceeding the critical concentration at which there is predicted molecular domain overlap [9, 17, 18]. The analysis is thus best suited to investigations of entanglement and intermolecular chain-chain association, as these would be concentration dependent and would be strongly favoured at high concentration, whereas electrostatic interactions and hydrogen bonding of adjacent disaccharides occur at all concentrations.

Concentration dependence of hyaluronan self diffusion

In characterising the general behaviour of hyaluronan solutions [10] it was shown that the lateral translational self-diffusion coefficients of hyaluronan showed a progressive fall with increasing concentration as it approached and exceeded the predicted critical concentration for domain overlap (Fig. 1). Similar smooth transitions between dilute, semi-dilute and concentrated regimes have been observed experimentally in many polymer / solvent systems [19–21], including HA solutions [20]. The lateral self-diffusion coefficients reduced steeply with concentration in a manner consistent with phenomenological descriptions of polymer self-diffusion in terms of a universal scaling equation [20].

$$D = D_0 \exp^{-\alpha c^\nu} \qquad (1)$$

Where D_0 is the polymer free self-diffusion defined in the limit of zero concentration and α and ν are empirically derived. The parameter α describes the strength of inter-polymer hydrodynamic interactions, and the deviation of ν from unity arises from chain contraction at high concentrations. Data were fitted to Equation (1) using a non-linear least squares fit (non-weighted). Analysis of data from results

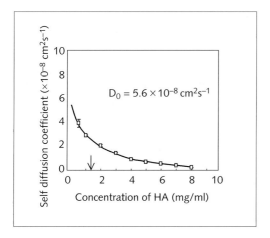

Figure 1
Concentration dependence of the lateral translational self-diffusion coefficients of hyaluronan. The lateral translational self-diffusion was determined by confocal-FRAP for hyaluronan (830 kDa) (□) at 0.5–8.0 mg/ml in PBS at 25°C. The vertical arrow marks the critical overlap concentration c*. The solid line shows the data fitted to Equation (1) and extrapolated to zero concentration to give D_0. (Data from [10]). Confocal-FRAP technique described in [17, 18].

with hyaluronan (830 kDa) gave $D_0 = 5.6 \times 10^{-8}$ cm^{2}s^{-1}, $\alpha = 0.63$ ml/mg and $\nu = 0.74$. These measurements were insensitive to pH over a broad range (pH 4–8) and also to temperature between 5°C and 60°C, when changes in solvent viscosity had been accounted for. There was thus no evidence for the thermal dissociation of ordered structures in this temperature range.

Effects of electrolytes on hyaluronan solution properties

Investigation of the effect of increasing electrolyte concentration on hyaluronan solution properties [10] showed that the self-diffusion coefficient of hyaluronan was very low in the absence of any supporting electrolyte, but increased dramatically with small increases in NaCl concentration, showing a 2.8-fold increase in lateral self-diffusion coefficient from zero to 100 mM (Fig. 2). This is consistent with increased electrostatic shielding resulting in polyanion coil contraction, and, as this was largely complete at 100 mM NaCl, the contribution of electrostatic effects to macromolecular stiffness under physiological conditions of ionic strength and pH is suggested to be small.

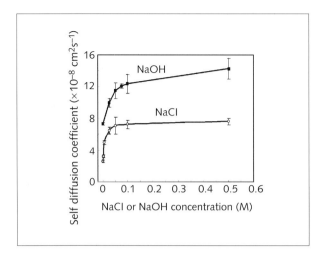

Figure 2
Effect of increasing concentrations of NaCl and NaOH on hyaluronan self-diffusion.
The lateral translational self-diffusion was determined for hyaluronan (500 kDa, 0.2 mg/ml) in increasing concentration of NaCl (○), or NaOH (■). Ionic strength of solutions in NaOH was maintained at 500 mM by the addition of NaCl. All measurements at 25°C. (From [10]).

The effects of different counterions on the self-diffusion of hyaluronan showed that Ca^{2+} caused a significant increase compared with Na^+, with less increase with Mn, Mg and K (Fig. 3), [11]. At all concentrations up to 10 mg/ml, the self-diffusion coefficients of HA were greater in $CaCl_2$ than in NaCl, although the difference became smaller at high concentration. The intrinsic effects of these counterions on the conformation of hyaluronan in dilute solution was investigated by gel filtration and multi-angle laser light scattering (MALLS) analysis. With HA (930 kDa) solutions in different electrolytes, but at similar ionic strength (Tab. 1), the Rg's in $CaCl_2$ (and $MnCl_2$) were significantly lower than in KCl ($p < 0.001$) and NaCl ($p < 0.001$). As the peak concentration of HA was < 30 µg/ml, this measurement provided a comparison of the Rg of individual chains and showed that there was a direct effect of Ca^{2+} in contracting the free solution domain of hyaluronan. The equivalent Stokes sphere radius (RH) (see Equation (2)) for HA (830 kDa) was 43 nm in 0.5 M NaCl, and 36 nm in 0.5 M $CaCl_2$.

The effect of Ca^{2+} on the properties of HA were also found to be dominant over the effects of Na^+, as in the presence of 0.15 M NaCl, the addition of $CaCl_2$ at low concentration (< 10 mM) caused a further significant increase in the self diffusion of HA (Fig. 4). The differences between the self-diffusion of HA in Ca^{2+} and in Na^+ were also accompanied by differences in the tracer diffusion of FITC-dextran

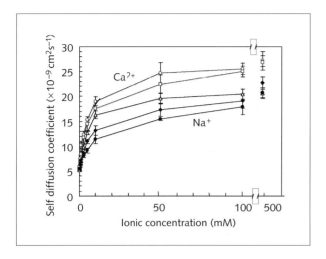

Figure 3
Effect of increasing cation concentration on hyaluronan self-diffusion.
Self diffusion coefficients of HA (830 Kda, 2 mg/ml) were determined as a function of salt concentration for NaCl (●), KCl (♦), MgCl$_2$ (ρ), MnCl$_2$, (□) and CaCl$_2$ (○). Hyaluronan was most mobile in CaCl$_2$ and least mobile in NaCl. All measurements at 25°C. (From [11].)

Table 1 - Weight averaged radius of gyration calculated from multi-angle laser light scattering (MALLS) analyses of 930 kDa HA in 100 mM ionic strength electrolyte solutions chromatographed on an S-1000 size exclusion column. Error values represent the SEM (N = 3). (From [11].)

Electrolyte	Ionic strength	R_g (± SE) (nm) (weight averaged)
NaCl	0.10	98.9 ±1.0
KCl	0.10	97.7 ± 2.2
MgCl$_2$	0.10	92.9 ± 2.0
CaCl$_2$	0.10	88.0 ± 1.3
MnCl$_2$	0.10	87.4 ± 3.2

(2000 kDa) in solutions of HA at up to 20 mg/ml and tracer mobility in 150 mM NaCl was also increased by the addition of CaCl$_2$ [11].

The contraction of the HA domain in calcium solutions suggested that Ca^{2+} increased the flexibility of the chain by promoting a greater range of movement at each glycosidic bond. This may be caused by Ca^{2+} ions perturbing the co-ordination

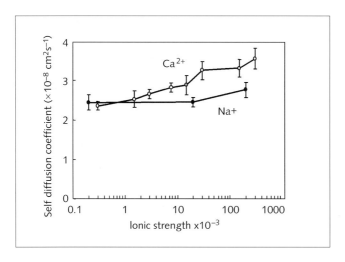

Figure 4
Effect of low concentrations of $CaCl_2$ on hyaluronan self-diffusion in NaCl (0.15 M). The self-diffusion coefficient of HA (830 kDa, 2 mg/ml) was determined in 150 mM NaCl with increasing concentrations of $CaCl_2$ (○) or NaCl (●) at 25°C. (From [11].)

of water molecules with HA chains, thereby destabilising some hydrogen bonds involving water bridges. In the presence of Ca^{2+}, the range of hydrogen bonds that bridge adjacent sugars in these linkages may thus be reduced [22, 23]. Overall, the changes in HA properties caused by Ca^{2+} are small compared to the effects of strong alkali (see below) [10]. However, it may be speculated that HA de-stiffening by Ca^{2+} may have a role in cell mediated matrix re-modelling processes.

The effects of alkali pH on hyaluronan self diffusion and on tracer diffusion in hyaluronan solutions

The effect of high pH in NaOH also contracted the domain size of hyaluronan, but the effect (Fig. 5) was much greater than the reduction found due to electrostatic shielding (Fig. 2). This effect is consistent with previously reported reductions in Rg and intrinsic viscosity [24]. Changes in the R_H and hydrodynamic volume of hyaluronan with NaCl and NaOH were calculated using the Stoke's Einstein approximation for the behaviour of a sphere:

$$D_0 = \frac{\kappa T}{6 \pi \eta R_H} \qquad (2)$$

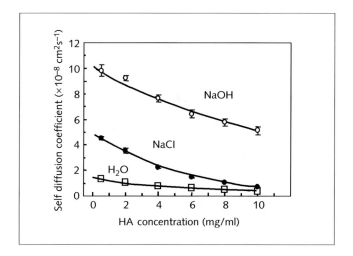

Figure 5
Comparison of the concentration dependance of hyaluronan self-diffusion in NaOH, NaCl and water. The concentration dependence of the lateral translational self-diffusion coefficient of hyaluronan (830 kDa) at 0.5–10 mg/ml was determined in, 0.5 M NaOH (○), 0.5 M NaCl (●), and de-ionised water (□). All measurements at 25°C. (From [10].)

where κ is Boltzmann's constant, T is temperature, and η is the solvent. If it is assumed that the self-diffusion coefficient at 0.2 mg/ml is approximately equal to the free diffusion coefficient (see Fig. 1), then for hyaluronan of 500 kDa, from the Stoke's Einstein equation, R_H contracted from 95 nm to 33.5 nm, in going from de-ionised water to 0.5 M NaCl, reducing further to 17.5 nm in 0.5 M NaOH (Fig. 6). These results show that in going from 0.5 M NaOH into de-ionised water, the apparent domains of hyaluronan chains are increased by more than 100 times and this is most likely to result from increased electrostatic interactions and hydrogen bond formation [10]. In the most compact configuration in alkali, the hydrodynamics of hyaluronan (500 kDa) became similar to those of the partly branched FITC-Dextran (2000 kDa, R_H = 19 nm), which is neither charged, nor is predicted to form comparable hydrogen bonds. For hyaluronan (500 kDa) in 0.5 M NaOH (Fig. 2), domain overlap is predicted to occur at 37 mg/ml. This implies that at 2–10 mg/ml, solutions are well below c* and this is entirely consistent with the comparatively greater network mobility observed in self-diffusion experiments, including those with higher molecular weight HA (830 kDa, Fig. 5). These effects in alkali were reversible and caused no significant de-polymerisation under the conditions used.

Tracer diffusion results at low hyaluronan concentrations (1–4 mg/ml) (Fig. 7) [10] show analogous behaviour to the changes in self diffusion (Fig. 5). The network

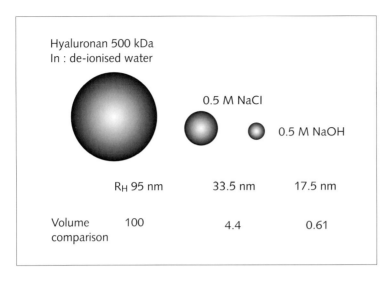

Figure 6
Comparison of the hydrodynamic radius (R_H) of hyaluronan (500 kDa) in de-ionised water and in salt and alkaline solutions.

is both more permeable and more mobile in 0.5 M NaCl than in de-ionised water, and this supports a model involving contraction of the hyaluronan chain conformation in the presence of increasing electrolyte. However, as the concentration of hyaluronan approached 20 mg/ml (Fig. 7), tracer mobility became progressively independent of salt concentration, indicating that at high concentrations chain density is the major determinant of matrix permeability. The lack of a salt effect at high hyaluronan concentration is interesting, as it suggests a lack of hydrophobic interactions between chains, as by analogy with RNA and DNA, high salt would be expected to favour chain-chain association.

The tracer studies also provide a measure of the major changes induced by NaOH. In hyaluronan (930 kDa) at 20 mg/ml (Fig. 7), the FITC-dextran translational diffusion is independent of NaCl concentration, but not of NaOH concentration. This reflects, as noted above, that for 930 kDa hyaluronan, 20 mg/ml is likely to represent a semi-dilute regime in the presence of NaOH, whereas it is clearly a concentrated, entanglement dominated regime, both in NaCl and in de-ionised water. These results strongly suggest that the solution properties at higher concentration in various solvents are directly related to the hydrodynamic volumes of single chains in the same solvent. Results at high pH, showing high mobility and permeability of hyaluronan, clearly reveal the degree to which, at neutral pH, intrachain hydrogen bonds profoundly affect chain stiffness, chain entanglement and inter-chain hydrodynamic interactions.

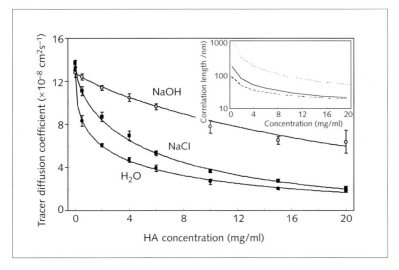

Figure 7
Comparison of tracer diffusion in hyaluronan solutions of different concentration, in NaOH, NaCl and water. Lateral translational diffusion coefficients of FITC dextran (2000 kDa) were determined in hyaluronan solutions (930 kDa) 0–20 mg/ml, in de-ionised water (●), 0.5 M NaCl (■), and 0.5 M NaOH (○). Inset shows the correlation length parameter (ξ) versus hyaluronan concentration for: de-ionised water (long dash), 0.5 M NaCl (solid line), and 0.5 M NaOH (short dash). All measurements in PBS at 25°C. (From [10]).

Effects of urea on hyaluronan solution properties

In concentrated polymer solutions, if the network properties are determined solely by chain entanglements, then they should be independent of agents that disrupt other associative mechanisms. To investigate whether HA properties were determined by hydrophobic chain-chain interactions, self-diffusion properties were investigated in the presence of urea, a potent disrupter of hydrophobic association [11]. Initially, the effect of urea on individual chain hydrodynamics was investigated by analysing HA self-diffusion at low polymer concentrations. In a dilute solution of HA, if there is association between segments of chains, this will tend to contract its hydrodynamic domain, whereas in concentrated solution it might serve to additionally make linkages between adjacent molecules. However, the self-diffusion of HA (500 kDa) in the presence of 6 M urea was consistently higher than in de-ionised water (Fig. 8). This showed that in urea, the polymer domain of hyaluronan became smaller and thus showed no evidence for the disruption of intramolecular chain-chain association. The increased free diffusion coefficient of hyaluronan in dilute

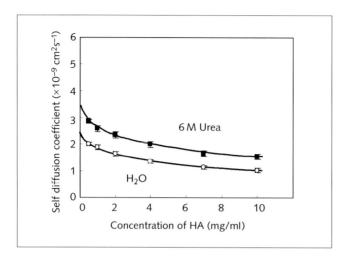

Figure 8
Effects of urea on hyaluronan self-diffusion. The concentration dependence of self-diffusion of HA (500 kDa) was determined 0.5–10 mg/ml for solutions in 6 M urea (■) and de-ionised water (○). All data at 25°C and corrected for solvent viscosity. Solid lines show data fitted to the polymer scaling equation. (From [11].)

solution in the presence of urea is therefore inconsistent with the presence of intramolecular chain associations. On the contrary, the reduced hydrodynamic domain size shows that urea increases the flexibility of HA chains.

The primary intramolecular chain stiffening mechanism for HA arises from hydrogen bonding, and the effect of urea may therefore be to reduce the hydrogen bonding between adjacent saccharides. However, the de-stiffening caused by urea is substantially less than that caused by 0.5 M NaOH, [10]. Therefore, urea may, for example, disrupt only a sub-fraction of H-bonds, such as those involving a water bridge. Interestingly urea had no detectible effect on HA diffusion in the presence of 0.5 M NaCl as supporting electrolyte, which suggested that it did not affect the chain-stiffening hydrogen bonds present in 0.5 M salt, but could affect those additionally present in de-ionised water.

Urea appeared to have little effect on intermolecular interactions between HA molecules, as the concentration dependence of self-diffusion, which is a measure of intermolecular interaction, follows a very similar form in urea and in de-ionised water (Fig. 8). These results therefore suggest that there are no chain-chain associations in HA in aqueous solution that are sensitive to this chaotropic agent. The absence of chain-chain interactions was also suggested by the inability of hyaluro-

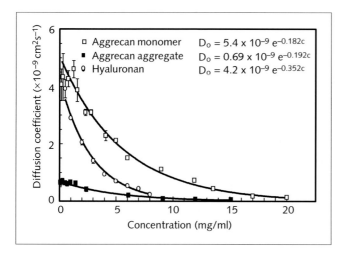

Figure 9
Comparison of the concentration dependance of self-diffusion for hyaluronan, aggrecan and aggrecan aggregate. The lateral translational self-diffusion was determined by confocal-FRAP for Aggrecan (□) (Mw 2.6 MDa), Aggrecan Aggregate (■) (Mw 75 MDa) and hyaluronan (900 kDa) (○) at 0.5–20 mg/ml in PBS at 25°C. The solid lines show the data fitted to Equation (1) and extrapolated to zero concentration to give D_0. (from [10].)

nan oligosaccharides (HA_{20-26}) to affect the self-diffusion of HA (2 mg/ml), even at an oligosaccharide concentration twice that of the full-length HA. These results therefore also suggest that intermolecular chain-chain associations are not important in determining the concentrated solution properties of HA [11].

Comparison of hyaluronan with aggrecan

As hyaluronan and aggrecan are both high molecular weight polyanions and chondroitin sulphate is a glycosaminoglycan with a structure closely related to HA, it is interesting to compare some of their physical properties. Aggrecan has a more compact structure and smaller hydrodynamic domain than HA of comparable mass, because it has a branched structure and chondroitin sulphate does not have the same chain stiffening as HA. Comparison of the concentration dependence of aggrecan self diffusion with hyaluronan, thus shows that aggrecan of 2.6 million average molecular weight is comparable to HA of 0.9 million molecular weight (Fig. 9). Tracer studies also showed aggrecan solutions in the semi-dilute range of concentrations to be more permeable than HA, even when compared at similar chain concentrations

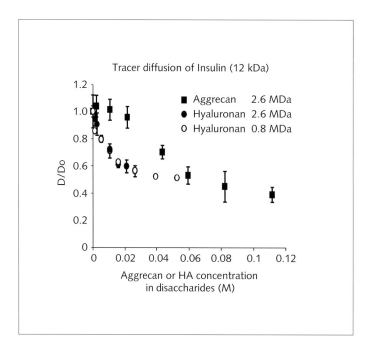

Figure 10
Comparison of tracer diffusion in aggrecan and hyaluronan solutions of different concentration. Lateral translational diffusion coefficients of FITC-insulin (12 kDa) were determined in solutions of hyaluronan (○) (Mw 800 kDa), hyaluronan (●) (Mw 2,600 kDa) and in aggrecan (■) (Mw 2,500 kDa). Concentrations are represented as disaccharide content in order to compare HA and aggrecan preparations on the basis of their glycosaminoglycan chain concentration. Disaccharide 0.1 M is ~40 mg/ml for HA and ~50 mg/ml for aggrecan. All measurements in PBS at 25°C. (Gribbon, Heng and Hardingham, unpublished results.)

(Fig. 10). This may be caused by the chondroitin sulphate chains being more flexible and/or dynamic than HA chains, but may also be because aggrecan's branched structure will form inhomogeneous solutions that will only tend to homogeneity at very high concentrations. Indeed, this difference in permeability became small at higher concentrations (~20 mg/ml) and was then more closely related to chain density in the solution. Tracer diffusion in concentrated HA solutions was also independent of HA molecular weight (Fig. 10).

These simple comparisons show aggrecan to be a more compact polyanion, perhaps better suited to the very high concentration found in cartilaginous tissues, which is essential for the compressive resilience of the tissue. As chondroitin sulphate is more highly charged than HA because of its sulphation, its compactness is also complemented by a higher charge density and therefore greater osmotic effect. The extracellular assembly of aggrecan into aggregates bound to HA creates a higher order structure, which further reduces its diffusion, although comparison between aggrecan aggregates and monomers shows a very similar form of concentration dependence. This suggests that aggregation does not result in a more compact or less permeable structure, but a structure with enhanced physical properties due to the linkage between many monomers. As with hyaluronan solutions, concentrated aggrecan aggregate solutions do not form a gel. Aggregates remain mobile within the solution, albeit with much lower diffusion coefficients than monomers. Aggrecan may thus be seen to fulfil a rather similar role to hyaluronan, but it has perhaps evolved for its function in the cartilage ECM, to provide a matrix that retains permeability to small macromolecules, in spite of its very high concentration.

Conclusions

The data presented in these studies [10, 11] suggest that intramolecular hydrogen bonds and polyanionic properties of hyaluronan both contribute to provide a highly expanded macromolecular conformation. However, under physiological conditions of ionic strength, the results predict the electrostatic effects to be modest. The major contribution to the large hydrodynamic volume of hyaluronan and, hence its other important non-Newtonian viscoelastic properties, are due to hydrogen bonding between adjacent saccharides. This restricts rotation and flexion at the glycosidic bonds and creates a stiffened polymer chain. The flexibility and permeability properties of the hyaluronan network can then be accounted for in terms of interchain hydrodynamic interactions of this extended structure, with entanglement being especially important at elevated concentrations. However, even at high concentrations, under physiological conditions, individual hyaluronan chains remain mobile and at no stage do hyaluronan solutions undergo transition to a gel-like state. These observations are incompatible with any significant degree of intermolecular association that is stable or co-operative. The results suggest that even at high concentrations, the properties can be directly predicted from the behaviour in dilute solution and a simple hydrodynamic model, with chains stiffened by hydrogen bonds and with domain overlap and entanglement at high concentration, accounts for the major solution properties of hyaluronan.

Biology on hyaluronan

How do these properties relate to the known biology of hyaluronan? The average turnover of HA in the body is reported to be quite fast, which shows that most hyaluronan is not bound in "long lasting structures", and examples, such as the aggrecan aggregates with HA in cartilage, must be an exception rather than the rule. The interactions of hyaluronan with protein ligands are highly specific and biologically important, but they may not dominate most HA function most of the time. From these studies it may be proposed that hyaluronan functions in most ECMs to favour disorder, and in this presentation, measurements of self-diffusion and tracer diffusion reveal that even at high concentration it forms a dynamic, mobile and permeable network, but does not self-associate, or form a cross-linked gel. This property of hyaluronan of non-interaction has its own functional significance, for it is only against a background of non-interaction that specificity and fidelity is achieved between interacting ECM partners to fulfil biological functions. Non-interaction amongst ECM components thus has an important part to play in matrix organisation, and I would suggest that amongst the properties of hyaluronan that have favoured its evolutionary selection as a component of ECMs, in addition to its inherent rheological and viscoelastic features, must be included its non-interactive properties; its significant non-engagement with most other ECM proteins and glycoconjugates, and particularly its lack of association with itself.

Hyaluronan is the Tween 20 of the ECM! It helps to hydrate and provide a disordered, dynamic and permeable matrix compartment that facilitates cellular functions, involving nutrient and waste product exchange, local and systemic communication between cells through cytokines, growth factors and hormones and provides the environment for the ordered assembly of fibrillar structures that form the structural scaffold for tissues.

Acknowledgement
We are very grateful to The Wellcome Trust (UK) and Seikagaku Corporation (Tokyo, Japan) for support for these studies.

References

1. Balazs EA, Gibbs DA (1970) The rheological properties and the biological function of hyaluronic acid. In: EA Balazs (ed): *Chemistry and molecular biology of the intercellular matrix*. Academic Press, London, New York, 1241–1254
2. Laurent TC (1995) Structure of the extracellular matrix and the biology of hyaluronan. In: RK Reed, NG McHale, JL Bert, CP Winlove, GA Laine (eds): *Intersitium, connective tissue and lymphatics*. Portland Press, London, 1–12
3. Lapcik L Jr, Lapcik L, De Smedt S, Demeest, J, Chabrecek P (1998) HA: preparation, structure, properties and applications. *Chemical Rev* 98: 2663–2684
4. Morris ER, Rees DA, Welsh EJ (1980) Conformation and dynamic interactions in HA solutions. *J Mol Biol* 138: 383–400

5   Wik KO, Comper WD (1982) Hyaluronate diffusion in semi-dilute solutions. *Biopolymers* 21: 583–599
6   Sheehan JK, Arundel C, Phelps CF (1983) Effect of the cations sodium, potassium and calcium on the interactions of hyaluronate chains: a light scattering and viscometric study. *Int J Biol Macromol* 5: 222–228
7   Reed CE, Li X, Reed WF (1989) The hydrodynamic scaling model for polymer self-diffusion. *Biopolymers* 28: 1981–2000
8   Almond A, Sheehan JK, Brass A (1997) Molecular dynamics simulations of the disaccharides of hyaluronan in solution. *Glycobiology* 7: 597–604
9   Hardingham TE, Gribbon P, Heng, BC (1999) New approaches to the investigation of hyaluronan networks. *Biochem Soc Trans* 27: 124–127
10 Gribbon P, Heng B C, Hardingham TE (1999) The molecular basis of the solution properties of hyaluronan investigated by confocal fluorescence recovery after photobleaching. *Biophys J* 77: 2210–2216
11 Gribbon P, Heng BC, Hardingham TE (2000) The analysis of intermolecular interactions in concentrated hyaluronan solutions suggest no evidence for chain-chain association. *Biochem J* 350: 329–335
12 Engel J (1989) Figure1 and Discussion comment. The biology of hyaluronan. *Ciba Foundation Symposium* 143: 18–19
13 Scott JE, Cummings C, Brass A, Chen Y (1991) Secondary and tertiary structures of HA in aqueous solution, investigation by rotary shadowing electron microscopy and computer simulation. *Biochem J* 274: 699–705
14 Scott JE, Heatley F (1999) Hyaluronan forms specific stable tertiary structures in aqueous solution: A C-13 NMR study. *Proc Natl Acad Sci USA* 96: 4850–4855
15 Glabe CG, Harty PK, Rosen SD (1983) Preparation and properties of fluorescent polysaccharides. *Anal Biochem* 130: 287–294
16 Kubitscheck H, Wedekind P, Peters R (1994) Lateral diffusion measurements at high spatial resolution by scanning microphotolysis in a confocal microscope. *Biophys J* 67: 946–965
17 Gribbon P, Hardingham TE (1998) Macromolecular diffusion of biological polymers measured by confocal fluorescence recovery after photobleaching. *Biophys J* 75: 1032–1039
18 Hardingham TE, Gribbon P (2000) Confocal-FRAP analysis of ECM molecular interactions. In: C Strueli, M Grant (eds): *Methods in molecular biology*, Vol 139, Extracellular matrix protocols. Humana Press, Totowa, NJ, 83–93
19 Callaghan PT, Pinder DN (1984) Influence of multiple length scales on the behaviour of polymer self-diffusion in the semidilute region. *Macromolecules* 17: 431–437
20 Phillies DJ (1989) The hydrodynamic scaling model for polymer self-diffusion. *J Phys Chem* 93: 5029–5039
21 Imhoff A, Van Blaadren A, Maret G, Mallema J, Dhont JKG (1994) A comparison between the long time self diffusion of and low shear viscosity of concentrated dispersions of charged colloidal silica spheres. *J Chem Phys* 100: 2170–2181

22　Almond A, Brass A, Sheehan JK (1998) Deducing polymeric structure from aqueous molecular dynamics simulations of oligosaccharides: predictions from simulations of hyaluronan tetrasaccharides compared with hydrodynamic and X-ray fibre diffraction data. *J Mol Biol* 284: 1425–1437
23　Almond A, Brass A, Sheehan JK (1998) Dynamic exchange between stabilized conformations predicted for hyaluronan tetrasaccharides: comparison of molecular dynamics simulations with available NMR data. *Glycobiology* 8: 973–980
24　Ghosh S, Khobal I, Zanette D, Reed WF (1993) Conformational contraction and hydrolysis of hyaluronate in sodium hydroxide solutions. *Macromol* 26: 4684–4691

Lectin domains in hyaluronan-binding proteoglycans

Anders Aspberg

Connective Tissue Biology, Lund University, BMC, C12, SE-221 84 Lund, Sweden

The large aggregating proteoglycans aggrecan, versican, neurocan or brevican are important components of extracellular matrices (ECM). These molecules all have a globular domain in each end joined by an extended central region. These structures have different functions. The N-terminal globular domain (G1) anchors the molecule to the long filaments of hyaluronan, The central region carries the glycosaminoglycan chains. The C-terminal G3 domain shows several interactions. It consists of several structural motifs: EGF-like repeat(s), a C-type lectin repeat, and a complement regulatory protein-like (CRP) motif. C-type lectin repeats often mediate molecular interactions, as, for example, in the selectins, and this is also the case for the proteoglycans. We originally found that the C-type lectin of versican [1] and of the other family members [2] binds the ECM glycoprotein tenascin-R with varying affinity. These interactions are calcium-dependent and mediated through direct protein-protein binding to FnIII-repeat 4 of tenascin-R. They are independent of carbohydrates [2]. Other interactions of the proteoglycan C-type lectins, however, involve carbohydrate ligands on sulfated glycolipids [3].

Tenascin-R expression is restricted to the central nervous system, but aggrecan and versican is not. In searching for novel interaction partners at other locations we identified fibulin-1 [4]. This ECM protein binds both versican and aggrecan with high affinities, but does not interact with the brain-specific proteoglycans brevican or neurocan. In recent work we found that the lectins of aggrecan, versican and brevican bind fibulin-2 with even higher affinities than fibulin-1 and tenascin-R [5]. The binding site was mapped to similar calcium binding EGF repeats as in fibulin-1. We were also able to actually demonstrate cross-linking of hyaluronan/aggrecan aggregates by fibulins through negative staining EM of the protein complexes (Fig. 1). All these ligands exist as dimers or multimers in the tissue, suggesting that the proteoglycan G3 domains mediate cross-linking of hyaluronan/proteoglycan complexes through binding tenascins or fibulins. This may be of functional importance for the organization of newly produced extracellular matrix in development and tissue repair.

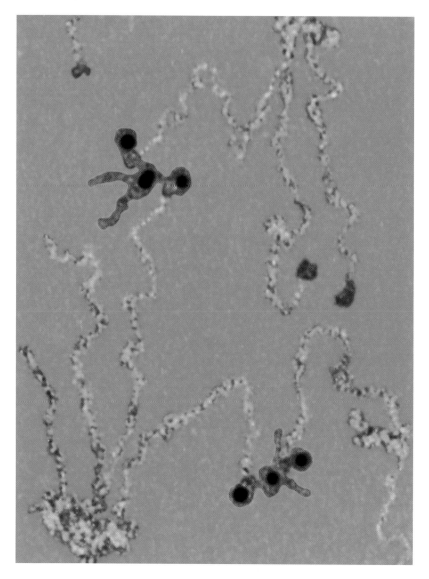

Figure 1.
Fibulin-2 cross-links hyaluronan/aggrecan complexes. This pseudocolored negative staining electron micrograph shows fibulin-2 dimers (colored red) binding the G3 domains of aggrecan core proteins (colored blue) extending from short aggregates purified from rat chondrosarcoma under native conditions. The black dots in the globular domains of fibulin2 molecules are from the thiocyanate gold labeling. Note that the fibulin-2 molecules bind aggrecan core proteins from different aggregates, forming a mesh-work. The image was prepared by Dr Matthias Mörgelin (for details see [5]).

There is coexpression of hyaluronan synthase-2 (HAS-2), proteoglycan and G3 ECM ligands such as tenascins and fibulins during heart formation and endochondral ossification. The HAS-2 knock-out [6] and the *heart defect* versican null mice [7] show similar failure to form endocardial cushion tissue during heart development, leading to death at embryonic day 10.5. The transient strong expression of fibulins 1 and 2 in this tissue, as well as in the cartilage template during endochondral ossification, suggests involvement in organization of the aggregates.

Several questions remain unanswered regarding the functions of the G3 domains. Most fundamentally, the *in vivo* relevance of the ECM molecule cross-linking of proteoglycan/hyaluronan aggregates has yet to be confirmed. We are addressing this question through deletion of the aggrecan lectin domain by gene targeting. The resulting mice should produce aggrecan, albeit lacking the lectin domain.

Both fibulin-1 and -2 are strongly expressed in embryonic cartilage, but absent from adult cartilage. It is possible that other proteins interacting with the proteoglycan lectin domain will be found in the adult cartilage matrix. In the mature cartilage, however, aggrecan is frequently proteolytically modified. This produces truncated proteoglycans, lacking the G3 domain but remaining anchored to the hyaluronan in the tissue. The fraction of full-length aggrecan molecules in the tissue decreases with age [8]. This suggests that the primary role of these interactions is in the organization of the forming extracellular matrix. It will be interesting to investigate whether these components are upregulated in response to damage, e.g., in wound healing, atherosclerosis and arthritis.

Interestingly, the aggrecan C-type lectin repeat is constitutively expressed, whereas the flanking EGF and CRP repeats are subjected to alternative splicing. It is not known whether the alternatively spliced repeats influence the interactions of the lectin. The flanking modules may of course have completely different functions – it has, for example, been reported that recombinant versican EGF-like repeats have effects on chondrocyte proliferation [9] and mesenchymal chondrogenesis [10] in explant cultures.

References

1 Aspberg A, Binkert C, Ruoslahti E (1995) The versican C-type lectin domain recognizes the adhesion protein tenascin-R. *Proc Natl Acad Sci USA* 92: 10590–10594
2 Aspberg A, Miura R, Bourdoulous S, Shimonaka M, Heinegård D, Schachner M, Ruoslahti E, Yamaguchi Y (1997) The C-type lectin domains of lecticans, a family of aggregating chondroitin sulfate proteoglycans, bind tenascin-R by protein-protein interactions independent of carbohydrate moiety. *Proc Natl Acad Sci USA* 94: 10116–10121
3 Miura R, Aspberg A, Ethell IM, Hagihara K, Schnaar RL, Ruoslahti E, Yamaguchi Y (1999) The proteoglycan lectin domain binds sulfated cell surface glycolipids and promotes cell adhesion. *J Biol Chem* 274: 11431–11438

4   Aspberg A, Adam S, Kostka G, Timpl R, Heinegård D (1999) Fibulin-1 is a ligand for the C-type lectin domains of aggrecan and versican. *J Biol Chem* 274: 20444–20449
5   Olin AI, Mörgelin M, Sasaki T, Timpl R, Heinegård D, Aspberg A (2001) The proteoglycans aggrecan and versican form networks with fibulin-2 through their lectin domain binding. *J Biol Chem* 276: 1253–1261
6   Camenisch TD, Spicer AP, Brehm-Gibson T, Biesterfeldt J, Augustine ML, Calabro A, Jr., Kubalak S, Klewer SE, McDonald JA (2000) Disruption of hyaluronan synthase-2 abrogates normal cardiac morphogenesis and hyaluronan-mediated transformation of epithelium to mesenchyme. *J Clin Invest* 106: 349–360
7   Mjaatvedt CH, Yamamura H, Capehart AA, Turner D, Markwald RR (1998) The Cspg2 gene, disrupted in the hdf mutant, is required for right cardiac chamber and endocardial cushion formation. *Dev Biol* 202: 56–66
8   Dudhia J, Davidson CM, Wells TM, Vynios DH, Hardingham TE, Bayliss MT (1996) Age-related changes in the content of the C-terminal region of aggrecan in human articular cartilage. *Biochem J* 313: 933–940
9   Zhang Y, Cao L, Yang BL, Yang BB (1998) The G3 domain of versican enhances cell proliferation via epidermial growth factor-like motifs. *J Biol Chem* 273: 21342–21351
10 Zhang Y, Cao L, Kiani CG, Yang BL, Yang BB (1998) The G3 domain of versican inhibits mesenchymal chondrogenesis via the epidermal growth factor-like motifs. *J Biol Chem* 273: 33054–33063

The matrilins: A novel family of extracellular adaptor proteins

Mats Paulsson, Andreas R. Klatt, Birgit Kobbe, D. Patric Nitsche and Raimund Wagener

Institute for Biochemistry, Medical Faculty, University of Cologne, Joseph-Stelzmann-Str. 52, D-50931 Cologne, Germany

Introduction

Matrilins are characterised by their modular structure made up from one or two von Willebrand Factor A-like domains (vWFA), a variable number of epidermal growth factor-like domains (EGF) and a coiled-coil α-helical domain, allowing the assembly of three or four subunits into a complete molecule (Fig. 1). Four matrilins have been described, and from analysis of the preliminary version of the human genome data base and from the consideration that matrilin genes are found on portions of the human genome that have undergone two duplications [1], we can be certain that these represent all members of this family.

Structure

Matrilin-1 (cartilage matrix protein, CMP) was the first family member to be discovered [2] and is by far the best studied matrilin. It typically forms homotrimers of 52 kDa subunits [3] that can be discerned in the electron microscope as three ellipsoids connected at a central point (Fig. 2) [4]. Matrilin-2 is the largest of all matrilins with subunits of 104 kDa. Its two vWFA domains are connected by ten EGF domains, and it additionally contains a unique segment between the C-terminal vWFA domain and the coiled coil [5]. When recombinantly expressed in human cells or when isolated from tissues, a complex spectrum of molecular forms from monomers to tetramers are found [6]. This heterogeneity is likely to be due to a limited proteolysis of an original tetramer (see below). Each matrilin-2 subunit has the shape of a doughnut, presumably due to interactions between the two vWFA domains causing the EGF domains to form a loop structure (Fig. 2). The unique region forms a flexible hinge connecting the rest of the subunit to the central tetrameric coiled coil. Matrilin-3 is structurally unique in containing only the N-terminal vWFA domain [7, 8], giving the 49 kDa subunit a tadpole shape (Fig. 2). On its own it assembles into tetramers, but in tissues it may also coassemble with matrilin-1 to yield mixed tetramers of varying stoichiometry [9, 10]. Also trimers

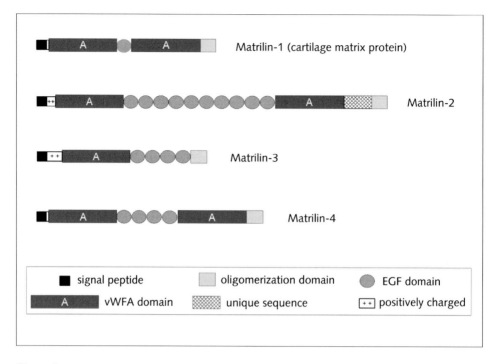

Figure 1
Comparison of the domain structures of the members of the matrilin family (modified from [25]).

are found, but these may be products of proteolytic processing. Finally, the 66 kDa matrilin-4 subunits assemble into trimers which have a more extended shape than those of other matrilins [11], presumably because of the vWFA domains in this matrilin being less prone to self-interactions (Fig. 2). Further variability of matrilins is possible as many alternative spliced mRNAs have been detected.

Proteolytic processing

A striking feature of all matrilins except matrilin-1 is that when recombinantly expressed in human cell lines or when extracted from tissues, a complex mixture of oligomeric forms is found. This could be due either to the formation of coiled-coil α-helices from a variable number of strands or to a proteolytic cleavage close to the assembly domain, releasing an almost complete subunit from the parent molecule. The notion of the same polypeptide forming coiled coils with different numbers of strands goes against prior experience of such structures, which tend to

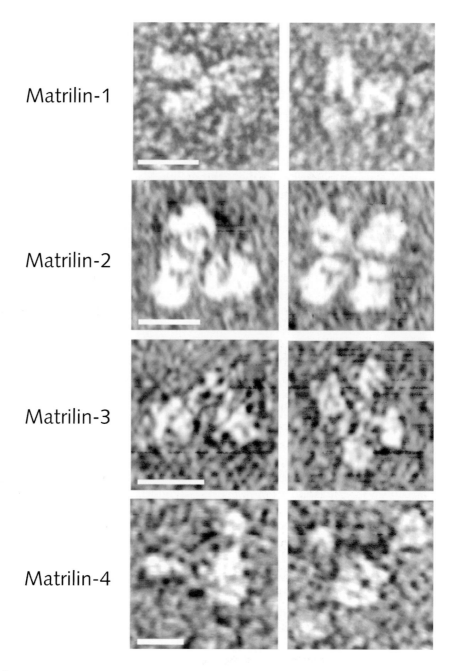

Figure 2
Electron micrographs of the matrilins obtained after negative staining. The bar represents 10 nm.

be highly specific in their assembly. Final proof for the occurrence of proteolytic cleavage came from studies of matrilin-4 [11]. In recombinantly produced material, a ladder of trimers, monomers and dimers was seen by SDS-PAGE or MALDI-TOF mass spectrometry. However, closer analysis showed that dimers had 7 kDa more mass than expected and monomers 14 kDa more than the subunit mass. On reduction this additional mass was released from the molecule in the form of 7 kDa fragments. Further, an endogenous 21 kDa fragment representing the trimeric coiled-coil domain could be isolated that gave the N-terminal sequence GIGAGTELRS. This sequence is found about 7 kDa from the C-terminus of the matrilin-4 subunit, just before the disulfide bonds that stabilise the coiled coil at its N-terminal end. It could be concluded that proteolytic cleavage occurs at this site, releasing the rest of the subunit, but leaving a complete trimeric coiled coil. The cleavage may affect one, two or three subunits while some molecules are completely intact. Directly at the N-terminal side of the cleavage site, two glutamic acid residues occur. This EE motif is conserved in all matrilins, and in matrilin-2, which is highly prone to degradation, it occurs a second time in the unique hinge region. We suspect that this motif is a recognition signal for cleavage in all matrilins. The distance between this sequence and the disulfide bonds varies between matrilins and is the shortest in matrilin-1. The stability of matrilin-1 to proteolysis may be due to sterical hindrance decreasing the efficiency of cleavage. In the case of matrilin-4, degradation products of very similar size are found in tissues and in cell culture media from cells expressing recombinant protein, indicating that the same mechanism operates *in vivo*. The cleavage appears to take place at or closely after secretion, indicating that the so far unidentified protease is localised in the secretory pathway or at the cell surface. The partial degradation of matrilins may affect their overall avidity for ligands through a loss of cooperativity, even though the affinity for a binding site contained in a single subunit will not be affected. Our hypothesis is that the proteolytic processing of matrilins may be a physiological mechanism for the modulation of matrix assembly.

Expression

Even though matrilin-1 was isolated from cartilage and is mainly expressed in this tissue [2, 12], other matrilins have a wider distribution. Matrilin-2 is found in tendons and ligaments, in many kinds of loose connective tissue and also associated with basement membranes [6]. In cartilage, it is weakly expressed as compared to other matrilins and mainly found in the proliferative and upper hypertrophic layers of the epiphyseal growth plate. Matrilin-3 has an expression in the skeletal system which is similar, but not identical to that of matrilin-1 [10]. Fine differences seen in differentiating cartilages show that synthesis of these two matrilins is separately regulated, despite the fact that they can form heterooligomers [13]. Matrilin-4 is the most ubiquitous matrilin, being found both inside and outside the skeletal system and being present in every tissue where another matrilin is expressed.

The temporal regulation of matrilin expression has not yet been fully explored, even though it is clear that matrilins take on their final expression domain first after birth. For example, in mouse embryos, matrilin-1 is transiently expressed in heart tissue and also detected at tendon insertions, in the pericartilaginous fibrous stroma and in the dermis [14]. Matrilin-2 on the other hand is more abundant in the skeletal system of embryos than in newborns. While matrilin-3 is highly expressed in cartilage of embryos and newborns, it is scarce in later life when matrilin-1 becomes the much predominating cartilage matrilin [9, 13]. The consequences of this differential regulation of matrilin genes is unclear, but it is closely correlated with events in skeletal development and maturation and may well be important to provide correct matrix organisation at each stage and location.

Supramolecular assembly

Matrilin-1 was first identifed as a protein tightly associated with aggrecan [2]. Later studies showed that it is bound to distinct sites of the core protein in its chondroitin sulfate attachment region and that this linkage may even become covalent [15]. Further, matrilin-1 was shown to be associated with cartilage collagen fibrils, but also to be found in a collagen independent fibrillar network [16, 17]. The affinity of matrilins for both collagens and proteoglycans has led to the assumption that they may have an adaptor function and mediate interactions between these major classes of matrix macromolecules. Also matrilin-3 is found in these two fibrillar systems in cartilage [10]. The collagen-free matrilin-3 filaments are pericellular, at least in cell culture, and appear to form a basket structure around the chondrocyte (Fig. 3). Upon cell division, the network connects the two cells formed as they move away from each other [10]. Matrilin-1 has been proposed as a ligand for integrin $\alpha1\beta1$ [18] and cell association of other matrilins may also be integrin mediated.

Genetic approaches to the study of matrilin function

Gene targeting of the matrilin-1 gene has been performed in mouse [19, 20]. Neither of the two mouse lines produced showed gross morphological changes, while in one case [20] fine disturbances in cartilage collagen fibril formation was reported. In the light of the discovery of three additional matrilins with related molecular properties and overlapping expression domains, this result is not surprising and is likely to reflect a redundancy among matrilins. Possibly, two or more matrilin genes will have to be inactivated in the same animal before phenotypes will be observed. Clearer evidence for an important function of matrilins has recently come from

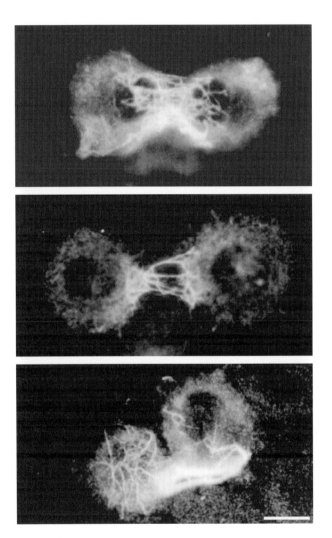

Figure 3
Immunofluorescence microscopy of matrilin-3 containing filaments in the pericellular matrix of cultured Swarm rat chondrosarcoma cells. The bar represents 5 μm.

human genetic studies. Families with multiple epihyseal dysplasia were recently shown to carry mutations in the gene coding for matrilin-3 [21]. In other families with similar phenotypes, mutations in COMP and collagen IX have been shown. The related disease pseudoachondroplasia is caused by COMP mutations and it appears that these may affect COMP conformation as well as interactions between COMP and collagens [22–24]. The fact that mutations in matrilin-3, the thrombospondin family member COMP, and the FACIT collagen type IX have similar consequences may possibly reflect common roles in mediating interactions between collagen fibrils and molecules in the surrounding ground substance.

Acknowledgements

Our work on matrilins is supported by grants from the "Deutsche Forschungsgemeinschaft" and the "Köln Fortune" program of the Medical Faculty, University of Cologne.

References

1   Gibson TJ, Spring J (2000) Evidence in favour of ancient octaploidy in the vertebrate genome. *Biochem Soc Trans* 28: 259–264
2   Paulsson M, Heinegård D (1979) Matrix proteins bound to associatively prepared proteoglycans from bovine cartilage. *Biochem J* 183: 539–545
3   Paulsson M, Heinegård D (1981) Purification and structural characterization of a cartilage matrix protein. *Biochem J* 197: 367–375
4   Hauser N, Paulsson M (1994) Native cartilage matrix protein (CMP). A compact trimer of subunits assembled via a coiled-coil α-helix. *J Biol Chem* 269: 25747–25753
5   Deák F, Piecha D, Bachrati C, Paulsson M, Kiss I (1997) Primary structure and expression of matrilin-2, the closest relative of cartilage matrix protein within the von Willebrand factor type A module superfamily. *J Biol Chem* 272: 9268–9274
6   Piecha D, Muratoglu S, Mörgelin M, Hauser N, Studer D, Kiss I, Paulsson M, Deák F (1999) Matrilin-2, a large, oligomeric matrix protein is expressed by a great variety of cells and forms fibrillar networks. *J Biol Chem* 274: 13353–13361
7   Wagener R, Kobbe B, Paulsson M (1997) Primary structure of matrilin-3, a new member of a family of extracellular matrix proteins related to cartilage matrix protein (matrilin-1) and von Willebrand factor. *FEBS Lett* 413: 129–134
8   Belluoccio D, Trueb B (1997) Matrilin-3 from chicken cartilage. *FEBS Lett* 415: 212–216
9   Wu JJ, Eyre DR (1998) Matrilin-3 forms disulfide-linked oligomers with matrilin-1 in bovine epiphyseal cartilage. *J Biol Chem* 273: 17433–174338
10 Klatt AR, Nitsche DP, Kobbe B, Mörgelin M, Paulsson M, Wagener R (2000) Molecular structure and tissue distribution of matrilin-3, a filament-forming extracellular matrix protein expressed during skeletal development. *J Biol Chem* 275: 3999–4006
11 Klatt AR, Nitsche DP, Kobbe B, Macht M, Paulsson M, Wagener R (2001) Molecular structure, processing and tissue distribution of matrilin-4. *J Biol Chem* 276: 17267–17275
12 Paulsson M, Heinegård D (1982) Radioimmunoassay of the 148-kilodalton cartilage protein. Distribution of the protein among bovine tissues. *Biochem J* 207: 207–213

13 Klatt AR, Paulsson M, Wagener R (2002) Expression of matrilins during maturation of mouse skeletal tissues. *Matrix Biol* 21: 289–296
14 Segat D, Frie C, Nitsche DP, Klatt AR, Piecha D, Korpos E, Deák F, Wagener R, Paulsson M, Smyth N (2000) Expression of matrilin-1, -2 and -3 in developing mouse limbs and heart. *Matrix Biol* 19: 549–555
15 Hauser N, Paulsson M, Heinegård D, Mörgelin M (1996) Interaction of cartilage matrix protein (CMP) with aggrecan. Increased covalent crosslinking with maturation. *J Biol Chem* 271: 32247–32252
16 Winterbottom N, Tondravi MM, Harrington TL, Klier FG, Vertel BM, Goetinck PF (1992) Cartilage matrix protein is a component of the collagen fibril of cartilage. *Dev Dyn* 193: 266–276
17 Chen Q, Johnson DM, Haudenschild DR, Tondravi MM, Goetinck PF (1995) Cartilage matrix protein forms a type II collagen-independent filamentous network: analysis in primary cell cultures with a retrovirus expression system. *Mol Biol Cell* 6: 1743–1753
18 Makihira S, Yan W, Ohno S, Kawamoto T, Fujimoto K, Okimura A, Yoshida E, Noshiro M, Hamada T, Kato Y (1999) Enhancement of cell adhesion and spreading by a cartilage-specific noncollagenous protein, cartilage matrix protein. *J Biol Chem* 274: 11417–11423
19 Aszódi A, Bateman JF, Hirsch E, Baranyi M, Hunziker EB, Hauser N, Bösze Z, Fässler R (1999) Normal skeletal development of mice lacking matrilin 1: redundant function of matrilins in cartilage? *Mol Cell Biol* 19: 7841–7845
20 Huang X, Birk DE, Goetinck PF (1999) Mice lacking matrilin-1 (cartilage matrix protein) have alterations in type II collagen fibrillogenesis and fibril organization. *Dev Dyn* 216: 434–441
21 Chapman KL, Mortier GR, Chapman K, Loughlin J, Grant ME, Briggs MD (2001) Mutations in the region encoding the von Willebrand factor A domain of matrilin-3 are associated with multiple epiphyseal dysplasia. *Nat Genet* 28: 393–396
22 Chen H, Deere M, Hecht JT, Lawler J (2000) Cartilage oligomeric matrix protein is a calcium binding protein and a mutation in its type 3 repeats causes conformational changes. *J Biol Chem* 275: 26538–26544
23 Maddox BK, Mokashi A, Keene DR, Bächinger HP (2000) A cartilage oligomeric matrix protein mutation associated with pseudoachondroplasia changes the structural and functional properties of the type 3 domain. *J Biol Chem* 275: 11412–11417
24 Thur J, Rosenberg K, Nitsche DP, Pihlajamaa T, Ala-Kokko L, Heinegård D, Paulsson M, Maurer P (2001) Mutations in cartilage oligomeric matrix protein (COMP) causing pseudoachondroplasia and multiple epiphyseal dysplasia affect binding of calcium and collagen I, II, and IX. *J Biol Chem* 276: 6083–6092
25 Deák F, Wagener R, Kiss I, Paulsson M. (1999) The matrilins: a novel family of oligomeric extracellular matrix proteins. *Matrix Biol* 18: 55–66

Superficial zone protein (SZP) is an abundant glycoprotein in human synovial fluid with lubricating properties

Thomas M. Schmid[1], Jui-Lan Su[2], Kathie M. Lindley[2], Vitaliy Soloveychik[1], Lawrence Madsen[1], Joel A. Block[3], Klaus E. Kuettner[1] and Barbara L. Schumacher[4]

[1]Department of Biochemistry and [3]Section of Rheumatology, Rush Medical College at Rush-Presbyterian-St. Luke's Medical Center, 1653 W. Congress Parkway, Chicago, IL 60612, USA; [2]Department of Protein Sciences, GlaxoSmithKline Inc., Five Moore Drive, Research Triangle Park, NC 27709, USA; [4]Department of Bioengineering, University of California at San Diego, 9500 Gillman Dr., La Jolla, CA 92093, USA

Superficial zone protein (SZP) is a 345 kDa glycoprotein which contains chondroitin sulfate and keratan sulfate chains [1]. The molecule is synthesized by the superficial zone chondrocytes of articular cartilage, but not by middle and deep zone chondrocytes. It is also made by synovial cells and is present in synovial fluid. SZP shows a high degree of homology with megakaryocyte stimulating factor (MSF) [2], camptodactyly-arthropathy-coxa vara-pericarditis protein (CACP) [3] and lubricin [4].

SZP concentration in synovial fluid and serum

A panel of monoclonal antibodies to human SZP have been generated [5]. A sandwich ELISA has been developed to measure the concentration of SZP in body fluids. Human synovial fluids were collected from donor joints through a collaboration with the Regional Organ Bank of Illinois (ROBI) and also from rheumatology patients after arthrocentesis procedures with IRB approval. The sandwich ELISA utilizes the binding of SZP to peanut lectin-coated plates followed by detection with anti-SZP monoclonal antibodies. The assay was able to measure SZP concentrations from 30–2000 ng/ml. An analysis of over 50 human synovial fluids gave a range of SZP concentrations from about 100 µg/ml to 600 µg/ml. Therefore SZP is an abundant glycoprotein in synovial fluid. The average concentration of SZP was substantially higher than reported estimates of lubricin concentration based on purification protocol yields. The monoclonal antibodies also detected forms of SZP in plasma and serum. The concentration of SZP in these fluids was substantially lower than in synovial fluid.

Cartilage-derived SZP has lubricating properties

SZP purified from human articular cartilage and synovial fluid was tested for its ability to lubricate a cartilage-glass interface using a test device similar in design to that of Swann et al. [6]. SZP was purified from serum-free culture medium of human articular cartilage organ cultures. Superficial slices of human tali were maintained in DME medium plus insulin, transferrin and selenium. The medium was collected every 2 days for 6 weeks. SZP was purified from the conditioned medium by a combination of DEAE chromatography, ultracentrifugation on a cesium chloride gradient and gel filtration chromatography [1]. The lubrication test device measured the frictional force exerted on an adult rabbit phalangeal bone as its articular cartilage surface moves on a glass plate revolving on a modified record player. The forces are measured at several different speeds. PBS was used as a negative control; it showed large (> 5 g) frictional forces at intermediate and low speeds. Human synovial fluid was used as a positive control. It showed good lubrication, as measured by low frictional forces at intermediate and low speeds. Hyaluronan (3.5 mg/ml) and BSA (1 mg/ml) showed poor lubrication qualities similar to the PBS control. Purified SZP at 40 μg/ml showed good lubrication with low frictional forces at intermediate and low speeds. Dilutions of human synovial fluid as high as 1/10 fold still showed good lubrication.

The lubrication activity of a synovial fluid dilution was tested after incubation with anti-SZP monoclonal antibodies. A 1/10 dilution of human synovial fluid in PBS was incubated with an anti-SZP monoclonal antibody (250 μg/ml) in solution or with the same antibody coupled to Sepharose beads. The solid phase antibody beads were separated by centrifugation and the supernatant tested. As shown in Figure 1, the 1/10 dilution of human synovial fluid (solid circles) showed good lubrication activity with low frictional forces at low speeds. PBS (open squares) and PBS plus the soluble monoclonal antibody (open triangles) did not lubricate well and showed high frictional forces at the low speeds of 20 and 40 rpm. The solid phase antibody beads (open circles, solid line) removed the lubrication activity from the synovial fluid. Even the soluble antibody (open circles, dashed line) incubated in solution with the synovial fluid generated an intermediate lubrication activity between the synovial fluid and PBS samples. Thus cartilage-derived and synovial fluid SZP demonstrated lubrication activity, and incubation with an anti-SZP antibody altered this activity.

Acknowledgement
This work was supported in part by NIH grant AR39239 and a collaborative research agreement with GlaxoSmithKline Corporation.

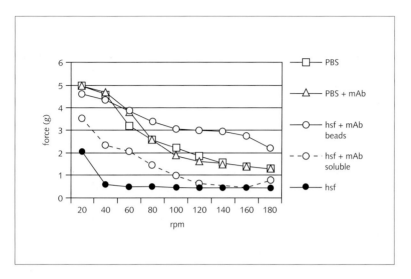

Figure 1
Effect of anti-SZP monoclonal antibody on the lubrication activity of human synovial fluid.

References

1 Schumacher BL, Block JA, Schmid TM, Aydelotte MB, Kuettner KE (1994) A novel proteoglycan synthesized and secreted by chondrocytes of the superficial zone of articular cartilage. *Arch Biochem Biophys* 311: 144–152
2 Flannery CR, Hughes CE, Schumacher BL, Tudor D, Aydelotte MB, Kuettner KE, Caterson B (1999) Articular cartilage superficial zone protein (SZP) is homologous to megakaryocyte stimulating factor precursor and is a multifunctional proteoglycan with potential growth-promoting, cytoprotective, and lubricating properties in cartilage metabolism. *Biochem Biophys Res Commun* 27: 535–541
3 Marcelino J, Carpten JD, Suwairi WM, Gutierrez OM, Schwartz S, Robbins C, Sood R, Makalowska I, Baxevanis A, Johnstone B et al (1999) CACP, encoding a secreted proteoglycan, is mutated in camptodactyly-arthropathy-coxavara-pericarditis syndrome. *Nature Genetics* 23: 319–322
4 Jay GD, Britt DE, Cha C-J (2000) Lubricin is a product of megakaryocyte stimulating factor gene expression by human synovial fibroblasts. *J Rheumatol* 27:594–600
5 Su J-L, Schumacher BL, Lindley KM, Soloveychik V, Burkhart W, Triantafillou JA, Kuettner K, Schmid T (2001) Detection of superficial zone protein in human and animal body fluids by crosspieces monoclonal antibodies specific to superficial zone protein. *Hybridoma* 20: 149–157
6 Swann DA, Hendren RB, Radin EL, Sotman SL, Duda EA (1981) The lubricating activity of synovial fluid glycoproteins. *Arthritis Rheum* 24: 22–30

Establishment of *in vitro* cell culture models for the investigation of the pathogenesis of cartilage diseases

Frank Zaucke, Robert Dinser, Patrik Maurer and Mats Paulsson

Institute for Biochemistry II, Medical Faculty, University of Cologne, Joseph-Stelzmann-Str. 52, D-50931 Cologne, Germany

Stabilisation of the chondrocyte phenotype

The investigation of cartilage abnormalities *in vitro* is hampered by the limited amount of available tissue. Thus, the chondrocytes have to be isolated and amplified in cell culture prior to a detailed biochemical analysis. However, it is well-known that chondrocytes dedifferentiate during extensive monolayer culture, losing not only their typical morphology but also changing their biosynthetic repertoire. The cells start to express collagen type I instead of the cartilage specific collage type II and shift to a more fibroblast-like phenotype [1, 2]. In addition to these well-established molecular markers for the chondrocyte phenotype, we have recently demonstrated that collagen type IX as well as non-collagenous matrix proteins, such as the cartilage oligomeric matrix protein (COMP), are downregulated very rapidly [3]. Although the function of some of these marker proteins remain to be elucidated, their presence might be of critical importance for the cartilage specific behaviour of the cells. To overcome the downregulation or even loss of chondrocyte-specific proteins, a variety of three-dimensional matrices, e.g., agarose, alginate or other complex polymers which stabilise or even re-induce the chondrocytic phenotype after dedifferentiation have been established [3–5]. The influence of a surrounding three-dimensional matrix on functional aspects is underscored in recent transplantation experiments in rabbits using chondrocytes embedded in alginate and agarose [6, 7]. For the development of our *in vitro* cell culture model, we have chosen the following approach (Fig. 1): directly after isolation, the primary articular chondrocytes were provided with the gene of interest and either maintained in a short-term monolayer culture or embedded directly after isolation into an alginate matrix and cultured for several weeks.

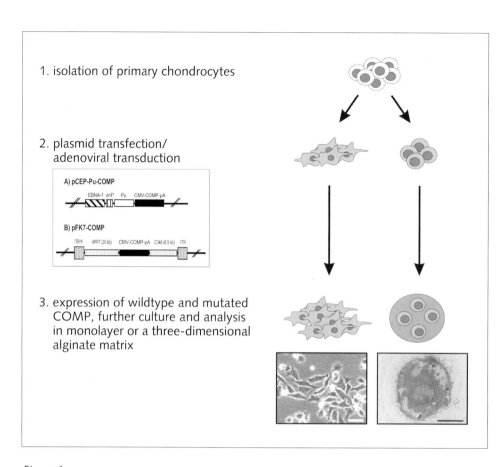

Figure 1
Approach to investigate cartilage diseases in an in vitro cell culture model. After isolation, the chondrocytes were immediately transfected/transduced with the vectors depicted schematically in the box (A, plasmid; B, adenovirus; described in detail in [10]). After gene transfer, two different culture conditions were used. The pictures below are showing a classical monolayer culture (light microscopy) and a chondrocyte encapsulated in alginate (transmission electron micrograph) after 1 week of culture (the bar represents 10 µm).

Transfection/transduction of primary chondrocytes

Genetically modified chondrocytes could serve as a useful tool in studies of genes involved in articular cartilage damage and repair and as a potential delivery method for therapeutic genes 4 [8–10]. In the present study, we wanted to investigate the effect of patient-derived mutations in the cartilage oligomeric matrix protein

(COMP) [11] in articular chondrocytes. It has been previously shown that mutations in COMP cause the dominantly inherited skeletal disorder pseudoachondroplasia (PSACH) which is characterised by disproportionate dwarfism, a waddling gait, joint laxity and severe premature osteoarthritis [12, 13]. Ultrastructural analysis of biopsies from affected patients revealed that COMP is retained in characteristic granular or lamellar inclusions within the endoplasmatic reticulum of chondrocytes [14, 15]. Before expressing different COMP constructs, we comparatively evaluated plasmid-based and adenoviral methods using enhanced green fluorescent protein as a reporter to optimize transfection/transduction efficiency in primary cells [10]. With these two approaches for gene transfer, we finally achieved long-term expression of wild type and mutated forms of COMP.

Analysis of chondrocyte cultures expressing PSACH associated mutations

Using the alginate-based cell culture model (Fig. 1) and the transgenic expression of mutated COMP in primary chondrocytes, we could reproduce the known features of PSACH *in vitro*, including the accumulation of COMP in the endoplasmatic reticulum and the co-retention of other matrix proteins. Moreover, the observations made in this model allow some additional conclusions: (1) the overexpression of mutated COMP affected the cell viability in a dose-dependent manner; (2) the secretion of COMP was not totally abolished, but was markedly delayed; (3) the extracellular matrix is composed of disrupted collagen fibrils and amorphous aggregates containing mutated COMP. In conclusion, the alginate bead culture in combination with an adenoviral transduction approach represents a promising *in vitro* model to investigate the pathogenic mechanisms of other cartilage diseases.

References

1 Abbott J, Holtzer H (1966) The loss of phenotypic traits by differentiated cells. III. The reversible behavior of chondrocytes in primary cultures. *J Cell Biol* 28: 473–487

2 Benya PD, Padilla SR, Nimni ME (1978) Independent regulation of collagen types by chondrocytes during the loss of differentiated function in culture. *Cell* 15: 1313–1321

3 Zaucke F, Dinser R, Maurer P, Paulsson M (2001) Cartilage oligomeric matrix protein (COMP) and collagen IX are sensitive markers for the differentiation state of primary articular chondrocytes. *Biochem J* 358: 17–24

4 Häuselmann HJ, Fernandes RJ, Mok SS, Schmid TM, Block JA, Aydelotte MB, Kuettner KE, Thonar EJ (1994) Phenotypic stability of bovine articular chondrocytes after long-term culture in alginate beads. *J Cell Sci* 107: 17–27

5 Benya PD, Shaffer JD (1982) Dedifferentiated chondrocytes reexpress the differentiated collagen phenotype when cultured in agarose gels. *Cell* 30: 215–224

6   Rahfoth B, Weisser J, Sternkopf F, Aigner T, von der Mark K, Brauer R (1998) Transplantation of allograft chondrocytes embedded in agarose gel into cartilage defects of rabbits. *Osteoarthritis Cartilage* 6: 50–65

7   Fragonas E, Valente M, Pozzi-Mucelli M, Toffanin R, Rizzo R, Silvestri F, Vittur F (2000) Articular cartilage repair in rabbits by using suspensions of allogenic chondrocytes in alginate. *Biomaterials* 21: 795–801

8   Baragi VM, Renkiewicz RR, Qiu L, Brammer D, Riley JM, Sigler RE, Frenkel SR, Amin A, Abramson SB, Roessler BJ (1997) Transplantation of adenovirally transduced allogeneic chondrocytes into articular cartilage defects *in vivo*. *Osteoarthritis Cartilage* 5: 275–282

9   Madry H., Trippel SB (2000) Efficient lipid-mediated gene transfer to articular chondrocytes. *Gene Ther* 7: 286–291

10 Dinser R, Kreppel F, Zaucke F, Blank C, Kochanek S, Paulsson M, Maurer P (2001) Comparison of long-term transgene expression after non-viral and adenoviral gene transfer into primary articular chondrocytes. *Histochem Cell Biochem* 116: 69–77

11 Hedbom E, Antonsson P, Hjerpe A, Aeschlimann D, Paulsson M, Rosa-Pimentel E, Sommarin Y, Wendel M, Oldberg Å, Heinegård D (1992) Cartilage matrix proteins. An acidic oligomeric protein (COMP) detected only in cartilage. *J Biol Chem* 267: 6132–6136

12 Briggs MD, Hoffman SM, King LM, Olsen AS, Mohrenweiser H, Leroy JG, Mortier GR, Rimoin DL, Lachman RS, Gaines ES et al (1995) Pseudoachondroplasia and multiple epiphyseal dysplasia due to mutations in the cartilage oligomeric matrix protein gene. *Nat Genet* 10: 330–336

13 Hecht JT, Nelson LD, Crowder E, Wang Y, Elder FF, Harrison WR, Francomano CA, Prange CK, Lennon GGM, Deere M, Lawler J (1995) Mutations in exon 17B of cartilage oligomeric matrix protein (COMP) cause pseudoachondroplasia. *Nat Genet* 10: 325–329

14 Stanescu R, Stanescu V, Muriel MP, Maroteaux P (1993) Multiple epiphyseal dysplasia, Fairbank type: morphologic and biochemical study of cartilage. *Am J Med Genet* 45: 501–507

15 Maynard JA, Cooper RR, Ponseti IV (1972) A unique rough surfaced endoplasmic reticulum inclusion in pseudoachondroplasia. *Lab Invest* 26: 40–44

Discussion

Electron microscope studies of collagen fibril formation in cornea, skin and tendon: implications for collagen fibril assembly and structure in other tissues

Q: *Ian Nieduszynski*: Can you identify if decorin is one of the molecules at the fibril surface?

A: *Karl Kadler*: We need to go back and do that. We plan to do immunolabeling combined with 3D AET to identify the surface molecules. We have a number of candidates. Decorin is going to be there and most probably lumican, also.

Q: *Robin Poole*: I am intrigued about the fusion of the C-tips. Do you have an explanation why this fusion occurs at the C-tip, as opposed to the N-end-tip? Is there some special chemical or structural property?

A: *Karl Kadler*: We are attempting to determine the high-resolution structure of the C-tips using tomography. I suspect that the recognition site for fusion is exposed at the surface of the C-tip. I think that the recognition site most probably involves more than one collagen molecule. The site might be at the surface of the microfibrils. There are unresolved questions about the microfibrils, e.g., what is the length of the microfibrils, what is their structure and conformation, and how are they exposed at the fibril surface? My guess is that the microfibrils have a different conformation at the N- and C-ends of fibrils, and this has something to do with why C-ends are needed for fusion.

Q: *Robin Poole*: If the literature is still correct, the N-propeptide tends to persist longer than the C-propeptide with respect to fibril formation, so that there is perhaps some influence of the persisting N-propeptide.

A: *Karl Kadler*: Yes.

Q: *Alan Grodzinsky*: Is it known how such large structures are secreted from the cell?

A: *Karl Kadler*: The very earliest complex that is identifiable as a collagen fibril is about one micron in length. We have tomography data of such early collagen fibrils lying within the boundary of the cell, and I think the most exciting future work on ECM assembly will be to understand how proteins are trafficked to the ECM post-Golgi. It is unknown how collagens, thrombospondins, proteoglycans, and integrins are trafficked to the ECM and are assembled. Nothing is known about what happens between proteins leaving the Golgi and arriving in the ECM. Researchers studying endoplasmic reticulum and Golgi trafficking are starting to use matrix molecules as tools to find out what those mechanisms are. Some people speculate that tubules link the Golgi to the matrix. I personally think that post-Golgi trafficking is a very dynamic process involving vesicles and tubules that transport proteins from the Golgi to the ECM. And my guess is this is where the early collagen fibrils are made.

Q: *Eugene Thonar*: What about branching in the cornea?

A: *Karl Kadler*: We see branching within tendon, and we see branching in the skin, but we don't see any branching within the cornea. Within the cornea, we only see bipolar fibrils. So there are no unipolar fibrils within the cornea and no fusion going on there.

Hyaluronics and aggrecanics

Q: *Vincent Hascall*: If the network were stabilized by hydrogen bonds, how would calcium affect it?

A: *Tim Hardingham*: Calcium loves water – it competes for water, it forms highly hydrated ions. Its effects on hyaluronan conformation compared with sodium are thus probably because of its more dominant effects on the water surrounding hyaluronan and its disruption of some of the hydrogen bonds stabilizing hyaluronan that bridge through water molecules.

Q: *Vincent Hascall*: What is the mechanism of the alkaline pH effect?

A: *Tim Hardingham*: It is unclear, but at high pH other groups on HA may become ionized, even hydroxyl groups, or the N-acetamido. This could disrupt some hydrogen bonds and thus reduce the stabilizing effect between adjacent disaccharides. This would reduce chain stiffening and cause HA properties to become similar to dextran.

Q: *Cheryl Knudson*: Do you think calcium might help to release the hyaluronan from the synthase?

A: *Tim Hardingham*: It depends what local calcium concentrations are achieved. It could be responsible for making the chain more mobile and perhaps competing with its interaction with other components such as the synthase.

Q: *Ted Oegema*: When you look at movement are you looking at one segment of the molecule moving relative to another segment or is it a bulk transfer you measure?

A: *Tim Hardingham*: This sort of technique has a big advantage because it measures the long-term lateral translational diffusion, which means you are getting bulk movement of the polymer. Other techniques such as dynamic light scattering are sensitive to segmental motion, which can lead to artifacts and misinterpretation. But this technique is not sensitive to these effects, and what we measure is movement of the polymer or molecules through the polymer.

Q: *Ted Oegema*: When you look at the movement of the chondrocyte in the territorial and interterritorial matrices, could you calculate whether the movement observed could be accounted for merely by diffusion or does it require some other mechanism?

A: *Tim Hardingham*: This can be calculated, but allowance has to be made for the fibrillar networks, which in cartilage accounts for a considerable fraction of the tissue volume. My measurements are in fibril-free glycosaminoglycan systems, which provide molecular size-dependent filtration and predict diffusion in the solute phase. The glycosaminoglycans (aggrecan, hyaluronan) form a solution phase network that is between the fibrils in cartilage.

Q: *Dick Heinegard*: Do you think there is a role for small basic proteins in binding or linking this network?

A: *Tim Hardingham*: There is an enormous concentration of these glycosaminoglycan chains in cartilage, and of course small basic proteins will bind to them. But it is my view that the system is dynamic and most of the chains are free most of the time. I remember that Dennis Tortia showed by NMR that chondroitin sulfate in cartilage was as mobile as chondroitin sulfate in solution, which suggested that most chains were not constrained by tight binding.

Q: *Dick Heinegard*: Could it be that these proteins could provide some channels for diffusion?

A: *Tim Hardingham*: There may be channels, but they might be dynamic rather than permanent, or fixed over a long time.

Lectin domains in hyaluronan-binding proteoglycans

Q: *Ian Nieduszynski*: Is there any evidence that the aggrecan C-type lectin repeat in the G3 domain binds carbohydrate?

A: *Anders Aspberg*: Yes, for example the interactions with sulfated glycolipids. The interactions with extracellular matrix glycoproteins, however, generally seem to be mediated through protein-protein interactions.

Q: *Robin Poole*: Could you use a fragment of a lectin ligand as a competitive inhibitor for the interaction? This could be an alternative to gene targeting approaches for studying the *in vivo* relevance of these interactions.

A: *Anders Aspberg*: We are doing such experiments right now, using a fragment of tenascin-R to inhibit the versican lectin interactions in an atherosclerosis model.

Q: *Robin Poole*: Is there any evidence for specific glycosylation within the fibrillar domains of the lectin ligands you have identified?

A: *Anders Aspberg*: These proteins certainly are glycosylated, but the carbohydrates do not seem to be involved in the lectin binding. For example, bacterially expressed tenascin-R fragments that were confirmed to lack glycosylation by mass spectrometry bind strongly to the proteoglycan lectins. We also used a small fragment of fibulin-1 with only one N-linked glycan, and saw no effect on the proteoglycan lectin affinity after deglycosylation. As you know, the proteoglycan lectin repeats were named on the basis of sequence homology, and there are in fact several other C-type lectins with protein ligands.

The matrilins: A novel family of extracellular adaptor proteins

Q: *John Couchman*: You showed that there were two distinct distributions of matrilins; one associated with fibrillar structures, and you've shown interactions with collagens, for example. The other interaction was at the cell surface. Is the cell surface pool also interacting with matrix molecules there, or are there direct interactions with cell surface receptors too?

A: *Mats Paulsson*: Previous work has shown an interaction between matrilin-1 and α1β1 integrins. So there is good reason to believe that there are cell surface receptors for matrilins. We have not done any work on these ourselves. I do not know if these networks around the cells are pure matrilin networks or if they contain other proteins. But if so, these other proteins are probably not collagens as they do not require ascorbate for their synthesis.

Q: *Ernst Hunziker*: I remember well that the matrilin expression in the adult articular cartilage is fairly low compared to the fetal and early postnatal articular cartilage. Do you have an interpretation for this?

A: *Mats Paulsson*: In an adult articular cartilage, at least in the superficial and central areas of the cartilage, there is very little matrilins. I assume that this bears a relation to the mechanical properties required for that particular cartilage. Matrilins are expressed throughout the cartilage during development up to birth. Loss of matrilin expression may be part of a modulation of matrix structure to provide optimal features for weight bearing. Matrilin-1 remains in other non-weight bearing cartilages, like in the trachea, where it becomes a very abundant protein with aging.

Q: *Ernst Hunziker*: Have you done some experiments or do you know any experiments where matrilin expression in immobilized joints has been studied?

A: *Mats Paulsson*: That has not been done to my knowledge. I agree that it would be interesting to see if mechanical loading regulates matrilin gene expression.

Q: *Robin Poole*: I notice that one of the matrilins had a distribution in the spine, and also in the annulus and the spinal ligaments. It struck me as being identical to that of versican, which we've been looking at. So the question is since versican and aggrecan both have a G3 domain; is there any evidence that this matrilin would interact with versican at the same site as for aggrecan and would it be the G3 domain?

A: *Mats Paulsson*: The interaction with aggrecan is not over the G3 domain, it's actually in the chondroitin sulfate carrying domain and, further, to distinct areas within that domain where it can be mapped by electron microscopy. I absolutely agree that matrilin-2 and versican show a marked co-distribution. I would not be surprised if there are complexes between versican and matrilin-2 in analogy to those between aggrecan and matrilin-1, but the work has just not been done.

Q: *Dick Heinegard*: Do you know anything about cooperativity of binding of matrilins to their ligands?

A: *Mats Paulsson*: We certainly know that monomeric arms can show the same specificity of binding as the whole molecule. These proteins are very "sticky" and show low solubility. Therefore, we have been limited to ELISA style binding tests and Biacore assays in our analyses. Even with these methods, we have had difficulties in obtaining reliable quantitative data. Even if we do see apparent differences in affinity between monomeric and tri- or tetrameric molecules, I personally don't want to interpret those data in a quantitative manner.

Superficial zone protein (SZP) is an abundant glycoprotein in human synovial fluid with lubricating properties

Q: *Hari Reddi*: Have you analyzed the distribution of SZP in tissues besides synovial fluid, plasma, and serum using the five different antibodies? Do you find SZP in intestine or some other sites that might explain the high concentration in plasma?

A: *Thomas Schmid*: We haven't really done a lot of tissue screening with these monoclonal antibodies. In previous studies with polyclonal antibodies, SZP was detected in tendon, where signals were stronger on the edge of the tendons. It was present throughout synovial tissue, again with very strong staining in the synovial-lining layer. Barbara Schumacher's work with monoclonal antibodies for bovine SZP made by Bruce Caterson showed very nice staining of synovial lining cells in fetal synovium. We haven't looked at a lot of other tissues. Matt Warman presented Northern blot data, which showed SZP mRNA signals in liver and heart, so we expect those tissues to contribute to the levels of SZP in the circulation. In our assays, we have tried to analyze SZP concentrations in serum, and it depends on which antibody we use. We think the levels in plasma are lower than the levels in synovial fluid, but they are still substantial.

Q: *Hari Reddi*: Eugene Thonar has analyzed keratan sulfate as a marker in various disease states. I was thinking that many of those samples must be in the freezer. Have you looked for SZP or find any differences in these tissues?

A: *Thomas Schmid*: We hope to do those types of studies, probably to compare SZP levels with CRP and HA concentrations. What happens in inflammation, does SZP go up or down? We don't know yet.

Q: *Alice Maroudas*: How does your story compare with the work of the Australian group, I think it is Hill and others, where they find the main synovial lubricant is lipid, and lipid associated with a lubricin-like protein? Also, do you find any relationship between your protein and the protein that Rita Stanescu described on the cartilage surface?

A: *Thomas Schmid*: My impression of the literature is that the Australian group showed that phospholipase destroyed the lubricating activity of synovial fluid and thus they attribute lubrication to lipid. Gregory Jay has shown that phospholipase preparations like those used by Hill and coworkers contained proteolytic activity, which was capable of degrading lubricin. Jay concluded that lubricin, and not phospholipid, was the primary boundary lubricant in synovial fluid. Our work shows that SZP binds to many macromolecules. If SZP binds albumin, and albumin binds lipid, then you have lipid bound to a SZP complex. I think some of these observations need more clarification.

Rita Stanescu did a very nice series of experiments looking at the lamina splendens. She found cationized ferritin bound to molecules at the cartilage surface. These molecules were not digested with hyaluronidase or collagenase, suggesting they were some other type of negatively charged macromolecules. Barbara Schumacher and Joel Block showed SZP contained chondroitin sulfate and keratan sulfate chains and also that SZP/lubricin is rich in sialic acid. These chemical groups would give SZP a high negative charge and make it likely to be the molecule that was binding cationized ferritin in Rita Stanescu's experiments.

Q: *Ernst Hunziker*: This protein has a great potential for analysis of the polarity of secretory activity in chondrocytes. You have shown before that it's secreted by synovial lining cells into the synovial cavity and these are clearly polarized cells. In the superficial zones of the cartilage, did you look at higher resolution in your immunohistochemistry to see if these cells secrete differentially towards the joint cavity or do they secrete SZP in all directions? Does the SZP molecule migrate through the cartilage?

A: *Thomas Schmid*: We have examined the distribution of SZP around chondrocytes at high magnification. At high magnification you can find cells where SZP forms a pericellular halo completely around a single cell. In other specimens, SZP forms a column-like structure in the matrix between superficial zone chondrocytes and the cartilage surface. So you can find SZP in both situations, where it might be secreted in a polarized fashion toward the synovial cavity, but at other times where it looks like it is trapped around the entire cell without polarity. So the distribution of SZP may depend on the time frame when a tissue is examined. We are in the process of trying to look at a number of tissues from patients and donors and trying to draw conclusions from these data, but we see a large variety of patterns. One complication is that as the surface layer gets damaged, you would expect it to become more porous, and now SZP will be able to diffuse into the cartilage from the synovial fluid. So at any particular time is SZP coming into the tissue or going out? We need to do some careful experiments to test these possibilities.

Q: *David Howell*: At the time David Swann did his work, the study was completely focused on cartilage, and your work is showing the possibility that SZP is being secreted by synovial cells. Eric Radin working with David Swann postulated that capsule lubrication is caused by hyaluronan, while the lubricin was strictly inside the joint. These new findings open up the possibility that even the capsule lubrication could be from the synovial cells and caused by lubrication with lubricin. Does SZP in the synovial fluid come from the superficial chondrocytes and synovial cells?

A: *Thomas Schmid*: I think it is a distinct possibility that both of these cell populations contribute to the pool of SZP in synovial fluid, and that how much each contributes may depend on the state of each tissue.

Establishment of *in vitro* cell culture models for the investigation of the pathogenesis of cartilage diseases

Q: *Karl Kadler*: You said that the secretion of collagen II is normal in your mutant COMP cells. Why then would fibril formation be affected? I'd be alert to whether the processing of collagen is normal and whether you see procollagen in the medium.

A: *Frank Zaucke*: We have analyzed cell culture supernatants only by pulse chase experiments and Western blotting. We detected similar secretion kinetics and amounts of collagen II when we compared wild-type and mutant transfected COMP cells, but we have not further characterized the collagen forms that were secreted.

Q: *Karl Kadler*: Do you know whether the collagens are secreted as mature collagens or procollagens?

A: *Frank Zaucke*: No, we have not determined that.

Q: *David Howell*: If the ER is diffusely full with these retarded molecules, I think this would interfere with synthesis and passage of other molecules like decorin and type II collagen. Wouldn't the secretion rate of these other proteins be slowed by this situation?

A: *Frank Zaucke*: Until now, we have done pulse-chase experiments only for collagen II and could not detect a delayed secretion. Immunofluorescence staining demonstrates that COMP is retained in the ER, whereas the further transport of collagen II and, quite likely decorin is not interfered with, and these molecules can be detected in the supernatant. Nevertheless, I think the precise molecular composition of the inclusions should be analyzed in more detail by immunogold labeling and electron microscopy.

Hyaluronan and joint biology

Introduction

Vincent C. Hascall

Orthopaedic Research Center, Department of Biomedical Engineering, The Cleveland Clinic Foundation, 9500 Euclid Avenue, Cleveland, OH 44195, USA

This session focuses on the role of hyaluronan in normal and pathological joint biology. The articles by Roger Mason et al. and by Endre Balazs and Charles Weiss describe the properties and functions of hyaluronan in the synovial fluid and adjacent synovial tissues. The high molecular weight distribution of hyaluronan in the small volume of synovial fluid in a normal knee joint has viscoelastic properties that contribute dynamically to distributing load and regulating drainage into the lymphatics. A careful assessment of the physical properties of the normal hyaluronan concentration and size distribution combined with measurements of the flow rates of drainage in joints subjected to different load led Mason et al. to conclude that reflectance of high molecular weight hyaluronan by the synovium increases local hyaluronan concentrations and inhibits flow. This property allows a dynamic control of flow rate in response to joint loading mechanics with high load and reflectance lowering flow while low load with diffusion of hyaluronan from the synovium allows higher flow. The viscoelastic properties of hyaluronan are ideally suited for distribution of load and protection of the synovial tissues as discussed by Balazs and Weiss. Further, they provide evidence that the concentration of hyaluronan in synovial fluid is maintained constant by finely tuned synthesis, catabolism and drainage, and that the synoviocites have a major role in this process. In joint pathology, effusion into the joint can increase the synovial fluid volume up to 50 fold or more while maintaining the same concentration of hyaluronan over long periods of time. However, in patients with effusions, the molecular weight distribution changes such that the majority of species are of lower molecular weight than present normally. The consequent changes in viscoelasticity and the impact on joint biology and function are discussed by Balazs and Weiss as is the beneficial effects of replacing the effusion with viscosupplements of appropriate hyaluronan preparations.

    Mason et al. note that normally, synovial fluid is perfused by a capillary filtrate that is balanced by the flow rate of drainage, but that in inflammation, leakage from the capillaries can overwhelm the drainage and lead to the effusions observed in OA and RA patients. A consequence of this is the exposure of serum proteins to the hyaluronan in the synovial fluid. That this occurs and, indeed, can be used as a marker for the disease status is described by Wannarat Yingsung et al. Their studies show that hyaluronan in serum can contain covalently bound proteins they named Serum-derived HA-binding Proteins (SHAPs). SHAPs were then shown to be the heavy chains that are covalently bound to chondroitin sulfate on the serum proteoglycan inter-alpha-trypsin inhibitor (ITI). The process of trans-esterifying SHAPs from ITI onto hyaluronan occurs when synovial fluid is exposed to serum. Drainage through the lymphatics can therefore introduce some of this structure into serum where it can be detected. The studies by Yingsung et al. show a strong correlation between hyaluronan concentrations and SHAP in serum from patients with OA and RA indicating the possible usefulness of SHAPs as a biomarker for individuals at risk for OA.

The metabolism of hyaluronan in most tissues is dynamic with both synthesis and catabolism influenced by a variety of environmental and chemical stimuli. This is particularly relevant to the development and function of the tissues of the synovial joint. Janet Lee et al. describe aspects of the biology of hyaluronan synthase 2 (HAS2), the major workhorse of the eukaryotic hyaluronan synthase family. They provide evidence that growth/differentiation factor 5 (Gdf5), a member of the transforming growth factor superfamily, regulates HAS2 transcription during cartilage and joint development. Because the HAS2 mouse knockout is embryonic lethal, Lee et al. discuss progress toward creating a conditional knockout in order to study the importance of this enzyme for a variety of tissue functions. CD44 is a major cell surface hyaluronan-binding protein that is involved in: (a) retention of hyaluronan and associated aggrecan molecules in the cell-associated matrix around the chondrocyte, and (b) the catabolism of hyaluronan with loss of the cell-associated matrix. Cheryl Knudson et al. describe the various factors that impact on CD44 that affect the balance between retention and catabolism of this cell-associated matrix. They discuss how IL-1 and OP-1 (BMP7) both upregulate CD44, but with opposite outcomes, i.e., IL-1 increases catabolism while OP-1 increases matrix retention. Several lines of experiments provide clues to possible underlying mechanisms for this ambivalent reaction, including demonstrations that the loss of cell associated hyaluronan matrix in chondrocytes and cartilage explants occurs: a) by disruption of the cytoskeleton, or b) by inhibition of phosphorylation of the cytoplasmic tail of CD44 via inhibition of CKII, or c) by overexpression of CD44 without its cytoplasmic tail, which acts as a dominant negative in disrupting the interactions with the cytoskeleton.

All of these exciting avenues of research are defining further roles for the function of hyaluronan in synovial biology far beyond its space filling attributes that not so long ago were considered its primary if not only role in connective tissue biology.

Role of hyaluronan in regulating joint fluid flow

Roger M. Mason[1], Peter J. Coleman[2], David Scott[2] and J. Rodney Levick[2]

[1]Cell and Molecular Biology Section, Division of Biomedical Sciences, Faculty of Medicine, Imperial College, Sir Alexander Fleming Building, Exhibition Road, London, SW7 2AZ, UK;
[2]Department of Physiology, St. George's Hospital Medical School, Cranmer Terrace, London, SW17 0RE, UK

Introduction

Hyaluronan, a glycosaminoglycan composed of alternating N-acetyl-D-glucosamine and D-glucuronate residues, linked β(1–4) and β(1–3) respectively, was first isolated from synovial fluid by Karl Meyer [1]. It is present in concentrations between 2–4 mg/ml in human and rabbit synovial fluids and has an average molecular mass between 2000 (rabbit) and 7000 (man) kDa [2–4].

Synovial fluid is formed in part from an ultrafiltrate of plasma across the walls of the fenestrated capillaries that form a dense network in the synovium, a thin layer of tissue surrounding the articular joint cavity. Ultrafiltration takes place during periods of low intra-articular pressure such as occur during extension of the joint, with the ultrafiltrate flowing through the interstitial matrix of the synovium to reach the cavity. Other components of the synovial fluid, notably hyaluronan and the glycoprotein, lubricin, are synthesized by fibroblast-like cells (B-type synoviocytes) resident in the synovium and are actively secreted into the joint cavity. At raised intra-articular pressure, such as occurs in joint flexion, the capillary pressure is opposed, and the direction of fluid flow is reversed across the synovium. This allows drainage of synovial fluid components into the subsynovial lymphatics. It provides a mechanism whereby diffusible metabolites such as proteoglycan fragments, which have passed out of articular cartilage in the joint into the synovial fluid, can be cleared from the joint cavity [5–7].

Very large anionic hyaluronan molecules adopt an expanded coiled configuration in an aqueous environment which encompasses a large volume of water and has a roughly spherical domain of radius ~100 nM, depending on chain length and counter ion concentration. Due to the massive size of this hydrated domain, adjacent hyaluronan domains overlap when their concentration exceeds ~1 mg/ml, so synovial fluid is characterized by a quasi-continuous network of chains [8]. This results in a highly viscous, non-Newtonian fluid whose main functions have been proposed to be: (1) to act as a lubricant for the joint surfaces under low load conditions, and (2) to provide a highway to transport nutrients (including oxygen) from the synovium to the avascular articular cartilage and metabolic products from the latter to the synovial capillaries. This requires the synovial fluid to be in a state of continual turnover driven by capillary pressure on the one hand and by changes in intra-articular pressure dependent on joint movement on the other. Under normal physiological conditions, the volume of synovial fluid remains remarkably constant, even during prolonged periods of joint flexion and raised intra-articular pressure. This indicates that mechanisms must be present which regulate fluid loss from the joint cavity under such conditions. We describe below two such mechanisms, one dependent on the hyaluronan component of the synovial interstitial matrix and the other on the hyaluronan present in the synovial fluid. We also discuss the secretion of hyaluronan into the synovial fluid and its conservation therein.

Synovial interstitial hyaluronan and fluid flow

The B-type synoviocytes synthesize an extracellular matrix around themselves as well as secreting hyaluronan and other products to the synovial fluid. We investigated the composition and structure of this matrix in samples of synovium microdissected from rabbit knee joints and using biochemical and ultrastructural methods of analysis [3, 9]. Collagens occupy much of the synovial interstitial space (120 mg/ml) with hyaluronan and proteoglycans residing with other macromolecules in the extrafibrillar compartment of this matrix. Since the volume occupied by collagen fibrils could be measured by morphometric methods [9], we were able to determine the extrafibrillar concentration of the various glycosaminoglycans present (Fig. 1A). Hyaluronan is a relatively minor component (~0.8 mg/ml), even in comparison to other glycosaminoglycans. However, this hyaluronan has an important role in regulating the hydraulic permeability of the synovium. Permeability is measured by infusing a Ringer solution into the joint space of the rabbit knee, *in vivo*, and measuring its rate of drainage across the synovium under a series of controlled intra-articular pressures [10].

Hydraulic conductance is defined by $d\dot{Q}_s/dP_j$ under steady-state conditions, where $\dot{Q}_s$ is the flow out of the joint at a series of different pressures, P_j (2–25 cm H_2O). We measured hydraulic permeability in control knee joints and in contralateral joints in which the synovium was depleted of hyaluronan by prior treatment with protease-free *Streptomyces* hyaluronidase or other hyaluronidases [11]. Following such treatment, the hydraulic permeability of the synovium increased by up to five-fold over controls (Fig. 1B). This was much greater than the ~1.2-fold increase predicted for loss of hydraulic drag (estimated for hyaluronan at a concentration of 0.8 mg/ml in the extrafibrillar space). Histochemistry [11] indicated that the synovial hyaluronan was fairly uniformly distributed in this extrafibrillar space, so the unexpectedly high increase in hydraulic permeability on removing hyaluro-

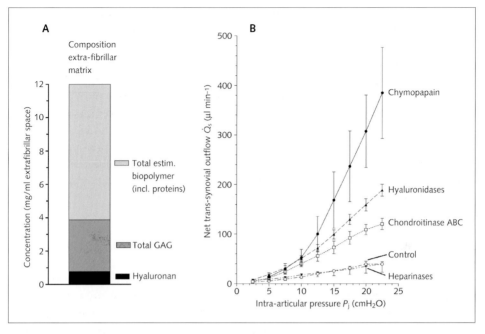

Figure 1
Synovial interstitial hyaluronan and fluid flow. (A) Concentration of hyaluronan and other biopolymers, including sulphated glycosaminoglycans and proteins, in the extrafibrillar space of the synovium. (B) The effect on trans-synovial outflow at different intra-articular pressures (hydraulic conductance) of selective removal of hyaluronan, other glycosaminoglycans, and glycoproteins, from the extrafibrillar space of the synovium. Biopolymers were removed by digestion in vivo *with specific hyaluronidases, chondroitinase ABC, heparinases, or chymopapain. For other details see the text and [11, 12].*

nan is not due to its selective concentration into a narrow zone of the tissue. Moreover, although different glycosaminoglycans have similar resistivities, they clearly do not contribute as much on a weight for weight basis to opposing synovial hydraulic permeability as hyaluronan does [12]. We propose that one explanation for the prominent role of interstitial hyaluronan in the control of synovial permeability may be that it functions as an organiser of the synovial matrix as well as contributing hydraulic drag. Thus, removal of hyaluronan may result in collapse of the matrix and loss of resistance to fluid flow through it. Ultrastructural analysis following digestion of the synovium with hyaluronidase supports this concept (P.J. Coleman, unpublished observations). It is noteworthy that non-collagenous proteins contribute significantly to synovial resistance to fluid flow (Fig. 1B), and it is possible that some of these are released from the tissue when hyaluronan is depleted.

Secretion of hyaluronan into the joint and reflection by the synovium

Measuring the amount of hyaluronan in the synovial joint space is not straightforward because the viscosity of the synovial fluid makes it difficult to aspirate it on a quantitative basis. However, using repeated washout of the joint, it is possible to ensure complete recovery of hyaluronan, and with this approach, we found an average total hyaluronan content of 182 ± 9.9 μg in the rabbit knee joint [13].

Since the concentration of hyaluronan in the synovial fluid was 3.62 ± 0.19 μg/μl, the volume of fluid is ~50 μl. Using the same washout method, we also estimated the rate of secretion of hyaluronan into the rabbit knee joint. This is 4.2 ± 0.77 μg/h and is due to an active process rather than to diffusion of pre-existing hyaluronan from the synovial interstitial matrix, since secretion ceases *post mortem* and after biochemical inhibition of the ATP supply. Interestingly, if the joint was expanded by injecting a 2.0 ml volume of Ringer solution, hyaluronan secretion increased to 5.80 ± 0.84 μg/h, indicating that the control of synovial hyaluronan synthesis is, in part, regulated by mechanical factors [13]. Protein kinase C-dependent pathways are potentially involved in the mechanotransduction pathway [14] since PMA stimulates HA secretion >two-fold.

The water content of the synovial fluid comes from capillary filtration and leaves the joint by lymphatic drainage, as discussed above. In our estimates the rate is fast, at about 28 μl/h [15]. Thus, given the relatively low rate of hyaluronan secretion into the joint, we predicted that its high concentration in the synovial fluid must be maintained by partial reflection by the synovial interstitial matrix [13]. We estimate the effective pore radius in this matrix to be ~30–90 nm [5], whilst synovial fluid hyaluronan molecules of ~2200 kDa will have a radius of gyration (Rg) of 107–180 nm [16, 17]. To test whether reflection occurs, joints were infused with solutions containing 3.6 mg/ml of hyaluronan of different sizes, ~2200, ~530, ~300, ~90 kDa. The concentration of hyaluronan was then measured again after a period of trans-synovial drainage from the cavity at controlled incremental intra-articular pressures. Large increases in intra-articular concentration confirmed that hyaluronan is indeed reflected by the synovium and that the degree of reflection is dependent on its molecular mass and therefore on its Rg. For hyaluronan 2200 kDa, the reflected fraction was 79% [17]. It is noteworthy that neutral dextran of ~2000 kDa with a similar molecular mass has a much smaller molecular radius (~31 nm) and was only reflected 31% by the synovium [18]. Albumin (mol. wt. 67 kDa), a component of the synovial fluid *in vivo*, has a much smaller radius (3.6 nm) and is not significantly reflected by the synovial interstitial matrix [19].

Outflow buffering across the synovium by synovial fluid hyaluronan

Trans-synovial flow can be measured after infusing the joint with Ringer's solution and raising the intra-articular pressure in a series of graded steps, as discussed above. McDonald and Levick [20] observed that opposition to fluid drainage increases with increasing pressure if hyaluronan is present in the solution, but not if it is absent. This indicated that hyaluronan in synovial fluid has a buffering effect on fluid loss from the joint cavity under conditions of raised intra-articular pressure. We investigated this effect further.

Rooster comb hyaluronan (3.6–4.0 mg/ml, ~2200 kDa), with a similar chain length to rabbit synovial fluid hyaluronan, caused the pressure-flow relationship to flatten as intra-articular pressure was raised, so that at 10–20 cm H$_2$O the slope d$\dot{Q}_s$/dP_j (0.05 ± 0.01 µl/min cm H$_2$O^{-1}) was almost a plateau [17]. In contrast, with Ringer's solution alone, the slope was 1.94 ± 0.01 µl/min/cm H$_2$O, or 39 times steeper than in the presence of hyaluronan (Fig. 2A). Bovine synovial fluid had a similar effect on the pressure-relationship as hyaluronan in Ringer's solution [16]. Mechanical perforation of the synovium or proteolytic damage to its matrix by prior digestion with chymopapain eliminated the buffering effect of intra-articular hyaluronan on trans-synovial flow, indicating that the effect depended on the ability of an intact interstitium to reflect the hyaluronan. The buffering effect of HA on outflow was reversible: if normal joints that had been treated with intra-articular hyaluronan were washed out to remove the glycosaminoglycan and then re-infused with Ringer's solution, the pressure-flow relationship returned to the same as that in control joints [16].

The effect of hyaluronan on buffering trans-synovial flow is dependent on the concentration of the hyaluronan and on its molecular mass. Low concentrations (0.2 mg/ml) of high molecular mass hyaluronan (~2100 kDa) were relatively ineffective, reducing outflow by ~50% relative to Ringer's solution and without flattening of the pressure-flow curve. Higher concentrations (2.0–4.0 mg/ml) flattened the curve. Viscometry indicated that concentrated solutions of this hyaluronan (1.35 mg/ml) underwent a change in state compatible with increased chain-chain interaction associated with overlapping of their molecular domains. It seems likely that flattening of the pressure-flow relationship occurs with hyaluronan solutions whose concentration is high enough for the overlap state to occur [21].

High concentrations (3.6 mg/ml) of low molecular mass hyaluronan (90–300 kDa) only partially reduced trans-synovial flow and did not flatten the pressure-flow curve (Fig. 2B). In contrast, solutions of hyaluronan of ~550 kDa and ~660 kDa markedly reduced flow and flattened the curve. As expected, very high molecular mass hyaluronan (~2200 kDa) had an even greater effect, reducing trans-synovial flow to ~8% of that with Ringer's solution alone at the highest pressure (Fig. 2A) [17]. The reflected fractions for hyaluronan – ~90, ~300, ~530, ~2200 kDa – were 0.12, 0.33, 0.25 and 0.79, respectively. These results show that decreas-

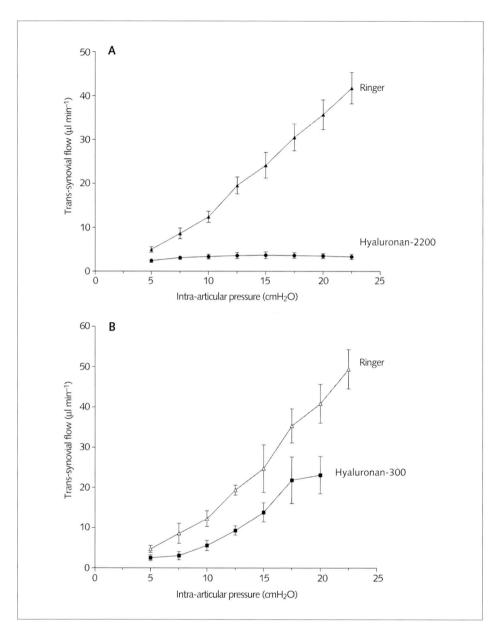

Figure 2
The effect of synovial fluid hyaluronan on trans-synovial flow. Trans-synovial flow was measured at graded increments of intra-articular pressure after infusing rabbit knee joints with either Ringer's solution, or with Ringer's solution containing 3.6 mg/ml hyaluronan of molecular weight 2200 kDa (A) or 300 kDa (B). For other details see the text and [17].

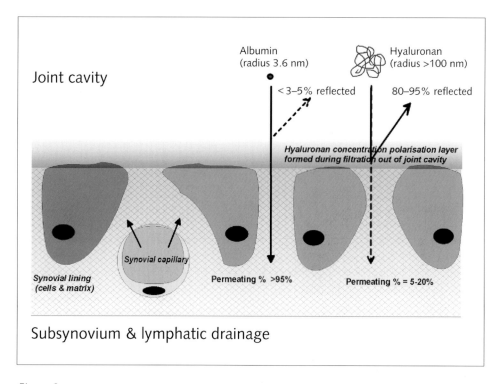

Figure 3
Outflow buffering across the synovium by synovial fluid hyaluronan. The figure shows a schematic representation of three compartments: the joint cavity containing synovial fluid, the synovium containing synoviocytes, capillaries and interstitial matrix, and the subsynovial connective tissue where lymphatics draining the joint are found. A high percentage of synovial fluid hyaluronan is reflected at the synovial interface during synovial fluid outflow because its molecular domain is larger than the effective pore size of the synovial interstitial matrix. We propose that the excluded hyaluronan forms a concentration polarization layer at the interface whose osmotic pressure opposes fluid outflow. A five-fold increase in hyaluronan concentration in this layer would be sufficient to account, osmotically, for the observed outflow buffering effect [16].

ing chain length is associated with reduced reflection of hyaluronan by the synovium, and they suggest that lower concentrations of the glycosaminoglycan would accumulate at the tissue interface during drainage. Thus the local osmotic pressure opposing fluid flow would be reduced with lower molecular mass hyaluronan and greater trans-synovial flow would occur compared with when high molecular mass hyaluronan is present [16, 17]. A concentration-polarization theory was developed on this basis (Fig. 3), and numerical solutions support the concentration-polarization explanation for the effect of high molecular mass hyaluronan on buffering trans-synovial flow at high pressure [16].

From a physiological point of view these observations indicate a key role for synovial fluid hyaluronan in opposing fluid outflow during periods of sustained joint flexion when intra-articular pressure rises. With extension, pressure falls, concentration polarization of hyaluronan at the synovial interface reduces due to back diffusion, and opposition to outflow decreases correspondingly.

Hyaluronan and trans-synovial flow in arthritis

Joint effusions occur frequently in arthritic conditions when capillary filtration due to inflammation exceeds the rate of drainage from the joint. They resolve when drainage exceeds filtration rate [22]. As we have shown above, trans-synovial flow is finely regulated in normal physiology by both synovial interstitial hyaluronan and by high concentrations of high molecular mass hyaluronan in the synovial fluid. Other factors may also be involved. However, changes occur in synovial fluid hyaluronan in arthritic states. Thus lower concentrations may occur in osteoarthritis [23] and both concentration and molecular mass are reduced in rheumatoid arthritis [2]. Such changes would be predicted to lead to enhanced fluid drainage from a joint, and this has been observed in practice [24]. Indeed, this may be nature's way of limiting a joint effusion. Increased drainage should also be taken into account when interpreting the results of synovial fluid or serum markers of disease activity in articular cartilage in arthritic disorders [6].

Acknowledgement
We are grateful to the Wellcome Trust for financial support.

References

1 Meyer K (1947) The biological significance of hyaluronic acid and hyaluronidase. *Physiol Rev* 27: 335–359
2 Dahl LB, Dahl IM, Engstrom-Laurent A, Granath K (1985) Concentration and molecular weight of sodium hyaluronate in synovial fluid from patients with rheumatoid arthritis and other arthropathies. *Ann Rheum Dis* 44: 817–822
3 Price FM, Levick JR, Mason RM (1996) Glycosaminoglycan concentration in synovium and other tissues of rabbit knee in relation to synovial hydraulic resistance. *J Physiol* 495: 803–820
4 Coleman PJ, Scott D, Ray J, Mason RM, Levick JR (1997) Hyaluronan secretion into

the synovial cavity of rabbit knees and comparison with albumin turnover. *J Physiol* 503: 645–656

5   Levick JR, Price FM, Mason RM (1996) Synovial matrix-synovial fluid system of joints. In: WD Comper (ed): *Extracellular matrix*. Harwood Academic Publishers, Amsterdam, 328–377

6   Levick JR, Mason RM, Coleman PJ, Scott D (1999) Physiology of synovial fluid and trans-synovial flow. In: CW Archer, B Caterson, M Benjamin, JR Ralphs (eds): *Biology of the synovial joint*. Harwood Academic Publishers, Amsterdam, 235–252

7   Mason RM, Levick JR, Coleman PJ, Scott D (1999) Biochemistry of the synovium and synovial fluid. In: CW Archer, B Caterson, M Benjamin, JR Ralphs (eds): *Biology of the synovial joint*. Harwood Academic Publishers, Amsterdam, 253–264

8   Laurent TC, Laurent UB, Fraser JR (1996) The structure and function of hyaluronan: An overview. *Immunol Cell Biol* 74: A1–A7

9   Price FM, Mason RM, Levick JR (1995) Radial organization of interstitial exchange pathway and influence of collagen in synovium. *Biophys J* 69: 1429–1439

10 Price FM, Levick JR, Mason RM (1996) Changes in glycosaminoglycan concentration and synovial permeability at raised intra-articular pressure in rabbit knee. *J Physiol* 495: 821–833

11 Coleman PJ, Scott D, Abiona A, Ashhurst DE, Mason RM, Levick JR (1998) Effect of depletion of interstitial hyaluronan on hydraulic conductance in rabbit knee synovium. *J Physiol* 509: 695–710

12 Scott D, Coleman PJ, Abiona A, Ashhurst DE, Mason RM, Levick JR (1998) Effect of depletion of glycosaminoglycans and non-collagenous proteins on interstitial hydraulic permeability in rabbit synovium. *J Physiol* 511: 629–643

13 Coleman PJ, Scott D, Ray J, Mason RM, Levick JR (1997) Hyaluronan secretion into the synovial cavity of rabbit knees and comparison with albumin turnover. *J Physiol* 503: 645–656

14 Anggiansah CL, Scott D, Poli A, Coleman PJ, James J, Houston A, Mason RM, Levick JR (2002) Control of hyaluronan secretion into joint fluid *in vivo*: role of protein kinase C (PKC). *Hyaluronan 2000; in press*

15 Levick JR, Coleman PJ, Scott D, Mason RM (2002) Secretory regulation *in vivo*, molecular reflection and hydraulic roles of synovial hyaluronan. *Hyaluronan 2000; in press*

16 Coleman PJ, Scott D, Mason RM, Levick JR (1999) Characterization of the effect of high molecular weight hyaluronan on trans-synovial flow in rabbit knees. *J Physiol* 514: 265–282

17 Coleman PJ, Scott D, Mason RM, Levick JR (2000) Role of hyaluronan chain length in buffering interstitial flow across synovium in rabbits. *J Physiol* 526: 425–434

18 Scott D, Coleman PJ, Mason RM, Levick JR (2000) Action of polysaccharides of similar average mass but differing molecular volume and charge on fluid drainage through synovial interstitium in rabbit knees. *J Physiol* 528: 609–618

19 Scott D, Coleman PJ, Mason RM, Levick JR (2000) Interaction of intraarticular

hyaluronan and albumin in the attenuation of fluid drainage from joints. *Arthritis Rheum* 43: 1175–1182

20 McDonald JN, Levick JR (1995) Effect of intra-articular hyaluronan on pressure-flow relation across synovium in anaesthetized rabbits. *J Physiol* 485: 179–193

21 Scott D, Coleman PJ, Mason RM, Levick JR (2000) Concentration dependence of interstitial flow buffering by hyaluronan in synovial joints. *Microvasc Res* 50: 345–355

22 Wallis WJ, Simkin PA, Nelp WB (1987) Protein traffic in human synovial effusions. *Arthritis Rheum* 30: 57–63

23 Praest RM, Greiling H, Kock R (1997) Assay of synovial fluid parameters: hyaluronan concentration as a potential marker for joint disease. *Clin Chim Acta* 266: 117–128

24 Wallis WJ, Simkin PA, Nelp WB, Foster DM (1985) Intra-articular volume and clearance in human synovial effusion. *Arthritis Rheum* 28: 441–449

Elastoviscous hyaluronan in the synovium in health and disease

Endre A. Balazs and Charles Weiss

Matrix Biology Institute, 65 Railroad Avenue, Ridgefield, NJ 07657, USA

Introduction

The vertebrate joint is composed of three structurally and functionally different tissues: cartilage, ligaments and synovium. The latter is the collective name for the synovial tissue covering all non-cartilaginous surfaces of the joint and for the synovial fluid imbibing them and filling the space where the movements of the joint occur. During the past century, cartilaginous tissues were investigated by many researchers. In contrast, research on the synovium and ligaments was less popular. In this paper, our current knowledge of the synovium is briefly summarized with the bias of our four decades of research interest in this field. The rationale behind the therapeutic modality of viscosupplementation, the intra-articular application of elastoviscous hyaluronan and its derivatives (hylan A and hylan B), is discussed briefly from the point of view of the biological role of hyaluronan in healthy and arthritic joints.

Hyaluronan

The dominant molecule of the extracellular matrix of the synovium is hyaluronan. It fills the space between the collagen networks of the synovial tissue and is the major macromolecular component of the synovial fluid, providing its elastoviscosity. In healthy tissues the average molecular weight of hyaluronan is approximately 6 million. As are all other hyaluronans in the intercellular matrix, synovial hyaluronan is also polydisperse. Molecules of up to 8 million molecular weight and as small as 2 million are present in the healthy human and other animal synovial fluids. This polydispersity is very important because in pathological conditions, the polydispersity increases as a result of the occurrence of molecules in the range of 250,000 to 2 million molecular weight [1, 2]. The occurrence of this low molecular weight fraction has been interpreted to be a result of partial intra-articular degradation (oxidative or free radical) or as a newly synthesized fraction of hyaluronan.

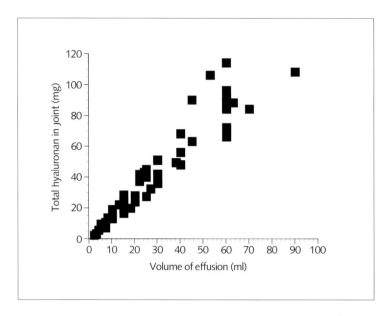

Figure 1
Relationship between volume and total hyaluronan content of 51 synovial effusions collected from 16 knees of 12 patients within a 43-month period. Each square represents one effusion.

Two interesting parameters of the synovial fluid in health and disease are its volume and its total hyaluronan content. While it is not easy to collect all the fluid from a healthy knee joint, during inflammation when the fluid volume increases (effusion) due to an influx of water, this task does not present a problem. There is a linear relationship between the volume of the collectable fluid from osteoarthritic joints and the total hyaluronan content. Figure 1 shows this relationship using 51 joint fluids collected from 16 knees analyzed during the course of a clinical study. It is important to note that this relationship is not dependent on the molecular weight distribution of the hyaluronan in the joint fluid. The average molecular weight of hyaluronan in these joints varied between 2 and 5 million, representing a broad distribution of molecules between 250,000 and 7 million molecular weight. Figure 2 demonstrates this broad molecular weight distribution in five synovial fluids obtained from one osteoarthritic knee of a patient during a period of 8 months. For comparison, the molecular weight distribution of synovial fluid from a healthy knee is also shown.

The polydispersity of the hyaluronan is fairly consistent in most of these patients, all of whom maintained a large effusion in the joint. The periodic removal of the fluid does not affect this picture. The important observation is that the relationship

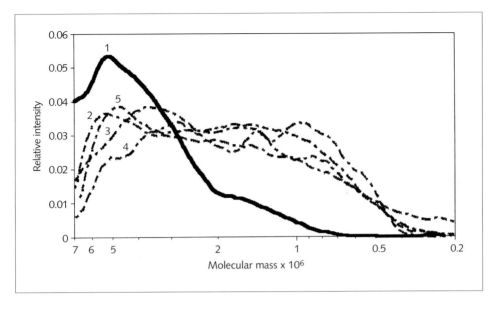

Figure 2
Molecular weight distribution of hyaluronan (HA) by gel electrophoresis in healthy synovial fluid (Curve 1) and in four effusions taken from the same osteoarthritic knee at various intervals (Curves 2–5). Curve 2, volume: 60 mL, HA concentration 2.2 mg/mL; Curve 3, volume 65 mL, HA concentration 2.1 mg/mL; Curve 4, volume 25 mL, HA concentration 1.1 mg/mL; Curve 5, volume 45 mL, HA concentration 2.0 mg/mL. Curve 2 represents the first arthrocentesis, Curve 2, 28 days later; Curve 3, 2 months later; Curve 4, 8 months later. The knee had no osteoarthritic pain during the 8-month period (gel electrophoresis according to the method described in [58]).

between volume and total hyaluronan content is fairly constant and does not depend on the variation in the mass of molecules contributing to the total hyaluronan content of the fluid. Similar observations were previously reported ([5]; Cowman et al., unpublished). This result indicates a feedback mechanism between volume and concentration. The volume of the fluid is regulated by turnover of water, and we know from Levick and his associates [3] that in rabbit joints, water turns over more rapidly than hyaluronan. It was shown that distending the synovial space in rabbit knees by injecting an excess volume of balanced salt solution temporarily increases the rate of synthesis, but not the concentration, of hyaluronan [3, 6]. It was also observed that the joint responds to the slightest injury to the synovial tissue or cartilage with a temporary increase in the rate of hyaluronan synthesis [6]. However, the concentration of hyaluronan in the injured joint never exceeds the normal level.

Hyaluronan in the superficial layer of the cartilage

This acellular, fibrillar structure covering the articular surface of the cartilage and menisci, called the superficial layer, consists of the lamina splendens and the acellular portion of the superficial zone. This layer contains a higher concentration of hyaluronan than either the underlying cartilage or the synovial fluid [4–6]. This superficial layer (1–10 μ thick, depending on the species) in healthy adults is easily recognizable in electron microscope sections because it has collagen fibrils different from those of the deeper zones [4, 8, 9]. Unlike the deeper cartilage layers, the superficial layer does not stain with cationic dyes, suggesting that the acidic groups of the hyaluronan and proteoglycans are not available for binding the positively charged dye molecules in this laminar structure [4, 6–8].

This superficial layer can be removed from the deeper layer of the cartilage and dissolved and dispersed by a hyaluronan-specific lysase ("bacterial hyaluronidase"), indicating that this structure is held together and anchored to the cartilage by hyaluronan [6]. The hyaluronan content of the superficial layer can be enhanced by injecting high concentrations of exogenous hyaluronan. In the rabbit joint, after 0.5 mL of 10 mg/mL high molecular weight hyaluronan was injected, the concentration of hyaluronan in the superficial layer was three times greater than the normal level [6]. By increasing the amount of hyaluronan injected (same concentration, greater volume), a saturation level can be reached, and by washing the joint with physiological saline solution, the hyaluronan content can be depleted to one-third of the unwashed control value [6]. From these experiments, Denlinger [6] calculated the concentration of hyaluronan in the interfibrillar space of the surface layer to be 15 mg/mL, which is approximately five times that of the concentration in the rabbit synovial fluid.

This hyaluronan-rich superficial layer also contains lubricin, a proteoglycan that was first identified by Swann and his coworkers [10] and found to contribute to the slipperiness of the cartilage surface. Lubricin is also present in the synovial fluid and was found to be homologous to the superficial zone protein [11, 12]. Surface-active phospholipids have also been identified in the lamina splendens, and it has been suggested that their hydrophobicity contributes to the reduction of friction on the cartilage surface [13]. The accumulation of fibronectin in the superficial zone was first demonstrated by immunofluorescence methods [6], and was found in the synovial fluid and in the synovial tissue, where it is apparently produced by the hyaluronan-synthesizing cells [14]. Fibronectin is also produced by macrophages [15]. Based on our work (see [16] for review) and that of other investigators, we believe that fibronectin counteracts the cell membrane stabilizing effect of elastoviscous hyaluronan solutions. The accumulation of fibronectin in the superficial layer where the highest hyaluronan concentration was found in the joint suggests that these two molecules represent a regulatory system for cells migrating on the surface of the cartilage. It is known that after cartilage surface injury, fibrocytes immediately migrate from the marginal zone, eventually covering the cartilage surface. We propose that this initial migration of cells and the later pannus formation are regulated by the hyaluronan-fibronectin system.

It was proposed that the superficial layer, due to the elastic properties of hyaluronan, provides mechanical protection to the cartilage cells and their intercellular matrix [5]. The superficial layers of the opposing cartilage surfaces and the elastoviscous synovial fluid trapped in the gap between them during joint loading represent a complex protective system. The importance of the superficial layer lies in its protective, elastic properties, its contribution to joint slipperiness, and its regulation of cell migration on the cartilage surface.

Hyaluronan in the synovial tissue

Cells that produce hyaluronan are localized in the intima of the synovial tissue. These cells in the surface layer of the tissue are embedded in a loose, randomly oriented collagen fibrous network. Since this tissue is constantly folding and unfolding with the movement of the joint, these cells are exposed to considerable mechanical stress and dislocation. Consequently, the morphology of the extracellular matrix and the exact shape and location of these cells as we see them in histological and electron microscopic sections will depend on whether the fixation was made on folded or on unfolded tissue (bent or stretched joint). In all published pictures of these cells, a considerable intercellular space, open toward the synovial fluid space, is evident. In other words, no structural separation exists between the intercellular matrix of the tissue and the synovial fluid. This is important because the hyaluronan molecules produced by these cells can flow freely during the movements of the joint or diffuse in the resting joint between the fluid (synovial fluid) and solid matrix (synovial tissue) compartments. The passage of hyaluronan between these two compartments is discussed in this volume in the chapter by Mason et al.

The synoviocytes, as these morphologically distinctive cells are called, are not fibrocytes, but are like the cells of the mononuclear phagocyte system (histocytes, monocytes and macrophages). They are structurally similar to the hyalocytes, the hyaluronan-producing phagocytic cells of the vitreus [17, 18]. Morphologists describing these cells distinguished between two or three types of cells (type A, B and C, or intermediate cells) [19, 20]. When the intima of the synovial tissue is cultured *in vitro*, the type A macrophage-like cells constitute only a small part of the cell population. The type B hyalocyte-like cells and the type C cells dominate the cultures [21, 22]. These authors concluded, as was confirmed by later investigators, that "the cells of the synovial intima belong to a single cell type, but that under appropriate environmental influence they can modulate into either of two very different sub-types". This is important because hyalocytes of the vitreus are very similar mononuclear phagocytes that, after intravitreal bleeding, phagocytose red blood cells and, in healthy conditions, produce and catabolize hyaluronan [23, 24]. We believe that the synoviocytes of the intima and the hyalocytes of the cortical layer of the vitreus are mononuclear phagocytes and are responsible for hyaluronan production in both tissues.

The cells of the intima of the synovial tissue produce hyaluronan [25–27]. The *in vitro* hyaluronan synthesis by cells cultured from synovial tissue of healthy and arthritic joints has been investigated by many authors. The cell cultures used in these studies do not represent a pure cell line, but a mixture of fibrocytes and synoviocytes. Using these mixed cell cultures originating from normal, osteoarthritic or rheumatoid human joints, it was shown that the rate of synthesis of hyaluronan is similar in all these cell cultures. When exogenous hyaluronan of different molecular weights and various concentrations was added to the cultures, significant stimulation of synthesis was observed only in the cultures originating from osteoarthritic joints and only by large molecular weight hyaluronan (avg. MW 3.8 and 4.7 million). This stimulation was concentration dependent. The highest concentration tested was 0.45 µg/mL, which is well below the concentration of hyaluronan surrounding these cells in healthy and osteoarthritic joints [28]. These data suggest that there are cells present in the cultures of synovial tissue of osteoarthritic joints that respond to high molecular weight exogenous hyaluronan with upregulation of hyaluronan synthesis.

The *in vitro* proliferation of cells of a rabbit synovial cell line obtained from healthy joints can be inhibited by elastoviscous hyaluronan solutions. The elastoviscosity depends on the concentration and molecular weight of the hyaluronan; consequently, the inhibitory effect also depends on these two parameters. Hyaluronan solutions of low concentration of any mix of molecular weights or high concentration solutions made of low molecular weight molecules have very low viscosity and are therefore ineffective inhibitors [29].

It is important to note that the presence of fibronectin in the intima has been demonstrated in a variety of joint diseases [30]. Fibronectin was also demonstrated in the hyalocytes (type B cells) of the intima using histochemical methods on biopsies from patients with traumatic, rheumatoid and osteoarthritis [31]. After the discovery that fibronectin was bound to collagen filaments in the hyaluronan-rich superficial layer of articular cartilage and in the cortical gel layer of the vitreus, it was assumed that the interaction between hyaluronan and fibronectin represents a biological regulatory mechanism [6, 56]. This was reinforced by the finding that the inhibitory effect of elastoviscous solutions (hyaluronan, DNA and gelatin) on the phagocytosis of peritoneal macrophages can be significantly counteracted by addition of plasma fibronectin. Since fibronectin itself in this system stimulates phagocytosis, an antagonist system between fibronectin and hyaluronan was suspected [16]. This antagonistic role of fibronectin and hyaluronan on the cartilage surface was demonstrated by Shizawa and coworkers in organ culture of synovial tissue cells overgrowing the cartilage surface. Hyaluronidase treatment of the surface promoted and hyaluronan coating inhibited, this pannus-like cellular overgrowth [53]. In an important series of studies, Homandberg, Kuettner and associates emphasized the role of fibronectin on the regulation of cartilage metabolism and, in osteoarthritis, the effect of hyaluronan in counteracting fibronectin-mediated chondrolysis [54–57]. They concluded that fibronectin fragments play an important role in the homeostasis of normal cartilage, and in osteoarthritic joints, they up-regulate metalloproteinase expression and enhance degradation of proteoglycans. We believe that this regulatory system has three elements: fibronectin of both plasma and tissue origin, hyaluronan of various molecular sizes and thin collagen filaments (as present in the vitreus and in the superficial layer of cartilage). The collagen network serves as an anchoring element for both fibronectin and hyaluronan. This system can regulate the movement of cells during the repair process after cartilage injury and during chronic inflammatory and degenerative processes. At the same time, this system serves as a matrix filter between the synovium and cartilage. The filter represents a molecular sieve as well as a selective binding system for noxious molecules. In the cortical layer of the vitreus the same system can regulate the invasion of cells from the retina in degenerative processes as well as in other injuries. The similarity of these two systems is that they both have a high hyaluronan concentration and fibronectin bound to the fine collagen filaments. The difference is that in the vitreus this layer contains cells (hyalocytes) that produce the hyaluronan, while the superficial layer is acellular and the hyaluronan originates from the synovial fluid.

Hyaluronan in the immobilized joint

The question of how mobility of the joint influences hyaluronan synthesis is a very important one, because arthritic pain in animals causes lameness and, in humans decreases the voluntary movements of the joint. Numerous animal studies have been performed during the past decades to answer this question. After immobilization of the rabbit knee joint, the hyaluronan concentration in the fluid does not increase, but the hyaluronan content does, due to an increase in the volume of the joint fluid [6]. Intra-articular injection of hyaluronan into the joint before immobilization significantly decreased the exudate formation during 7 days of immobilization, and also significantly decreased the protein content of the exudates. The hyaluronan content of the synovial tissue was also doubled after 7 days of immobilization (from 0.031 to 0.067% of wet tissue weight) [6].

After long immobilization (9 weeks), the hyaluronan content in the rabbit periarticular fibrous connective tissue (ligaments, capsule, synovium and fascia) significantly decreased [32]. The hyaluronan concentration in the synovial fluid also decreased in sheep joints after 12 weeks of immobilization [33]. In this study, the uridine diphosphoglucose dehydrogenase activity of the synovial intima cells significantly decreased. The immobilized rabbit joints (2–8 weeks) also demonstrated an increase of type B cells and a decrease in the hyaluronan-producing type A cells [34].

Based on all these data, one can conclude that the short-term (1 week) effect of immobilization results in an accumulation of hyaluronan in the joint. After longer immobilization, due to the increased amount of hyaluronan in the joint, a feedback mechanism down-regulates hyaluronan synthesis. This hypothesis is supported by a study in which rabbit knee joints became contracted and arthritic (cartilage degradation) during immobilization. When hyaluronan of various molecular weights (9.8×10^5, 1.73×10^6, 2.02×10^6) was injected, cartilage degeneration and joint contraction were significantly inhibited, and these effects were more pronounced with the higher molecular weight hyaluronan [35].

Studies on the hyaluronan content of immobilized human joints are not available. However, the effect of intra-articular injection of hyaluronan on the pain experienced in the remobilization of healthy joints after prolonged immobilization was studied. Elastoviscous hyaluronan was injected weekly into painful joints after remobilization of the joint [36]. Significant reduction of pain at rest and on movement, and of stiffness on flexion and extension was observed. This study confirmed the analgesic effect of hyaluronan in the healthy, but previously immobilized, human joint. A study on immobilized human knees following joint injury demonstrated that knees injected with hyaluronan (avg. MW 500,000) had significantly reduced pain and increased joint mobility post-immobilization [37]. If hyaluronan synthesis is down-regulated during immobilization, resulting in a decrease of its concentration in the synovial tissue, then less protection is available for the nociceptors and their sensitivity increases, resulting in pain on movement. The intra-articular injection of elastoviscous hyaluronan restores joint homeostasis and acts as an analgesic.

The analgesic effects of hyaluronan

The analgesic effects of elastoviscous solutions of hyaluronan were discovered in post-traumatic arthritis of racehorses in the early 1970s [38]. Highly purified (NIF-NaHA), high molecular weight (avg. MW 2–3 million) hyaluronan solutions (1%) prepared from human umbilical cord or rooster comb relieved the lameness when injected into the painful joint. A similar pain-relieving effect was found in dogs after experimentally produced cartilage damage [39, 40]. The analgesic effect of 1% hyaluronan solutions (avg. MW 2–3 million) on post-traumatic osteoarthritis was confirmed with objective, quantitative gait analysis in osteoarthritic racehorses [41] and, nearly two decades later, in experimental osteoarthritis of sheep produced by medial meniscectomy [42]. In this study, 1% hyaluronan solutions with increasing average molecular weight were used (0.89–2 million). The attenuation of the pain was greatest with solutions of 2 million average molecular weight, confirming that this effect is based on the elastoviscosity and not on the concentration of hyaluronan in the solution. Japanese investigators [43–45] used mouse and rat behavioral pain models to confirm the analgesic effects of hyaluronan. These pain models were produced by intra-articular injections of bradykinin, prostaglandin E_2 or monosodium urate crystals. These authors also concluded that the analgesic effect of hyaluronan is affected by the concentration and average molecular weight of the HA. The greatest and most lasting effect was obtained using 1% solutions with an average MW of 2.2 million.

The first neurophysiological studies on the effect of hyaluronan on arthritic pain were reported in 1996 using cat knee joints to determine the influence of elastoviscous and non-elastoviscous hylan A solutions (hylan A is a derivative of hyaluronan) on nerve responses from nociceptive afferent fibers of the medial articular nerve in healthy and arthritic animals [46]. In healthy animals, pain was produced by displacement of the joint beyond its normal range of motion. This noxious movement produced a sudden rise in the rate of neural discharge. Twenty to 40 min after the injection of elastoviscous hylan A solution, the noxious, movement-evoked discharges were significantly reduced to an average of 65% of control values (Fig. 3). Between the noxious stimuli, there were substantially smaller, but consistent, nerve discharges (ongoing discharges). The frequency of these ongoing discharges was also attenuated by the hylan A solutions. When this experiment was carried out either with non-elastoviscous hylan A solutions made of the same concentration, but with molecules of low molecular weight (avg. MW 30,000), or with physiological buffer solution (the solvent of hylan A), the discharge rate of both the ongoing and the noxious movement-evoked nerve activity was unaltered. These studies were also carried out on knee joints of anesthetized cats with experimentally induced, acute inflammatory arthritis produced by intra-articular injection of kaolin and carrageenan. A gradual build up of nerve discharges was observed, with an increasing rate of discharge reaching a constant level at 2–3 h after the injection. Then elastoviscous hylan A was injected into the joint, resulting in a decrease in both the ongoing and movement-evoked discharges, which eventually returned almost to the level that was observed before kaolin-carrageenan injection. As control, the above-described non-elastoviscous hylan A solution was also used, and it was found that both the ongoing and movement-evoked discharges were unaffected by the injection (Fig. 4) [47].

Studies on cat knees were extended to the rat knee model. Nociceptors from the afferent fibers (A-delta and C fibers) of the medial articular nerve responded in the same way to the elastoviscous and non-elastoviscous solutions as did fibers in the cat joints. These studies confirmed in a second animal model that elastoviscous solutions significantly reduce the responses of these primary afferent fibers in

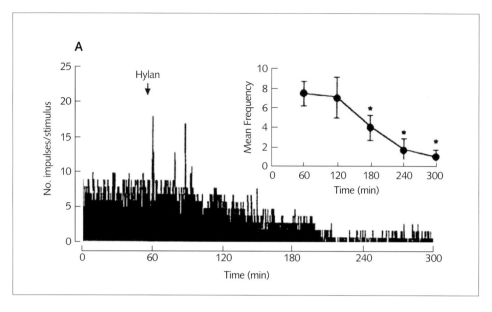

Figure 3
Inhibition of the response of joint afferent fibers to noxious movements in the healthy (intact) knee joint of cats by intra-articular injection of elastoviscous hyaluronan solutions (hylan A, 0.8%, avg. MW 8 million).
Histogram of the number of impulses per stimuli of an A-delta fiber elicited by an outward rotation of 5-s duration repeated at 2 min intervals continually during 5 h. An outward rotation beyond the normal range of notion of the joint, created a displacement of the joint in the noxious range, causing pain. Hylan A was injected (at the arrow) 1 h after the onset of the mechanical stimulus.
Insert: The average numbers of impulses during 1-h periods during the 5-h experiment, showing the mean frequencies before (60 min) and after the intra-articular injection of the elastoviscous solution (hylan A). The difference between the 60 min and the times marked with asterisks are significant at $p < 0.05$. (From [47].)

both healthy and experimentally-induced arthritic joints. These authors compared the analgesic effect of intra-articularly injected hylan in cats with systemically applied indomethacin in rats. They found that both depressed the ongoing and provoked activities of the nociceptive neuroterminals to essentially the same degree [48].

Recently completed work of Belmonte, Schmidt and coworkers substantially expanded the scope of this inquiry to a cell model using oocyte (*Xenopus laevis*) cell membranes. In the first set of experiments, the cell membrane of the intact oocyte

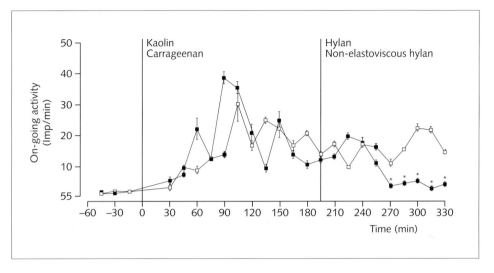

Figure 4
The build-up of ongoing neural activity of two A-delta fibers from two different knees after kaolin-carrageenan injection and the effect of intra-articularly injected elastoviscous (filled circle) and non-elastoviscous (open circles) hylan solutions. The mean frequency of the ongoing activity was measured during 10-min periods. The injection of kaolin-carrageenan (at 0 time) and the injection of hylan solutions (at 190 min) are marked. Vertical bars are standard error of means. Asterisks indicate significant differences at $p < 0.05$ levels between the effects of the two hylan solutions. (From [47].)

was deformed by gradually increased suction through a microelectrode pipette. The opening of the stretch-activated ion channels was blocked by using elastoviscous hylan solutions (hylan A or Synvisc, see Fig. 5). Non-elastoviscous hylan solutions made of the same concentration, but low average molecular weight (96,000) hylan did not have this blocking effect [49].

In another series of experiments, excised cell membrane patches from the oocytes were used, and the elastoviscous solutions or control fluids were applied on both the inside (cytoplasmic side) and outside (external side) of the cell membrane. In this case, the micropipette providing the suction was applied on the opposite side of the membrane. This experiment confirmed the results obtained on intact oocytes. The importance of this finding is that the elastoviscous solution prevented the opening of the channels when the stretch was applied on the opposite side of the membrane [50]. This finding suggests that both sides of the membranes have mechanoreceptors that interact with the elastoviscous molecular network, and that the stabilizing effect is not dependent on the chemical properties of the hyaluronan, but on its net-

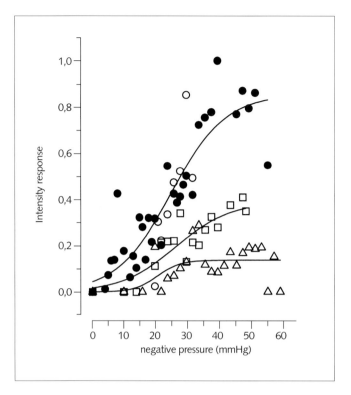

Figure 5
The effect of hylan solutions on the intensity response on stretch activated ion channels of oocyte membrane patches. The intensity response represents the normalized average current that is proportional to the opening probability of the ion channels triggered by increasing negative pressure applied on the cell membrane patches. Control Barth's solution (filled circles); non-elastoviscous hylan A solution (open circles); elastoviscous hylan A solution (open squares); hylan G-F 20 (open triangles; see text). (From [50].)

work-forming structure, which is the basis of elastoviscosity. We believe that the analgesic effect of elastoviscous solutions on the opening of mechanosensitive ion channels of cell membranes points to a general biological phenomenon: the mechanical decoupling effect of elastoviscous hyaluronan solutions in the immediate environment of the cell. This mechanical decoupling effect can be perceived as a stabilization or protection of the cell from noxious mechanical events. A similar "stabilization" effect of elastoviscous solutions was described by one of the authors (Balazs) on cells of the lymphomyeloid system, including stabilization of such functions as phagocytosis, cell migration, cell division and release of important signal molecules (for review see [16]).

The therapeutic use of hyaluronan in osteoarthritis

The injection of elastoviscous hyaluronan or hylan solutions into painful knee joints (viscosupplementation) has been shown to be effective in reducing pain and improving function in a broad spectrum of osteoarthritic patients. There are, however, numerous questions concerning its mechanism of action and, in particular, its prolonged analgesic effect, which may extend weeks, months and occasionally years beyond its residence time in the joint. To explain this prolonged analgesic effect, a hypothesis based on empirical observations on osteoarthritis, laboratory and clinical studies is presented.

Because articular cartilage is aneural, and nociceptors in the subchondral bone and ligaments are beyond the penetration of viscosupplements, we will focus on nociceptive terminals of synovium and the subsynovial capsule. In the healthy joint, nociceptive afferent units are stimulated by stretching the soft tissues beyond the normal range of movement of the joint. In chronic arthritic conditions or in acute inflammation (such as the experimental arthritis in cats and rats), the nociceptors with a low threshold become sensitized, which means they can be stimulated by the mechanical impulses within the normal range of motion of the joint. The chemical agents in elevated concentrations that sensitize the mechanoreceptors may also stimulate afferent fibers that are not sensitive to movement. The important point is that in the arthritic joint, the sensitization and stimulation by chemical agents and stimulation by movement are not separable events. The pivotal point of this hypothesis is that pain in the arthritic joint elicited by an increase of sensory inflow from articular nociceptors is the result of the change of the pericellular environment of afferent nerve endings. This means that the intercellular matrix around these nociceptive nerve endings is modified. This modification is the result of the decrease of both concentration and average molecular weight, and also of the availability of specific binding sites of the hyaluronan in the intercellular matrix that surrounds the nerve endings. The latter is important because it is stipulated that binding sites inactivate nociceptor-stimulating noxious agents. Thus both the chemical and physical protective environments are compromised. The protective and regulatory mechanism is based on the filtering and exclusion volume effects of the hyaluronan molecular network, as well as its elastoviscous properties and special binding characteristics.

Today, two very different therapeutic products for viscosupplementation are available to osteoarthritic patients worldwide. The first type contains a 1.0 to 1.5% solution of hyaluronan with an average molecular weight of 0.5 to 2 million, with elastoviscosity in the range of the pathological synovial fluid. The second type contains a combination of hylan A and hylan B (generic name: hylan G-F 20 (Synvisc), containing hylan A (0.8% solution) and hylan B gel (0.5% polymer content) mixed in an 8:2 volume ratio) with an elastoviscosity similar to that of healthy, young synovial fluid. The longest lasting analgesic effect of the low molecular weight solutions was observed after 5 to 10 weekly injections, while long lasting analgesia was obtained using the higher molecular weight hylan solutions after only 2 to 3 weekly injections (for review see [51, 52]). During the course of numerous clinical trials on both types of elastoviscous solutions, it became obvious that in order to obtain an optimal effect, the injections must be carried out under the strictest aseptic conditions, and the effusions must be carefully removed before injection of the viscosupplement. Since the treatment is an application of hyaluronan into a joint that often already has more than the normal amount of hyaluronan of both low and high molecular weight fractions, patient selection must address the amount of hyaluronan present in the joint. From this point of view, one can differentiate three different types of patients: (1) patients with no effusion at all ("dry joint", *arthrosis deformans*); (2) patients with moderate effusion volume; and (3) patients with chronic, large effusion volumes (> 20 mL). Clinical studies indicate that joints of all three groups can benefit from viscosupplementation, but with varying degrees of response, depending on other factors, such as the status of the cartilage damage and the instability of the joint, which can cause mechanical stress on ligaments, menisci and cartilage.

From the point of view of hyaluronan synthesis, two events can be distinguished in these knees. In all three groups of knees, a large amount (30 to 60%) of the hyaluronan is present in a lower-than-normal molecular size, which substantially decreases the elastoviscosity of the fluid and, therefore, its physical protective properties. Since the lower elastoviscosity means a less elastic, less coherent and less stable molecular network, it may also decrease its chemical protective (interactive) properties. Numerous tissue cultures and animal studies have shown that hyaluronan, depending on concentration and molecular size (elastoviscosity of the network), inhibits cellular activities that characterize the inflammatory process (release of prostaglandins, phagocytosis, etc.).

The second important event that occurs in all joints with effusion is the increased rate of hyaluronan synthesis. Both the low and high molecular weight fractions are overproduced and partially retained in the joint. We believe that this represents a protective mechanism designed to flood all tissues of the joint with increased amounts of hyaluronan to provide enhanced physical and chemical protection.

The short-term analgesic effect of viscosupplementation is also amplified by the removal of the effusion that contains nociceptor-stimulating agents and by the "washing out" and dilution effect of the injected viscosupplement. Long-term analgesic effects can be achieved only if the rate of increased hyaluronan synthesis is sustained until the chronic synovial and cartilage wounds heal. The healing of the chronic synovial wound means that the influx of leucocytes ceases, and the concentrations of inflammatory and pain-inducing chemicals decrease or are neutralized by hyaluronan. At this stage, the homeostasis is restored and the rate of hyaluronan synthesis returns to a normal level. Viscosupplementation with repeated exchanges of the pathological fluid with a pure hyaluronan solution acts as an analgesic, thereby restoring the mobility of the joint, which in turn increases the blood and lymph flow in the synovial tissue and the outflow of the synovial fluid, facilitating the healing of the intra-articular wounds. Thus, viscosupplementation can be regarded as an internal lavage of a chronic wound with a high concentration of elastoviscous hyaluronan or hylan that acts directly or indirectly as a molecular and free radical scavenger, reducing the pain and inflammation.

References

1. Balazs EA, Watson D, Duff IF, Roseman S (1967) Hyaluronic acid in synovial fluid. I. Molecular parameters of hyaluronic acid in normal and arthritic human fluids. *Arthritis Rheum* 10: 357–375
2. Balazs EA, Briller S, Denlinger JL (1981) Na-hyaluronate molecular size variations in equine and human arthritic synovial fluid and effect on phagocytic cells. In: JH Talbott (ed): *Seminars in arthritis and rheumatism*. Grune and Stratton, New York, 11: 141–143
3. Coleman PJ, Scott D, Ray J, Mason RM, Levick JR (1997) Hyaluronan secretion into the synovial cavity of rabbit knees and comparison with albumin turnover. *J Phys* 503(3): 645–646
4. Balazs EA, Bloom GD, Swann DA (1966) Fine structure and glycosaminoglycan content of the surface layer of articular cartilage. *Fed Proc* 25(6): 1813–1816
5. Balazs EA, Gibbs DA (1970) The rheological properties and biological function of hyaluronic acid. In: EA Balazs (ed): *Chemistry and molecular biology of the intercellular matrix*. Academic Press, London and New York, 1241–1254
6. Denlinger JL (1982) *Metabolism of sodium hyaluronate in articular and ocular tissues*. Ph.D. thesis, Université des Sciences et Techniques de Lille, Lille, France
7. Rosenburg L (1971) Chemical basis for the histological use of safranin-O in the study of articular cartilage. *J Bone Joint Surg* 53A: 69
8. Weiss C, Rosenberg L, Helfet AJ (1968) An ultrastructural study of normal young adult human articular cartilage. *J Bone Joint Surg* 50A: 663
9. Guerra D, Frizziero L, Losi M, Bacchelli B, Mezzadri G, Pasquali-Ronchetti I (1996) Ultrastructural identification of a membrane-like structure on the surface of normal articular cartilage. *J Submicrosc Cytol Pathol* 28(3): 385–393
10. Swann DA, Sotman S, Dixon M, Brooks C (1977) The isolation and partial characterization of the major glycoproteins (LGP-1) from the articulating lubricating fraction from bovine synovial fluid. *Biochem J* 161: 473–485
11. Jay GD, Britt DE, Cha C-J (2000) Lubricin is a product of megakaryocyte stimulating factor gene expression by human synovial fibroblasts. *J Rheum* 27: 594–600
12. Su J-L, Schumacher BL, Lindley KM, Soloveychik V, Burkhart W, Triantafillou JA, Kuettner K, Schmidt T (2001) Detection of superficial zone protein in human and animal body fluids by cross-species monoclonal antibodies specific to superficial zone protein. *Hybridoma* 20(3) 149–157

13. Hilla BA, Butler BD (1984) Identification of surfactants in synovial fluid and their ability to act as boundary lubricants *Ann Rheum Dis* 48: 51–57
14. Mapp PI, Revell PA (1985) Fibronectin production by synovial intimal cells. *Rheumatol Int* 5: 229–237
15. Alitalo K, Hori T, Vaheri A (1980) Fibronectin is produced by human macrophages. *J Exp Med* 151: 602–613
16. Balazs EA (1998) The viscoelastic intercellular matrix and control of cell function by hyaluronan. In: T Laurent (ed): *The chemistry, biology and medical applications of hyaluronan and its derivatives* (Proceedings of the Wenner-Gren Foundation International Symposium held in honor of Endre A. Balazs, September 18–21, 1996, Stockholm, Sweden), Portland Press, London, 185–204
17. Balazs EA, Denlinger JL (1984) The synovial cell. In: DG Scarpelli, G Migaki (eds): *Comparative pathobiology of major age-related diseases: current status and research frontiers*. Alan R. Liss, Inc., New York, 129–143
18. Bloom GD, Balazs EA (1965) An electron microscopic study of hyalocytes. *Exp Eye Res* 4: 249–255
19. Dobbie JW, Hind C, Meijers P, Bodart C, Tasiauz N, Perret J, Anderson JD (1995) Lamellar Body secretion: Ultrastructural analysis of an unexplored function of synoviocytes. *British J Rheum* 34: 13–23
20. Iwanaga T, Shikichi M, Kitamura H, Yanase H, Nozawa-Inoue K (2000) Morphology and functional roles of synoviocytes in the joint. *Arch Histol Cytol* 63(1): 17–31
21. Fell HB, Glauert AM, Barrat MEJ, Green R (1976) The pig synovium I. The intact synovium *in vivo* and in organ culture. *J Anat* 122(3): 663–680
22. Barratt MEJ, Fell HB, Coombs RRA, Glauert AM (1977) The pig synovium II. Some properties of isolated intimal cells. *J Anat* 123(1): 47–66
23. Österlin SE, Jacobson B (1968) The synthesis of hyaluronic acid in vitreous I. Soluble and particulate transferases in hyalocytes. *Exptl Eye Res* 7: 497–510.
24. Balazs EA (1961) Molecular morphology of the vitreous body. In: GK Smelser (ed): *The structure of the eye*. Academic Press, New York, 293–310
25. Castor CW, Dorstewitz EL (1966) Abnormalities of connective tissue cells cultured from patients with rheumatoid arthritis. I. Relative unresponsiveness of rheumatoid synovial cells to hydrocortisone. *J Lab Clin Med* 68: 300–313
26. Castor CW, Dorstewitz EL, Rowe K, Ritchie JC (1971) Abnormalities of connective tissue cells cultured from patients with rheumatoid arthritis. II. Defective regulation of hyaluronate and collagen formation. *J Lab Clin Med* 77: 65–77
27. Castor CW, Smith SF, Ritchie JC, Dorstewitz EL (1971) Connective tissue activation. II. Abnormalities of cultured rheumatoid synovial cells *Arthritis Rheum* 14: 55–66
28. Smith MM, Ghosh P (1987) The synthesis of hyaluronic acid by human synovial fibroblasts is influenced by the nature of the hyaluronate in the extracellular environment. *Rheumatol Int* 7: 113–122
29. Goldberg RL, Toole BP (1987) Hyaluronate inhibition of cell proliferation. *Arth Rheum* 30(7): 769–778

30 Mapp PI, Revell PA (1985) Fibronectin production by synovial intima cells. *Rheum Int* 5: 229–237
31 Mayston V, Mapp PI, Davies PG, Revell PA (1984) Fibronectin in the synovium of chronic inflammatory joint disease. *Rheum Int* 4: 129–133
32 Akeson WH, Woo S, Amiel D, Coutts RD, Daniel D (1973) The connective tissue response to immobility: biochemical changes in periarticular connective tissue of the immobilized rabbit knee. *Clin Ortho Rel Rsch* 93: 356–362
33 Pitsillides AA, Skerry, TM, Edwards, JCW (1999) Joint immobilization reduces synovial fluid hyaluronan concentration and is accompanied by changes in the synovial intimal cell populations. *Rheumatology* 38: 1108–1112
34 Finsterbush A, Friedman B (1973) Early changes in Immobilized rabbits knee joints: a light and electron microscopic study. *Clin Ortho Rel Rsch* 92: 305–319
35 Sakakibara Y, Miura T, Iwata H, Kikuchi T, Yamaguchi T, Yoshima T, Itoh H (1994) Effect of High Molecular weight sodium hyaluronate on immobilized rabbit knee. *Clin Orth Relat Rsch* 299: 282–292
36 Wigren A. (1981) Hyaluronic acid treatment of postoperative joint stiffness. *Acta Orthop Scand* 52: 123–127
37 Di Marco C, Letizia GA (1995) Hyaluronic Acid in the treatment of pain due to knee joint immobilization. *Clin Drug Invest* 10(4): 191–197
38 Rydell NW, Butler J, Balazs EA (1970) Hyaluronic acid in synovial fluid. VI. Effect of intra-articular injection of hyaluronic acid on the clinical symptoms of arthritis in track horses. *Acta Vet Scand* 11: 139–155
39 Rydell N, Balazs EA (1971) Effect of intra-articular injection of hyaluronic acid on the clinical symptoms of osteoarthritis and on granulation tissue formation. *Clin Orthop* 80: 25–32
40 Balazs EA, Denlinger JL (1985) Sodium hyaluronate and joint function. *J Equine Vet Sci* 5: 217–228
41 Gingrich DA (1981) Effect of exogenous hyaluronic acid on joint function in experimentally induced equine osteoarthritis: dosage titration studies. *Res Vet Sci* 30: 192–197
42 Ghosh P, Read R, Armstrong S, Wilson D, Marshall, R McNair P (1993) The effects of intraarticular administration of hyaluronan in a model of early osteoarthritis in sheep I. Gait analysis and radiological and morphological studies. *Sem in Arth Rheum* 22: 18–30
43 Miyazaki K, Gotoh S, Ohkawara H, Yamaguchi T (1984) Studies on analgesic and antiinflammatory effects of sodium hyaluronate (SPH). *Pharmacometrics* 28(6): 1123–1135
44 Gotoh S, Miyazaki K, Onaya J, Sakamoto T, Tokuyasu K Namiki O (1988) Experimental knee pain model in rats and analgesic effect of sodium hyaluronate (SPH). *Folia Pharmacol Japon* 92: 17–27
45 Aihara S, Murakami N, Ishii R, Kariya K, Azuma Y, Hamada K, Umemoto J, Maeda S (1992) Effects of sodium hyaluronate on the nociceptive response of rats with experimentally induced arthritis. *Folia Pharmacol Japon* 100: 359–365
46 Pozo MA, Balazs EA, Belmonte C (1997) Reduction of sensory responses to passive

movements of inflamed knee joints by hylan, a hyaluronan derivative. *Exp Brain Res* 116: 3–9

47   Belmonte C, Pozo MA, Balazs EA (1998) Modulation by hyaluronan and its derivatives (hylans) of sensory nerve activity signaling articular pain. In: T Laurent (ed): *The chemistry, biology and medical applications of hyaluronan and its derivatives* (Proceedings of the Wenner-Gren Foundation International Symposium held in honor of Endre A. Balazs, September 18–21, 1996, Stockholm, Sweden). Portland Press, London, 205–217

48   Pawlak M, Gomis A, Just S, Heppelmann B, Belmonte C, Schmidt RF (2002) Mechanoprotective actions of elastoviscous hylans on articular pain receptors. In: JF Kennedy, GO Phillips (eds): *Hyaluronan 2000*, vol. 2. Woodhead Publishing, Cambridge, UK, 341–352

49   De la Peña E, Sala S, Schmidt RF, Belmonte C (2002) Effect of elastoviscous solutions of hyaluronan derivatives on mechanotransduction. In: JF Kennedy, GO Phillips (eds): *Hyaluronan 2000*, vol. 2. Woodhead Publishing, Cambridge, UK, 407–418

50   De la Peña E, Sala S, Rovira JC, Schmidt RF, Belmonte C (2002) Elastoviscous substances with analgesic effects on joint pain reduce stretch-activated ion channel activity *in vitro*. *Pain; in press*

51   Peyron JG (1999) Viscosupplementation for the treatment of osteoarthritis of the knee with hyaluronan and hylans: rationale and state of the art. In: S Tanaka, C Hamanishi (eds): *Advances in osteoarthritis*. Springer-Verlag, Tokyo 213–236

52   Rosier RN, O'Keefe RJ (2000) Hyaluronic acid therapy. In: CT Price (ed): *Instructional Course Lectures*. American Academy of Orthopaedic Surgeons, Rosemont, IL 49: 495

53   Shiozawa S, Yoshihara R, Kuroki Y, Fujita T, Shiozawa K, Imura S (1992) Pathogenic importance of fibronectin in the superficial region of articular cartilage as a local factor for the induction of pannus extension on rheumatoid articular cartilage. *Ann Rheum Dis* 51(7)869

54   Homandberg GA, Hui F, Wen C, Kuettner KE, Williams JM (1997) Hyaluronic acid suppresses fibronectin fragment mediated cartilage chondrolysis: I. *in vitro*. *Osteo Cart* 5(5): 309–319

55   Williams JM, Plaza V, Hui F, Wen C, Kuettner KE, Homandberg GA (1997) Hyaluronic acid suppressed fibronectin fragment mediated cartilage chondrolysis: II. *in vitro*. *Osteo Cart* 5(4): 235–240

56   Kang Y, Eger W, Koepp H, Williams JM, Kuettner KE, Homandberg GA (1999) Hyaluronan suppresses fibronectin fragment-mediated damage to human cartilage explant cultures by enhancing proteoglycan synthesis. *J Orthop Res* 17(6): 858–869

57   Homandberg GA (1999) Potential regulation of cartilage metabolism in osteoarthritis by fibronectin fragments. *Front Biosci* 4: D713–730

58   Lee HG, Cowman MK (1994) An agarose gel electrophoretic method for analysis of hyaluronan molecular weight distribution. *Anal Biochem* 219: 278–287

The covalent complex formation of hyaluronan with heavy chains of inter-α-trypsin inhibitor family is important for its functions

Wannarat Yingsung[1,2], Lisheng Zhuo[1], Masahiko Yoneda[1], Naoki Ishiguro[3], Hisashi Iwata[3] and Koji Kimata[1]

[1]Institute for Molecular Science of Medicine, Aichi Medical University, Nagakute, Aichi, 480-1195, Japan; [2]Present address: Department of Biochemistry, Faculty of Medicine, Chiang Mai University, Amphure Muang, Chiang Mai, 50200 Thailand; [3]Department of Orthopedic Surgery, Nagoya University School of Medicine, Tsurumai, Nagoya, 466-8550, Japan

In culture, not all, but most cells form a hyaluronan (HA)-rich matrix around themselves. In a typical case, chondrocytes in serum-supplemented culture elaborate the matrix consisting of hyaluronan (HA), hyaluronan-binding chondroitin sulfate proteoglycans such as aggrecan and PG-M/versican, and the Serum-derived Hyaluronan-Associated Proteins, which we previously named SHAP [1]. We found that SHAP exactly corresponds to the heavy chains of the serum inter-alpha-trypsin inhibitor (ITI) family molecules, and is covalently bound to HA. The structure of the SHAP-HA complex and a possible pathway of its formation are shown in Figure 1 [2, 3]. ITI family molecules comprise a common subunit, bikunin and one or two of the three genetically distinct, but related heavy chains which are covalently attached to the chondroitin sulfate chain of bikunin *via* an ester bond of the same type as the one found in the SHAP-HA complex [4]. Accumulated evidence by us [1, 5, 6] and others [7] shows that the SHAP-HA complex is formed by replacing the chondroitin sulfate portion of ITI family molecules by HA with concomitant release of bikunin. The reaction needs at least intact forms of ITI family molecules and HA as substrates and a serum factor(s) as an enzyme. Since ITI family molecules are synthesized in the liver and secreted into the blood at the high concentrations (0.15–0.5 mg/ml in human), the SHAP-HA complex will be formed in any HA-containing tissue whenever it happens to encounter blood.

The synovial cavity of the articular joint can be considered an intercellular space because of the absence of any basement membrane in the synovial membrane. Therefore, the transit of macromolecules such as proteins from the blood stream to the joint cavity and their reverse transit are only limited by the presence of capillary walls. However, once the synovium suffers from injury or inflammation such as in osteoarthritis or rheumatoid arthritis, these characteristic features result in the

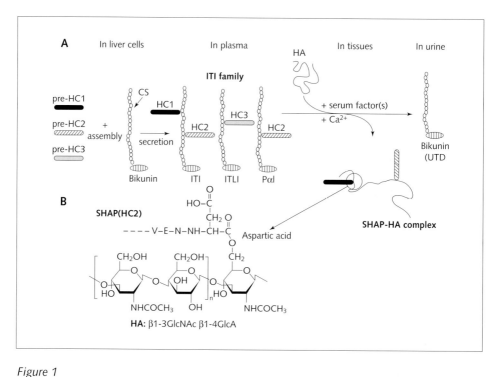

Figure 1
(A) The structures of inter-α-trypsin inhibitor (ITI) family molecules and their intracellular assembly, and the possible pathway of the SHAP (the heavy chains of the ITI family molecules)-hyaluronan (HA) complex formation. ITI family molecules are synthesized in liver by assembling processed heavy chains with bikunin. The covalent assembly is a prerequisite for the formation reaction of the SHAP-HA complex; HA displaces the chondroitin sulfate of bikunin and forms an ester bond with a heavy chain, accompanied by the release of bikunin. The reaction is catalyzed by an as yet uncharacterized enzymatic factor(s) in the presence of divalent cations. The released bikunin is secreted into urine as urinary trypsin inhibitor (UTI). (B) The structure of the linkage region of the SHAP-HA complex. SHAPs are linked to HA via an ester bond between the α-carboxyl group of the C-terminal aspartic acid of each SHAP and the C-6 hydroxyl group of an internal GlcNAc of HA.

almost free exchange of fluid and macromolecules in the blood and the synovial fluid, because the synovium is highly vascularized [8]. Synovial tissues and cartilage are rich in HA in the extracellular matrix. Therefore, the HA in such injured and diseased tissues could easily form covalent complexes with SHAP after the influx of ITI family molecules and serum factor(s) from the blood. In fact, we have purified the SHAP-HA complex from the synovial fluid of RA patients for further characterization of the complex [3].

We previously established ELISA methods to measure the HA content and the SHAP-HA content, respectively, in serum samples using the HA-binding region (HABP) of aggrecan for HA and commercially available anti-ITI antibodies for SHAP [9]. Using this method, we have measured the levels of the SHAP-HA complex in sera of more than a hundred RA patients. Figure 2A demonstrates that the levels of the complex are extremely high in most of RA sera, compared to those of normals. This was also the case with sera of OA patients, although the levels were not as high as those of RA sera [9]. We also measured the HA levels in sera of RA patients, and found a high correlation between the levels of the SHAP-HA complex and HA (Fig. 2B), suggesting that HA in sera of RA patients mostly exists as the SHAP-HA complex. Interestingly, in the RA sera examined, the levels of the SHAP-HA complex appeared to be more related to clinical variables for the degree of RA than to the levels of HA (data not shown). It is very likely that the observed high levels of the SHAP-HA complex in those sera are due to the abnormal formation and accumulation of the SHAP-HA complex in the synovial tissues of diseased joints and the subsequent release of the complex into the blood stream. Therefore, the serum levels of the SHAP-HA complex could reflect the degree of RA progression in these patients.

One may ask, then, what is the physiological significance of the formation of such a unique covalent complex in diseased tissues, and how different is the HA-rich matrix with SHAP from the one without SHAP. To help answer this question, we prepared knockout mice where the formation of the complex was abolished as a result of the inactivation of the bikunin gene. These mice contain no ITI family molecules and, therefore, have no chance to form the SHAP-HA complex. Details for the experimental strategy, procedures, and results are described elsewhere (see [6]). Briefly, the bikunin-deficient mice exhibit severe female infertility due to an abnormality in the expansion of the cumulus oophorus in preovulatory follicles. The cumulus cells are irregularly arranged around the oocyte, indicative of a defect in the formation of the cumulus HA-rich matrix. The results clearly showed that the formation of the SHAP-HA complex is essential for the structure and function of the HA-rich cumulus oophorus matrix.

At this time, how the SHAP-HA complex participates in the matrix formation has not been clarified. It may be necessary to characterize the interaction of SHAP with other matrix components such as PG-M/versican [5]. In addition, the product of the TSG-6 gene (tumor necrosis factor-stimulated gene-6), a 35 kDa glycoprotein with a link protein module domain, may be involved in the complex formation [10–12].

The SHAP-HA complex deficiency in the bikunin-gene knockout mice did not cause other significant abnormalities even in mice older than 1 year, suggesting that the SHAP-HA complex formation may be primarily involved in responses to abnormal conditions rather than being necessary in normal development and homeostasis. ITI family molecules and HA may distribute in different body compartments,

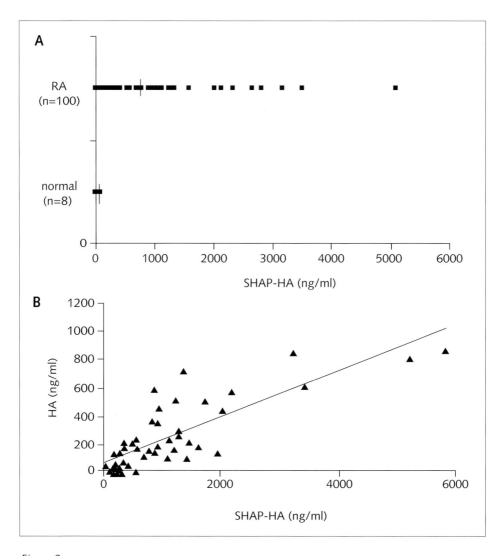

Figure 2
(A) Serum concentrations (ng/ml) of the SHAP-HA complex in patients with RA and in normals. (B) Regression analysis between serum HA concentration (ng/ml) and serum SHPA-HA complex concentration (ng/ml) in patients with RA. The experimental procedures are essentially the same as described previously [6]. Briefly, both the concentrations of the SHAP-HA complex (ng/ml) and HA (ng/ml) in the same sera were measured by ELISA using wells coated with HA-binding protein (HABP). For the SHAP-HA complex, bound SHAP was assayed using rabbit anti-human ITI antibodies and HRP-conjugated anti-rabbit IgG antibodies. For the HA, bound HA was measured using biotinylated HABP and HRP-streptavidin. A positive correlation in B was found ($r = 0.81$, $p < 0.001$, $n = 50$).

namely the blood circulating systems and connective tissues, respectively. If this is the case with every normal tissue, the convergence of serum and HA may be the rate-limiting step. Our previous data [9] and present data (Fig. 2), which indicate that the SHAP-HA complex is present in large amounts in both synovial fluid and sera of RA patients, but absent in normal sera, support the above hypothesis. Despite no direct evidence for this possibility, it is plausible that SHAP-binding affects the interaction of HA with specific receptors such as CD44 on inflammatory cells thereby modifying the CD44-HA interaction that mediates leukocyte activation and invasion. If so, the SHAP-HA complex formation may represent a new target for controlling some HA-related pathological events such as those that occur in OA and RA.

Acknowledgments
We thank Drs. Klaus Kuettner and Vincent C. Hascall for inviting us to contribute to this volume.

References

1   Yoneda M, Suzuki S, Kimata K (1990) Hyaluronic acid associated with the surfaces of cultured fibroblasts is linked to a serum-derived 85-kDa protein. *J Biol Chem* 265: 5247–5257
2   Huang L, Yoneda M, Kimata K (1993) A serum-derived hyaluronan-associated protein (SHAP) is the heavy chain of the inter alpha-trypsin inhibitor. *J Biol Chem* 268: 26725–26730
3   Zhao M, Yoneda M, Ohashi Y, Kurono S, Iwata H, Ohnuki Y, Kimata K (1995) Evidence for the covalent binding of SHAP, heavy chains of inter-alpha-trypsin inhibitor, to hyaluronan. *J Biol Chem* 270: 26657–26663
4   Salier JP, Rouet P, Raguenez G, Daveau M (1996) The inter-alpha-inhibitor family: from structure to regulation. *Biochem J* 315: 1–9
5   Yoneda M, Zhao M, Zhuo L, Watanabe H, Yamada Y, Huang L, Nagasawa S, Nishimura H, Shinomura T, Isogai Z, Kimata K (2000) Roles of inter-alpha-trypsin inhibitor and hyaluronan-binding proteoglycans in hyaluronan-rich matrix formation. In: G Abatangelo, PH Weigel (eds): *New frontiers in medical sciences: redefining hyaluronan*. Elsevier, Amsterdam, 21–30
6   Zhuo L, Yoneda M, Zhao M, Yingsung W, Yoshida N, Kitagawa Y, Kawamura K, Suzuki T, Kimata K (2001) Defect in SHAP-hyaluronan complex causes severe female infertility. A study by inactivation of the bikunin gene in mice. *J Biol Chem* 276: 7693–7696
7   Chen L, Zhang H, Powers RW, Russell PT, Larsen WJ (1996) Covalent linkage between proteins of the inter-alpha-inhibitor family and hyaluronic acid is mediated by a factor produced by granulosa cells. *J Biol Chem* 271: 19409–19414

8   Jee WSS (1988) *The skeletal tissues*. In: L Weiss (ed): Cell and tissue biology, a textbook of histology. Urban & Schwarzenberg, Baltimore, 211–254
9   Kida D, Yoneda M, Miyaura S, Ishimaru T, Yoshida Y, Ito T, Ishiguro N, Iwata H, Kimata K (1999) The SHAP-HA complex in sera from patients with rheumatoid arthritis and osteoarthritis. *J Rheumatology* 26: 1230–1238
10 Lee TH, Wisniewski HG, Vilcek J (1992) A novel secretory tumor necrosis factor-inducible protein (TSG-6) is a member of the family of hyaluronate binding proteins, closely related to the adhesion receptor CD44. *J Cell Biol* 116: 545–57
11 Carrette O, Nemade RV, Day AJ, Brickner A, Larsen WJ (2001) TSG-6 is concentrated in the extracellular matrix of mouse cumulus oocyte complexes through hyaluronan and inter-alpha-inhibitor binding. *Biol Reprod* 65: 301–308
12 Mukhopadhyay D, Hascall VC, Day AJ, Salustri A, Fulop C (2001) Two distinct population of tumor necrosis factor stimulated gene-6 protein in the extracellular matrix of expanded mouse cumulus cell-oocyte complexes. *Arch Biochem Biophys* 394: 174–181

In vivo investigation of hyaluronan and hyaluronan synthase-2 function during cartilage and joint development

Janet Y. Lee[1], Ryan B. Rountree[2], David M. Kingsley[2] and Andrew P. Spicer[3]

[1]Department of Cell Biology and Human Anatomy, University of California, School of Medicine, Davis, CA 95616, USA; [2]Department of Developmental Biology, Beckman Center, Stanford University, Palo Alto, CA 94305, USA; [3]Center for Extracellular Matrix Biology, Texas A and M University, Health Science Center, 2121 W. Holcombe Blvd., Houston, TX 77030, USA

Introduction

Hyaluronan (HA) is a major component of the extracellular and pericellular matrix. It is a simple, linear glycosaminoglycan composed of thousands of disaccharide repeats of glucuronic acid and N-acetylglucosamine [1]. Initially isolated from the vitreous humor, this macromolecule can be found in numerous tissues and body fluids of vertebrates, as well as in some bacteria [1]. In vertebrates, HA is synthesized by any one of three hyaluronan synthases (Has) designated Has1, Has2 and Has3, respectively [2, 3]. During embryogenesis, large amounts of HA are synthesized by many cell types, particularly those that are proliferating and migrating. Through specific inactivation of the mouse *Has2* gene [4], we have recently shown that HA is critical for embryogenesis and is required for the maintenance and possibly the creation of extracellular matrix-defined spaces throughout the embryo, in addition to the migration of endocardial cushion cells.

The roles of HA in tissue remodeling extend to disease states, where HA is implicated in osteoarthritis, rheumatoid arthritis, pulmonary fibrosis, wound healing, as well as numerous malignant cancers including lung, mammary, and hepatic carcinoma [5–7]. Although the presence of HA appears to be necessary in these biological systems, the actual roles of HA remain unclear.

Model for investigation of hyaluronan function

HA is proposed to have multiple functions that affect cell behavior. It may play a scaffolding role in the extracellular matrix by creating space through exerting hydrostatic pressure on neighboring structures [8], or by organizing the matrix through interactions with other ECM molecules such as link proteins and chondroitin sulfate proteoglycans [9, 10]. HA, in its diverse forms, may also act in a signaling role. Depending on the concentrations of HA, it can stimulate or inhibit cellular processes such as migration and differentiation [11–14].

The combination of these structural and signaling roles of HA and the balance of HA synthesis and removal may be critical during morphogenesis, when tissues are undergoing rapid proliferation, migration, and differentiation. A useful model for studying the potential roles of HA during morphogenesis is the dynamic process of cartilage and joint development.

In the limb, cartilage development is initiated by proliferation of a subset of mesenchymal cells that become committed to a chondrogenic fate. These cells will then aggregate to form a condensed area, wherein discrete regions will further develop into joints or bones [15, 16]. HA may be important in mediating the early proliferative, as well as adhesion and anti-adhesion events in chondrogenesis. Prior to condensation, HA is distributed throughout the limb mesenchyme at high levels. During condensation, a high level of HA remains at the limb periphery, but HA levels decrease within the condensing mesenchyme [17]. Treatment of prechondrogenic cells with hyaluronidase or blocking HA binding to cell surface proteins prevented normal aggregation and differentiation [18].

HA may play a structural role in the precartilage matrix by stabilizing and organizing the matrix proteoglycan aggregates that are essential for normal cartilage development [10, 19].

During joint development, HA may play a role in cavitation. HA, as well as CD44, is localized in the precavitating joint, and blocking CD44-HA binding led to a failure of joint cavitation [20].

To investigate the source of HA during limb development, we performed whole mount *in situ* hybridization on staged mouse embryos using probes for *Has1*, *Has2*, and *Has3*. We observed that Has2 is the major HA synthase involved. The expression pattern of *Has2* was coincident with that of HA, as expected (Fig. 1A, D, G).

Potential regulation by BMP/TGFβ family members

Based on these preliminary data, we began exploring potential regulatory mechanisms by comparing expression patterns of key players in cartilage and joint development. These key players included members of the BMP/TGFβ superfamily [21], and since several of the family members (BMP7, TGFβ, and growth/differentiation factor-9 [Gdf9]) have been reported to mediate *Has2* expression [10, 22, 23], it was not surprising to find that certain members had expression patterns that corresponded closely to that of *Has2*. These members are Gdf5 and Gdf6 [24].

Gdf5 is a coordinator for cartilage and joint development, mainly in the digits, while Gdf6 may have similar roles in the wrists and ankles [24]. Mutations in *Gdf5* cause brachypodism (*bp*) in mice, characterized by shorter bones in limbs, altered formation of bones and joint in the sternum, and reduced number of bones in the digits [25].

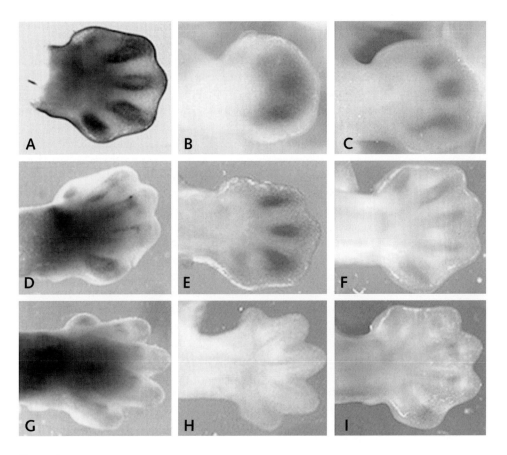

Figure 1
Spatial and temporal localization of Has2 *transcripts during limb development of wildtype and mutant mice.* Has2 *transcripts are detected using non-radioactive whole mount in situ hybridization. Panels A–C are dorsal views of embryonic mouse forelimbs at E12.5. Panels D–F are of E13.5, while G–I are of E14.5. Limbs from wildtype embryos are shown in panels A, D, and G, whereas limbs from Gdf5 null and Gdf6 null embryos are shown in panels B, E, and H, and C, F, and I, respectively. (A–C) E12.5 limbs express high levels of* Has2 *in the digital rays. (D–F) In E13.5 and (G, I) E14.5 limbs,* Has2 *transcripts become progressively restricted to the interzone region of the developing metacarpophalangeal and metatarsophalangeal joints, as well as the wrists and ankles. (H) In E14.5 Gdf5 mutant limb,* Has2 *transcripts are absent from the interzone region.*

To investigate the impact of *Gdf5* and *Gdf6* mutations on *Has2* expression during cartilage and joint development, we performed whole mount *in situ* hybridization using the *Has2* probe on homozygous mutant embryos at embryonic day (E)

12.5, 13.5, and 14.5. We observed that *Has2* transcript levels in E14.5 *Gdf5* null embryos are lower than in wild-type embryos. Conversely, E12.5 and E13.5 *Gdf5* mutants, as well as *Gdf6* mutants exhibited no obvious change in *Has2* expression (Fig. 1), after taking slight age variations into account. These preliminary data suggest that Gdf5, but not Gdf6, may regulate, in part, the expression of *Has2*. Further investigation is necessary to support these results, and to determine whether HA synthetic activity correlates with *Has2* expression levels in the mutants.

Conditional Has2 knock-out

Our most recent strategy for investigation of HA biosynthesis and function during vertebrate embryogenesis employs site-specific recombinases to create conditional or tissue-specific deficiencies of Has2 function during embryogenesis [26]. Conditional gene targeting relies upon the use of tissue-specific promoter/enhancer sequences to restrict expression of a Cre recombinase transgene, the activity of which is required for inactivation of the gene of interest. In order to investigate Has2 function in the tissue of interest, the Cre transgene must be expressed in either the cell that expresses *Has2* or its precursor. In our experiments, we will utilize *Col2a1-Cre* and *Gdf5-Cre* mouse lines in order to study the effect of cartilage- or joint-restricted deficiency of Has2-dependent HA synthesis on skeletal development.

To create a Has2 conditional knockout in the mouse, we cloned three contiguous mouse *Has2* genomic DNA fragments of 4.5 kb, 900 bp and 3.6 kb into a vector designed for allelogenic gene targeting. The 4.5 kb fragment represents the 5' arm of homology, the 900 bp fragment was cloned between two *loxP* sites, and the 3.6 kb fragment, representing the 3' arm of homology was cloned downstream of the selectable *PGKneo* cassette, which is flanked by two *frt* sites. A hypomorphic *Has2* allele is predicted to result from insertion of the *PGKneo* cassette into intron 3. Through crossing with transgenic mice expressing Flp recombinase, the *neo* cassette will be deleted, resulting in a functionally wild-type *Has2* allele [27]. By a similar strategy, crossing with transgenic mice expressing Cre recombinase will delete exon 3, converting the targeted allele or its flp-recombined descendants into a functionally null allele.

Conclusions

Has2 is the major HA synthase involved in HA biosynthesis during vertebrate embryogenesis, and is expressed in multiple tissues in a distinct temporal and spatial pattern, including the developing limb. Has2 may play a pivotal role in limb and joint development, as its expression pattern closely mimics that of Gdf5 and Gdf6, growth factors necessary for limb and joint development. Gdf5, but not Gdf6, may be involved in regulation of HA synthesis through Has2.

The conditional targeting approach should provide a particularly powerful tool in dissecting the role of Has2-dependent HA function during embryonic development and potentially through adult maturation.

Acknowledgements

This work was supported by research grants from the American Heart Association National Office AHA 0030184N and the March of Dimes Birth Defects Foundation #1-FY00-361 to APS.

References

1 Laurent TC (ed) (1998) *The chemistry, biology and medical applications of hyaluronan and its derivatives*. Wenner-Gren International Series, Vol 72, Portland Press, London
2 Spicer AP, McDonald JA (1998) Characterization and molecular evolution of a vertebrate hyaluronan synthase (HAS) gene family. *J Biol Chem* 273: 1923–1932
3 Itano N, Sawai T, Yoshida M, Lenas P, Yamada Y, Imagawa M, Shinomura T, Hamaguchi M, Yoshida Y, Ohnuki Y, Miyauchi S, Spicer AP, McDonald JA, Kimata K (1999) Three isoforms of mammalian hyaluronan synthases have distinct enzymatic properties. *J Biol Chem* 274: 25085–25092
4 Camenisch TD, Spicer AP, Brehm-Gibson T, Biesterfeldt J, Augustine ML, Calabro Jr. A, Kubalak S, Klewer SE, McDonald JA (2000) Disruption of hyaluronan synthase-2 abrogates normal cardiac morphogenesis and hyaluronan-mediated transformation of epithelium to mesenchyme. *J Clin Invest* 106: 349–360
5 Gerdin B, Hallgren R (1997) Dynamic role of hyaluronan (HYA) in connective tissue activation and inflammation. *J Int Med* 242: 49–55
6 Knudson W, Biswas C, Li X-Q, Nemec RE, Toole BP (1989) The role and regulation of tumour-associated hyaluronan. In: D Evered, J Whelan (eds): *The biology of hyaluronan*. Ciba Foundation Symposium 143. Wiley, Chichester, NewYork, 150–169
7 Delpech B, Girard N, Bertrand P, Courel M-N, Chauzy C, Delpech A (1997) Hyaluronan: fundamental principles and applications in cancer. *J Int Med* 242: 41–48
8 Fraser JRE, Laurent TC, Laurent UBG (1997) Hyaluronan: nature, distribution, functions and turnover. *J Int Med* 242: 27–33
9 Hardingham TE (1998) Cartilage: Aggrecan-link protein-hyaluronan aggregates. Glycoforum internet article http: //www.glycoforum.gr.jp/
10 Nishida Y, Knudson CB, Eger W, Kuettner KE, Knudson W (2000) Osteogenic protein 1 stimulates cells-associated matrix assembly by normal human articular chondrocytes: up-regulation of hyaluronan synthase, CD44, and aggrecan. *Arthritis Rheum* 43: 206–214

11. Goldberg RL, Toole BP (1987) Hyaluronate inhibition of cell proliferation. *Arthritis Rheum* 30: 769–778
12. Kosaki R, Watanabe K, Yamaguchi Y (1999) Overproduction of hyaluronan by expression of the hyaluronan synthase Has2 enhances anchorage-independent growth and tumorigenicity. *Cancer Res* 59: 1141–1145
13. West DC, Hampson IN, Arnold F, Kumar S (1985) Angiogenesis induced by degradation products of hyaluronic acid. *Science* 228: 1324–1326
14. Kujawa MJ, Carrino DA, Caplan AI (1986) Substrate-bonded hyaluronic acid exhibits a size-dependent stimulation of chondrogenic differentiation of stage 24 limb mesenchymal cells in culture. *Dev Biol* 114: 519–528
15. Gilbert SF (ed) (1997) *Developmental biology*, 5th ed. Sinauer Associates, Inc., Sunderland, Massachusetts
16. Hall BK, Miyake T (2000) All for one and one for all: condensations and the initiation of skeletal development. *BioEssays* 22: 138–148
17. Knudson CB, Toole BP (1985) Changes in the pericellular matrix during differentiation of limb bud mesoderm. *Dev Biol* 112: 308–318
18. Maleski MP, Knudson CB (1996) Hyaluronan-mediated aggregation of limb bud mesenchyme and mesenchymal condensation during chondrogenesis. *Exp Cell Res* 225: 55–66
19. Watanabe H, Yamada, Y, Kimata K (1998) Roles of aggrecan, a large chondroitin sulfate proteoglycan, in cartilage structure and function. *J Biochem* 124: 687–693
20. Dowthwaite GP, Edwards JCW, Pitsillides A (1998) An essential role for the interaction between hyaluronan and hyaluronan binding proteins during joint development. *J Histochem Cytochem* 46: 641–651
21. Zou H, Wieser R, Massagué J, Niswander L (1997) Distinct roles of type I bone morphogenetic protein receptors in the formation and differentiation of cartilage. *Genes Dev* 11: 2191–2203
22. Usui T, Amano S, Oshika T, Suzuki K, Miyata K, Araie M, Heldin P, Yamashita H (2000) Expression regulation of hyaluronan synthase in corneal endothelial cells. *Invest Ophthalmol Vis Sci* 41: 3261–3267
23. Elvin JA, Amander TC, Wang P, Wolfman NM, Matzuk MM (1999) Paracrine actions of growth differentiation factor-9 in the mammalian ovary. *Molec Endocrin* 13: 1035–1048
24. Storm EE, Kingsley DM (1999) GDF5 coordinates bone and joint formation during digit development. *Developmental Biol* 209: 11–27
25. Storm EE, Huynh TV, Copeland NG, Jenkins NA, Kingsley DM, Lee SJ (1994) Limb alterations in brachypodism mice due to mutations in a new member of the TGF-beta-superfamily. *Nature* 368: 639–643
26. Meyers EN, Lewandoski M, Martin GR (1998) An Fgf8 mutant allelic series generated by Cre- and Flp-mediated recombination. *Nature Genetics* 18: 136–141
27. Dymecki SM (1996) Flp recombinase promotes site-specific DNA recombination in embryonic stem cells and transgenic mice. *Proc Natl Acad Sci USA* 93: 6191–6196

CD44 and cartilage matrix stabilization

Cheryl B. Knudson, Kathleen T. Rousche, Richard S. Peterson, Geraldine Chow and Warren Knudson

Department of Biochemistry, Rush Medical College, 1653 W. Congress Parkway, Chicago, IL 60612, USA

Introduction

One prominent feature of osteoarthritic cartilage is an inherent failure to retain a proteoglycan-rich extracellular matrix. This lack of retention occurs even as the chondrocytes respond with episodes of pronounced matrix biosynthesis. Thus, osteoarthritis is likely manifested in part by the inability of chondrocytes to retain matrix at the cell surface. Another aspect of degenerative changes in osteoarthritis is enhanced catabolism, often termed chondrocytic chondrolysis. Although many facets of chondrocyte catabolism and attempted repair have been extensively investigated, the signals that initiate such metabolic changes remain unclear. Matrix receptors direct the assembly and retention of the pericellular matrix and provide a linkage to the cytoskeleton. CD44 is the principal receptor for hyaluronan (HA) expressed by chondrocytes [1–3]. The interactions between matrix components and receptors establish a cell-associated pool of extracellular matrix molecules. These associations have the potential to signal changes in cell behavior, such as proliferation, apoptosis and matrix biosynthesis or turnover. In addition to chondrocyte integrins [4], CD44 represents another class of receptors that can participate in these matrix-cell-cytoskeleton interactions [3, 5].

CD44 and chondrocyte matrix assembly and retention

Bovine articular cartilage was isolated from metacarpophalangeal joints and normal human articular cartilage was isolated from the talocrural joints of donors obtained from the Regional Organ Bank of Illinois (ROBI). Chondrocytes were cultured in alginate beads [1, 6] and matrix visualized on living cells by the particle exclusion assay [7, 8]. We have shown previously that the HA-proteoglycan aggregates are tethered to the chondrocyte cell surface [8] *via* multivalent interactions with CD44 [9]. This interaction can be blocked with anti-CD44 antibodies [9], or by competition with exogenous HA or HA hexasaccharides (HA6) [8]. Trypsin treatment removes all cell surface CD44 (as well as all bound HA and aggrecan). The intracellular precursor pool of CD44 is small, and thus new synthesis of receptor protein is required to replenish the chondrocyte cell surface over a period of 24 h [2]. Chondrocytes taken at earlier time intervals of recovery and thus, exhibiting varying levels of CD44, were fixed and tested for matrix assembly using exogenous HA and aggrecan [10]. Formaldehyde-fixed cells that express CD44 can serve as spheres for matrix assembly [10]. The correlation between the re-population of CD44 receptors and the capacity for matrix assembly was $r = 0.98$, while chondrocytes exhibiting only 25% of pre-trypsinization levels of CD44 had the capacity to assemble and retain a matrix [11]. Below this level of expression, the amount of HA bound is insufficient to maintain a visible cell-associated matrix. Thus, a minimum density of CD44 appears to be required.

In an attempt to selectively modulate CD44 expression, without directly altering the matrix itself (as with trypsin) or overall chondrocyte metabolism (with cytokines), an antisense approach was undertaken [12]. Antisense treatment of articular chondrocytes resulted in a 60–75% reduction of cell surface expression of CD44 with no change in DNA content, total protein content or $\beta1$ integrin protein expression. When tested for function, the CD44 antisense-treated chondrocytes displayed variability in their capacity to assemble endogenous cell-associated matrices. These results were not entirely unexpected as the level of CD44 inhibition by antisense was close to the minimum density of CD44 receptors determined by trypsin-recovery experiments as necessary for matrix assembly.

Treatment of bovine or human articular chondrocytes with IL-1 resulted in enhancement in the expression of CD44 mRNA and protein [1, 13]. Even with this large increase in CD44, there was a consistent decrease in the size of cell-associated matrices following IL-1 treatment, whereas the capacity for HA internalization and intracellular degradation increased [1, 13]. Treatment of human articular cartilage slices with IL-1 also enhanced HA synthase-2 (HAS-2), however the HA failed to accumulate within the tissue slices. This is significant because cartilage proteoglycan cannot be retained in the tissue if the HA is not retained. Our interpretation of these results is that, even with the increase in CD44, its function was modulated by IL-1 from a primary role in matrix retention to a more predominate role in receptor-mediated HA catabolism.

In the presence of the anabolic cellular mediator OP-1 (a.k.a. BMP-7) the mRNA copy number for CD44 and HAS-2 were again, significantly upregulated (increased CD44 protein measured by flow cytometry). However, unlike the stimulation induced by IL-1, articular chondrocytes treated with OP-1 exhibited extremely large cell-associated matrices – matrices that extend to a distance > 1.5 cell diameters. Intact cartilage tissue slices treated with OP-1 displayed prominent proteoglycan as well as enhanced HA retention. Thus, these results [14, 15] demonstrate that OP-1 stimulates not only the synthesis of the major cartilage extracellular matrix components aggrecan and type II collagen, but also two associated molecules necessary for the retention of aggrecan, namely hyaluronan and CD44. While an elevation of CD44 may augment or promote a particular function of CD44 (i.e., HA retention or HA catabolism), the increase in CD44 does not appear to regulate the choice of function. These results limit the ability to interpret increases in CD44 observed osteoarthritis samples (our preliminary observations and [16]). Does the increase in CD44 represent "attempted repair" or enhanced matrix turnover *via* CD44-mediated internalization of HA? Since changes in total cell surface CD44 do not appear to control whether CD44 participates primarily in HA retention, HA catabolism or some other function, additional mediators of CD44 function may be key factors, particularly mediators that act upon CD44 intracellularly.

Regulation from within: interactions of the CD44 cytoplasmic domain in chondrocytes

A differential membrane extraction method has been used by several investigators to separate and distinguish pools of transmembrane proteins, such as CD44. Extraction of cells with 0.2% NP-40 alone preferentially solubilizes transmembrane proteins that do not have a strong linkage to the underlying cytoskeleton. Following the NP-40 extractions, the membranes remain intact, and the residual transmembrane proteins are predominantly those linked to cytoskeleton-associated proteins. These residual proteins can be solubilized by subsequent extraction with agents that disrupt the actin cytoskeleton, such as DNAase-I and 1% EmphigenBB in 0.5% NP-40 [17]. Using matrix-intact bovine articular chondrocytes, we determined that approximately 50% of the total cell surface CD44 could be extracted with 0.2% NP-40 (Fig. 1). No additional CD44 could be solubilized even with higher concentrations of NP-40. The other 50% CD44 could be subsequently solubilized in the presence of the cytoskeleton-disrupting agents. After depletion of the matrix through the use of Streptomyces hyaluronidase, 0.2% NP-40 alone was able to extract >70% of the total CD44. These results suggest that the prevention of extracellular matrix binding to CD44 (i.e., reduced receptor occupancy) triggers an alteration in the association of CD44 with the underlying chondrocyte cytoskeleton. In reverse, matrix binding may affect a cytoskeletal-anchored stabilization of CD44.

Since changes in matrix assembly altered CD44: cytoskeletal associations, we explored the disruption of the cytoskeleton and matrix retention. Matrix assembly was monitored in the presence or absence of cytochalasin [18] or latrunculin [19]. Treatment of alginate bead cultures with these agents reduced chondrocyte cell-associated matrix assembly while cell viability remained >95%. Relative to control, untreated chondrocytes, there was a 58% decrease in the amount of bound FITC-HA [20] in the cytochalasin-treated cells and a 56% decrease with the latrunculin-treated cells [21]. However, these treatments did not alter CD44 expression detect-

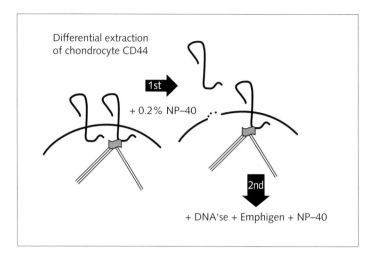

Figure 1
CD44 in articular chondrocytes exists in two cellular pools. The first pool is not associated with the cytoskeleton and can be extracted with 0.2% NP-40, while the second pool requires solubilization from the residual cytoskeleton.

ed by flow cytometry. This indicates that disruption of the chondrocyte cytoskeleton can affect the capacity of CD44 to retain extracellular matrix.

To expand these studies to cartilage explants, three approaches were used. Addition of HA_6 was performed with the hypothesis that the oligosaccharides would uncouple chondrocyte-HA interactions. Addition of antisense oligonucleotides was performed with the hypothesis that reduction in cell surface expression levels of CD44 would reduce CD44-HA interactions. Finally, latrunculin and cytochalasin treatments were evaluated with the hypothesis that HA-CD44 binding would be reduced upon cytoskeletal disruption. By 48 h of treatment with cytochalasin and 5 days of incubation with latrunculin, a decrease in safranin O staining was detected, indicating the loss of proteoglycan from these tissues as compared to cultured controls [21]. The HA_6 induced a state of chondrocytic chondrolysis, including a significant loss of proteoglycan from the tissue as observed by the absence of safranin O staining, activation of gelatinolytic activity detected by zymography, while the tissue exhibited aggrecan neoepitope (NITEGE) expression detected by immunohistochemistry [22] indicative of "aggrecanase" activity [23]. Although the final outcome of the chondrocytic chondrolysis elicited by the HA_6 is the loss of matrix proteoglycans, the chondrocytes attempt matrix repair with an upregulation of iNOS, aggrecan and HAS-2 mRNA. RITC-labeled oligonucleotides penetrated the chondrocytes within the cartilage explants as well as the alginate bead cultures [12]. Similar to osteoarthritic cartilage, antisense-treated cartilage slices displayed a nearly total loss of safranin O staining proteoglycan-rich matrix, as well as intense staining for the aggrecanase-generated neoepitope. Thus, CD44 expression, and HA-CD44 interactions are needed for maintenance of cartilage homeostasis.

CD44 may be a rheostat for chondrocytes, capable to "sense" and "respond to" matrix changes. Although a myriad of proteins (structural, kinases, signaling) have been described for the focal contacts of other cells [24], little is known about the chondrocyte matrix adhesion complex (CMAC) [25]. We found that in normal articular chondrocytes, both ankyrin and spectrin co-immunoprecipitate with CD44, but moesin does not [26]. Thus, although the cytoplasmic tail domain of CD44 contains putative binding sites for both the ezrin/radixin/moesin (ERM) family of actin-binding proteins [27, 28] and ankyrin [29], chondrocytes apparently use ankyrin to interact with the cytoskeleton. However, these lysates were derived from matrix-intact chondrocytes. It is still possible that in an altered metabolic state (e.g., osteoarthritis), chondrocytes switch from CD44:ankyrin to CD44:moesin and, in the process, alter the function of CD44. When HA binds to chondrocytes, there also is receptor-mediated internalization, along with HABR-decorated HA [20, 30]. No consensus binding sites for adaptins [31] are present in the CD44 cytoplasmic domain to facilitate clathrin coat formation, and no clathrin coats were associated with CD44 [9]. Since we have identified spectrin/ankyrin binding to chondrocyte CD44, this suggests that the ankyrin/spectrin/actin linkage [32] may facilitate vesicle formation for receptor-mediated internalization.

HA binding to chondrocytes upregulated the phosphorylation of CD44, which could influence its interaction with cytoskeletal components [11, 33]. Phosphorylation of CD44 from *Streptomyces* hyaluronidase-treated chondrocytes was reduced in comparison to CD44 from matrix-intact control chondrocytes although the total amount of CD44 was nearly equivalent (Fig. 2). Addition of HA to *Streptomyces* hyaluronidase-treated chondrocytes restored CD44 phosphorylation. Our model for this event is that the presence of an HA-rich extracellular matrix induces CD44 stabilization and the clustering/ordering of the receptor – a process that is further stabilized by intracellular phosphorylation leading to enhanced interaction with the cortical actin cytoskeleton. If correct, the reverse event would also be likely; namely, changes in CD44 phosphorylation may induce a destabilization of CD44-mediated cell-matrix interactions.

Two laboratories have recently reported that the enzyme casein kinase II (CKII) is likely responsible for the serine phosphorylation of CD44 [34, 35]. Our immunohistochemical analyses [36] localized CKIIα' in the chondrocytes of all zones of normal (Collins' Grade 0) human cartilage [37], with the superficial zone exhibiting the most intense staining. This staining was diminished throughout damaged (Grades 2, 3) cartilage, indicating a loss of CKIIα'. Although these results are preliminary, if diminished CKII is associated with disease progression, the lack of appropriate CD44 phosphorylation could result in failed HA/proteoglycan retention even in the presence of elevated matrix biosynthesis. Furthermore, CD44 is not likely the only

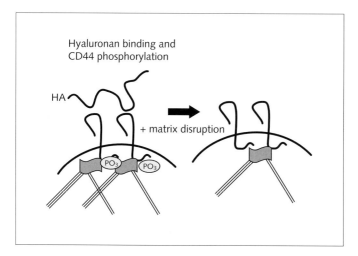

Figure 2
Phosphorylation of CD44 was reduced after matrix disruption in comparison to CD44 from matrix-intact chondrocytes. The actin-binding protein ankyrin may link CD44 to the cortical cytoskeleton.

phosphorylation target of CKII, and thus, diminished CKII may have pleiotropic effects on chondrocyte metabolism.

To investigate this further, we designed a phosphorothioate antisense oligonucleotide directed against the CKIIα' mRNA sequence [36]. Following antisense treatment, an enhancement in 0.2% NP-40 extraction of CD44 was observed indicating that the decrease in CKIIα' resulted in diminished CD44-cytoskeletal interactions. Also, chondrocytes treated with CKII antisense oligonucleotides exhibited a reduced capacity for cell-associated matrix assembly compared to scrambled control.

The CD44 gene consists of 20 exons with two principal sites of alternative splicing [38]. One site lies within the sequence coding for the extracellular domain of the protein. Alternative splicing usage of exons 6–15 (also known as variant exons 1–10) within this site gives rise to variant isoforms of CD44 as reported for human tumors [39] and murine embryonic tissues [40]. The second site for alternative splicing includes the sequence that codes for the C-terminal cytoplasmic domain. The predominant isoform of CD44 (CD44Hwt) contains no variant exons within the extracellular domain (CD44H) and expresses the "long" (72 amino acids) cytoplasmic tail domain (exon 20). However, alternative splicing usage of exon 19 instead of exon 20 results in a CD44 protein with a "short" cytoplasmic domain containing only 5 amino acids (CD44exon19) [3]. All normal human donor cartilages examined (n = 31) expressed both CD44exon19 as well as CD44exon20 mRNA.

The proportion of CD44exon19 to total CD44 mRNA varied from donor to donor with a mean of 21% ± 7 [41].

The protein generated from CD44exon19 mRNA lacks all of the signaling motifs and functional domains necessary for interaction with actin-binding or other cytosolic proteins or for phosphorylation. Given the implications that expression of this isoform may have on CD44 function, a "mock short-tail" expression mutant pCD44HΔ67 was created by the introduction of a stop codon (Cys^{295}-TGT→TGA-Stop codon) in exon 20 employing PCR-mediated site-directed mutagenesis. This CD44 construct yields a CD44 protein with a truncated cytoplasmic tail domain identical to the naturally occurring CD44exon19 isoform.

COS-7 cells do not express endogenous CD44 mRNA. The expressed pCD44*Hwt* was fully glycosylated by the COS-7 transfectants, exhibiting a molecular weight of ~85 kD by western blotting. The pCD44HΔ67 ran as a ~78 kD protein – the expected size protein with a 7 kD truncation. This demonstrates that the truncated tail isoform can be translated, properly glycosylated and transported correctly to the cell surface. COS-7 cells transfected with pCD44*Hwt* assembled pericellular matrices with exogenous HA and aggrecan as shown before [42]. However, the CD44HΔ67 transfectants did not exhibit the capacity for matrix assembly with exogenous aggrecan and HA [43]. These results suggest that interactions between the intact cytoplasmic tail domain and components of the cortical cytoskeleton are required for HA-dependent matrix assembly.

CD44 constructs were also expressed in bovine as well as human articular chondrocytes following transient transfection employing an optimized suspension protocol using FuGENE as a facilitator following *Streptomyces* hyaluronidase pre-treatment [44, 45]. The pTracer vectors constructs provide for the simultaneous expression of both green fluorescence protein (GFP) and inserted cDNA. Ten to fifteen percent of the chondrocytes transfected with the pTracer-CD44HΔ67 expressed the GFP marker 48 h post-transfection. Native matrices were observed on all GFP-negative chondrocytes. However, none of the GFP-positive chondrocytes, representing cells over-expressing short tail CD44HΔ67, exhibited pericellular matrices. As a control, GFP-positive chondrocytes transfected with the pCD44*Hwt* or no insert continue to exhibit large cell-associated matrices [43]. All of the chondrocytes express endogenous, native CD44*Hwt*. Thus, one explanation of these results is that the over-expression of pCD44HΔ67 acts as a dominant negative in these cells, inhibiting the activity of the endogenous CD44*Hwt* (Fig. 3).

Summary

In osteoarthritis, proteoglycan retention is poor [46], and proteoglycan degradation is increased [23] – resulting in impaired cartilage function. Chondrocytes may exhibit metabolic changes following modifications of the matrix and through the

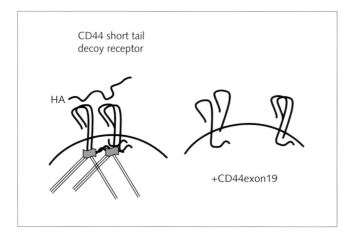

Figure 3
Human articular chondrocytes express varied levels of a natural truncated variant isoform of CD44, CD44exon19. Over-expression of the decoy receptor reduces the capacity for matrix retention.

interaction with the matrix initiate matrix remodeling or repair. The interaction of CD44 with the cytoskeleton and the matrix allow us to speculate that both inside-out and outside-in communication patterns are occurring within the chondrocytes *via* CD44. Uncoupling chondrocytes from their matrix by antisense oligonucleotides directed against CD44 or with HA oligosaccharides results in loss of cartilage homeostasis and induction of repair mechanisms. Also, our results indicate that destabilization of the cytoskeleton may bring about a change in the matrix. Human articular chondrocytes obtained from different donors express varied levels of mRNA coding for a natural truncated variant isoform of CD44 [41]. Over-expression of this decoy receptor in articular chondrocytes resulted in the loss of the capacity for matrix retention [44].

The integrity of the articular cartilage extracellular matrix is essential for the biomechanical properties of the tissue [47, 48]. HA is a key component of the extracellular matrix of cartilage, functioning as the central filament of the cartilage proteoglycan aggregate. Changes in matrix structure, due to faulty maintenance or repair, trauma, or possibly intrinsic age-related changes, give rise to degenerative disease states such as osteoarthritis [49, 50]. It has long been known that in cartilage, the extracellular matrix exerts effects on chondrocyte cell shape, phenotype and metabolism [51]. Matrix receptors, including CD44, direct the assembly and retention of the matrix and provide a linkage to the cytoskeleton and cytosolic components of signaling pathways. Stability of cell-matrix interactions promotes tissue homeostasis [5]. Maintenance of the interaction between hyaluronan and CD44 and the cytoskeleton sustains chondrocyte phenotype, whereas disruptions of these interactions initiate changes in the cartilage matrix. Chondrocytes may have the capacity to sense changes in the matrix, possibly early matrix damage or uncoupling of cell-matrix interactions, and attempt matrix modeling and repair. The chondrocyte response to matrix disruption or damage is likely mediated by cross-talk between specific receptors to matrix components and BMPs.

Acknowledgments
The contributions by former members of our laboratories, especially Drs. Y. Nishida, D. Aguiar and G. Nofal, cited in this chapter are worthy of special note. Collaboration with Dr. Allan Valdellon and his staff at the Regional Organ Bank of Illinois is gratefully acknowledged. This work is supported in part by NIH grants P50-AR39239, AR43384, AR39507 and the Arthritis Foundation.

References

1   Chow G, Knudson CB, Homandberg G, Knudson W (1995) Increased CD44 expression in bovine articular chondrocytes by catabolic cellular mediators. *J Biol Chem* 270: 27734–27741
2   Aguiar DJ, Knudson W, Knudson CB (1999) Internalization of the hyaluronan receptor CD44 by chondrocytes. *Exp Cell Res* 252: 292–302
3   Knudson W, Knudson CB (1999) Hyaluronan receptor, CD44. Glycoforum http://www.glycoforum.gr.jp/science/hyaluronan/HA10/HA10E.html: 1–30
4   Loeser RF (2000) Chondrocyte integrin expression and function. *Biorheology* 37: 109–116
5   Knudson CB, Knudson W (1993) Hyaluronan-binding proteins in development, tissue homeostasis and disease. *FASEB J* 7: 1233–1241
6   Hauselmann HJ, Aydelotte MB, Schumacher BL, Kuettner KE, Gitelis SH, Thonar EJ-MA (1992) Synthesis and turnover of proteoglycans by human and bovine adult articular chondrocytes cultured in alginate beads. *Matrix* 12: 130–136
7   Clarris BJ, Fraser JRE (1968) On the pericellular zone of some mammalian cells *in vitro*. *Exp Cell Res* 49: 181–193
8   Knudson CB (1993) Hyaluronan receptor-directed assembly of chondrocyte pericellular matrix. *J Cell Biol* 120: 825–834
9   Knudson W, Aguiar DJ, Hua Q, Knudson CB (1996) CD44-anchored hyaluronan-rich pericellular matrices: An ultrastructural and biochemical analysis. *Exp Cell Res* 228: 216–228
10 Knudson W, Knudson CB (1991) Assembly of a chondrocyte-like pericellular matrix on non- chondrogenic cells. *J Cell Sci* 99: 227–235

11 Knudson CB, Nofal GA, Pamintuan L, Aguiar DJ (1999) The chondrocyte pericellular matrix: a model for hyaluronan-mediated cell-matrix interactions. *Biochem Soc Trans* 27: 142–147
12 Chow G, Nietfeld J, Knudson CB, Knudson W (1998) Antisense inhibition of chondrocyte CD44 expression results in cartilage chondrolysis. *Arthritis Rheum* 41: 1411–1419
13 Nishida Y, D'Souza AL, Thonar JMA, Knudson W (2000) IL-1a stimulates hyaluronan metabolism in human articular cartilage. *Arthritis Rheum* 43: 1315–1326
14 Nishida Y, Knudson CB, Eger W, Kuettner KE, Knudson W (2000) Osteogenic protein-1 stimulates cell-associated matrix assembly by normal human articular chondrocytes. *Arthritis Rheum* 43: 206–214
15 Nishida Y, Knudson CB, Kuettner KE, Knudson W (2000) Osteogenic protein-1 promotes the synthesis and retention of extracellular matrix within bovine articular cartilage and chondrocyte cultures. *Osteoarthritis Cartilage* 8: 127–136
16 Ostergaard K, Salter DM, Andersen CB, Petersen J, Bendtzen K (1997) CD44 expression is up-regulated in the deep zone of osteoarthritic cartilage from human femoral heads. *Histopathology* 31: 451–459
17 Camp RL, Kraus TA, Pure E (1991) Variations in the cytoskeletal interaction and posttranslational modification of the CD44 homing receptor in macrophages. *J Cell Biol* 115: 1283–1292
18 Brown PD, Benya PD (1988) Alterations in chondrocyte cytoskeletal architecture during phenotypic modulation by retinoic acid and dihydrocytochalasin B-induced reexpression. *J Cell Biol* 106: 171–179
19 Spector I, Shochet N, Blasberger D, Kashman Y (1989) Latrunculins-novel marine macrolides that disrupt microfilament organization and affect cell growth. *Cell Motility Cytoskeleton* 13: 127–144
20 Hua Q, Knudson CB, Knudson W (1993) Internalization of hyaluronan by chondrocytes occurs *via* receptor-mediated endocytosis. *J Cell Sci* 106: 365–375
21 Nofal GA, Knudson CB (2000) Cytoskeletal disruption by latrunculin A decreases chondrocyte matrix retention. *Ortho Res Soc Trans* 25: 1005
22 Knudson W, Casey B, Nishida Y, Eger W, Kuettner KE, Knudson CB (2000) Hyaluronan oligosaccharides perturb cartilage matrix homeostasis and induce chondrocytic chondrolysis. *Arthritis Rheum* 43: 1165–1174
23 Lark MW, Bayne EK, Flanagan J, Harper CF, Hoerrner LA, Hutchinson NI, Singer II, Donatelli SA, Weidner JR, Williams HR, Mumford RA, Lohmander LS (1997) Aggrecan degradation in human cartilage. *J Clin Invest* 100: 93–105
24 Sastry SK, Burridge K (2000) Focal adhesions: a nexus for intracellular signaling and cytoskeletal dynamics. *Exp Cell Res* 261: 25–36
25 Svoboda KKH (1998) Chondrocyte-matrix attachment complexes mediate survival and differentiation. *Micro Res Tech* 43: 111–122
26 Nofal GA, Knudson CB (1999) Chondrocyte CD44 interacts with the cytoskeleton *via* ankyrin. *Ortho Res Soc Trans* 24: 130
27 Tsukita S, Oishi K, Sato N, Sagara J, Kawai A (1994) ERM family members as molec-

ular linkers between the cell surface glycoprotein CD44 and actin-based cytoskeletons. *J Cell Biol* 126: 391–401

28 Legg JW, Isacke CM (1998) Identification and functional analysis of ezrin-binding site in the hyaluronan receptor, CD44. *Current Biol* 8: 705–708

29 Lokeshwar VB, Fregien N, Bourguignon LYW (1994) Ankyrin-binding domain of CD44(GP85) is required for the expression of hyaluronic acid-mediated adhesion function. *J Cell Biol* 126: 1099–1109

30 Embry J, Knudson W (2001) Hyaluronan and the hyaluronan binding region of aggrecan are co-internalized by articular chondrocytes. *Orthop Res Soc Trans* 26: 401

31 Naux D (1992) The structure of an endocytosis signal. *Trends Cell Biol* 2: 189–192

32 Michaely P, Kamal A, Anderson RG, Bennett V (1999) A requirement for ankyrin binding to clathrin during coated pit budding. *J Biol Chem* 274: 35908–35913

33 Isacke CM (1994) The role of the cytoplasmic domain in regulating CD44 function. *J Cell Sci* 107: 2353–2359

34 Formby B, Stern R (1998) Phosphorylation stabilizes alternatively spliced CD44 mRNA transcripts in breast cancer cells: inhibition by antisense complementary to casein kinase II mRNA. *Mol Cell Biochem* 187: 23–31

35 Lesley J, Hyman R (1998) CD44 structure and function. *Frontiers Bioscience* 3: 616–630

36 Rousche KT, Knudson CB (2001) Casein kinase: a potential mediator of CD44 phosphorylation in human articular cartilage. *Ortho Res Soc Trans* 26: 323

37 Muehleman C, Bareither D, Huch K, Cole AA, Kuettner KE (1997) Prevalence of degenerative morphological changes in the joints of the lower extremity. *Osteoarthritis Cartilage* 5: 23–37

38 Screaton GR, Bell MV, Jackson DG, Cornelis FB, Gerth U, Bell JI (1992) Genomic structure of DNA encoding the lymphocyte homing receptor CD44 reveals at least 12 alternatively spliced exons. *Proc Natl Acad Sci USA* 89: 12160–12164

39 Gunthert U, Hofmann M, Rudy W, Reber S, Zoller M, Haussmann I, Matzku S, Wenzel A, Ponta H, Herrlich P (1991) A new variant of glycoprotein CD44 confers metastatic potential to rat carcinoma cells. *Cell* 65: 13–24

40 Sherman L, Wainwright D, Ponta H, Herrlich P (1998) A splice variant of CD44 expressed in the apical ectodermal ridge presents fibroblast growth factors to limb mesenchyme and is required for limb outgrowth. *Genes Dev* 12: 1058–1071

41 Jiang H, Knudson CB, Knudson W (2001) Antisense inhibition of alternatively spliced CD44 variant in human articular chondrocytes promotes hyaluronan internalization. *Arthritis Rheum* 44: 2599–2610

42 Knudson W, Bartnik E, Knudson CB (1993) Assembly of pericellular matrices by COS-7 cells transfected with CD44 homing receptor genes. *Proc Natl Acad Sci USA* 90: 4003–4007

43 Jiang H, Peterson R, Wang W, Bartnik E, Knudson CB, Knudson W (2002) A requirement for the CD44 cytoplasmic domain for hyaluronan binding, pericellular matrix

assembly and receptor-mediated endocytosis in COS-7 cells. *J Biol Chem* 277: 10531–10538
44 Peterson R, Knudson CB, Knudson W (2001) Expression of CD44 isoforms and truncation mutants by articular chondrocytes. *Trans Ortho Res Soc Trans* 26: 395
45 Madry H, Trippel SB (2000) Efficient lipid-mediated gene transfer to articular chondrocytes. *Gene Ther* 7: 286–291
46 Bayliss MT (1992) Metabolism of animal and human osteoarthritis cartilage. In: KE Kuettner, R Schleyerbach, JG Peyron, VC Hascall (eds): *Articular cartilage and osteoarthritis*. Raven Press, New York, 487–500
47 Muir H (1995) The chondrocyte, architech of cartilage: Biomechanics, structure, function and molecular biology of cartilage matrix macromolecules. *BioEssays* 17: 1039–1048
48 Knudson CB, Knudson W (2001) Cartilage proteoglycans. *Seminars Cell & Dev Biol* 12: 69–78
49 Kuettner KE (1994) Osteoarthritis: Cartilage integrity and homeostasis. In: JH Klippel, PA Dieppe (eds): *Rheumatology*. Mosby-Year Book Europe Limited, St. Louis, MO, 6.1–6.16
50 Reddi AH (1998) Role of morphogenetic proteins in skeletal tissue engineering and regeneration. *Nature Biotech* 16: 247–252
51 Daniels K, Solursh M (1991) Modulation of chondrogenesis by the cytoskeleton and extracellular matrix. *J Cell Sci* 100: 249–254

Discussion

Role of hyaluronan in regulating joint fluid flow

Q: *Vince Hascall:* I am interested in the difference between the dextran and the hyaluronan. If you increased the molecular weight of the dextran so that it had similar size, would you expect it to behave the same way since it's not a polyanion?

A: *Roger Mason:* The problem for us, of course, is getting commercially a high enough molecular weight dextran.

Q: *Vince Hascall:* What is the contribution of the carboxyl anionic groups on the hyaluronan?

A: *Roger Mason:* They must contribute, but I don't think I could comment quantitatively on that.

Q: *Warren Knudson:* Do you know if hyaluronan in synovial fluid is decorated with anything else such as proteoglycan remnants that may change some of its reflective properties, as opposed to when you do it in the model system with purified hyaluronan?

A: *Roger Mason:* We don't know that, but clearly if that were the case, then the hyaluronan would become effectively even larger. That could clearly have an effect on the system.

Q: *Ernst Hunziker:* I assume that if we remove certain tissues from the joint, we break the synovial membrane barrier. What is the critical percentage of synovial tissue we can remove in the joint before it gets completely depleted in synovial fluid?

A: *Roger Mason:* Very minimal! If we even just puncture the synovium with a catheter needle when we are setting up these experiments, then we do not see these effects, and there is a tremendous increase in transynovial flow as you increase pressure within the joint.

Q: *Ernst Hunziker:* Can we conclude that if transynovial flow is re-established that the repair process at this level is very efficient?

A: *Roger Mason:* I don't know whether the repair process establishes the same sort of matrix that is present in the normal synovium *in vivo*. So, I don't know what the effects would be other than to predict that any damage would be fairly adverse from a physiological point of view.

Q: *Klaus Kuettner:* Based on your data would you recommend lavaging of the synovial fluid in an inflammatory situation and replacing it by high molecular weight or low molecular weight hyaluronic acid?

A: *Roger Mason:* Well that would be a logical way to proceed. On the other hand, of course, experience is not always the same in practice as a theoretical prediction.

Q: *Jody Buckwalter:* When listening to your data, I think of the millions of artheroscopic procedures that are done per year with lavages. You have shown very convincing evidence that this disturbs the filtration mechanism profoundly. There must be a very rapid recovery or we would see a lot more complications with these procedures.

A: *Roger Mason:* In the physiological state, the synovium is not entirely stretched. It is folded and maybe the fact that it is folded actually helps to smooth the puncture site over more quickly.

Q: *Jody Buckwalter:* One other thought that occurred to me is that inflammatory disease is only one set of problems, but what about osteoarthritis? Do you think the same sort of phenomenon occurs?

A: *Roger Mason:* As far as I'm aware, in osteoarthritis, the main problem for changes in the synovial fluid hyaluronan is a decrease in concentration, and clearly if you fall below the overlap concentration, then you will alter the transynovial flow rate even though a high molecular weight hyaluronan may be present.

Elastoviscous hyaluronan in the synovium fluid in health and disease

Q: *Vince Hascall:* When pressure is put on the membrane, does it deform to the same extent in the presence of the viscoelastic material?

A: *Endre A. Balazs*: If we are talking about the cell membrane of the oocytes, the assumption is that when the negative pressure (suction) is applied in the presence of the elastoviscous media, the membrane will not deform to the same extent as in the less or non-elastoviscous media. Consequently, the activity of the ion channels is reduced. One must remember that, in this experimental model, the ion channels open by physical deformation of the membrane. The membrane, however, is a complex system, and if one assumes that it is covered by very small microcilia of molecular dimensions the opening of the ion channels can be mediated by the dislocation of the cilia. In this case, the elastoviscous medium would act as a mechanical stabilizer, not on the folding of the membrane, but of the movement of the cilia. The important aspect of these experiments is that it proves that the activity of a mechanical sensor system can be reduced by the elastoviscous properties of the fluid surrounding them.

Q: *Katalin Mikecz*: You told us about the properties of hyaluronan in normal and OA joints. Is there anything particular to synovial fluid hyaluronan in rheumatoid arthritis?

A: *Endre A. Balazs*: In rheumatoid arthritis, the molecular distribution of various molecular mass hyaluronan molecules is similar to that found in osteoarthritis. While we do not have extensive studies available on these fluids, in general, the lower molecular weight hyaluronan is more dominant.

Q: *David Howell*: I wonder how the hyaluronan might stimulate synovial cells to produce more hyaluronan throughout several months or at least a few days to a week. Do studies on the synovial cells in cultures show some evidence of stimulation to produce hyaluronan?

A: *Endre A. Balazs*: In *in vitro* studies made decades ago, it was shown that hyaluronan added to hyaluronan-producing cells stimulates hyaluronan synthesis. More recently, Peter Ghosh compared the hyaluronan synthesis using a radioisotope incorporation technique in cell cultures of normal, osteoarthritic and rheumatoid arthritic synovial tissue. When exogenous hyaluronan was added to these cultures, a significant stimulation of hyaluronan synthesis was observed only in the cultures obtained from osteoarthritic patients.

Q: *Tibor Glant*: Is the wide range of the molecular weight of HA in the synovial fluid of the patient that you showed the result of synthesis or degradation?

A: *Endre A. Balazs*: There are investigators who believe that the appearance of hyaluronan of lower than normal molecular mass is the result of degradation of the molecule in the fluid itself caused by oxidative or free radical events. I believe that this is not the case. The number of hyaluronan molecules with lower than 2 million molecular mass in the fluids of osteoarthritic patients varies from patient to patient and often, at various times. The volume of the exudates varies as well. We have not seen any pathological synovial fluid that did not contain a varying amount of hyaluronan with normal, high molecular mass. Furthermore, there are very few cells present in the fluid of these patients, and often inflammatory cells (leucocytes) are not present at all. Some investigators assume that the hyaluronan-degrading mechanism is associated with these types of cells. Most importantly, no hyaluronan degrading agents are present in the freshly collected osteoarthritic fluids. Consequently, I believe that the appearance of the lower-than-normal molecular mass hyaluronan molecules in these patients is the result of changes in hyaluronan synthesis.

In vivo investigation of hyaluronan and hyaluronan synthase-2 function during cartilage and joint development

Q: *Warren Knudson*: Does GDF5 occur only in periods of chondrogenesis or is there any GDF5 that may be expressed earlier that may make these mice lethal at an earlier stage?

A: *Janet Lee*: Actually, the GDF5 knockout in mice is not lethal. They survive and breed as homozygotes. So they really express the effect just in proximal digits.

Q: *Vince Hascall*: Can you tell us a little bit more about GDF5 function? How does it open up the Has2 gene? Is there anything known about GDF5 that would lead you to speculate as to what it is doing in that respect?

A: *Janet Lee*: In the GDF5 mice many cartilage markers such as collagen 2 are down regulated, so we think it may downregulate Has2 also. But according to our data, if there is no Has2 in the E14.5 limbs, GDF5 may have an opposite effect on Has2. So it's speculative right now.

Q: *Vince Hascall*: Is anything known about how GDF5 inhibits collagen 2 in terms of its synthesis?

A: *Janet Lee*: Not that I know. I didn't mention this earlier, but the Has2 gene contains SMAD boxes. Thus, members of that superfamily may regulate Has2.

CD44 and cartilage matrix stabilization

Q: *John Couchman*: Do you have an explanation for the result where you overexpressed the short tailed form of CD44? Maybe it is simply competing for hyaluronan on the cell surface. Yet you clearly deplete the ability to maintain a hyaluronan matrix. Is that the case, or in fact, is the short tail CD44 forming dimers and poisoning some sort of relationship on the cell surface?

A: *Cheryl Knudson*: Those are the two hypotheses that we are trying to work on now. I think that your second theory is our favorite, that it's acting somewhat as a dominant negative receptor on the cytoplasmic side, somehow altering the patterning, or dimerizing, or ligamerizing of the wild-type CD44 to inhibit its capacity to retain matrix on articular chondrocytes.

Q: *Vince Hascall*: It appears from your slides that when you use the oligosaccharides and disrupt the pericellular matrix, you eventually get a loss of the entire matrix. After disrupting the cytoskeleton some signaling pathways are probably induced that open up a different kind of biology. Do you know anything about the signaling pathways that might be involved?

A: *Cheryl Knudson*: No, unfortunately, we don't know. When you look at the longer-term response, we don't know if it's cytokine mediated. We don't know if there's IL1 production in the presence of the HA oligos. The binding of HA oligos increases the production of the iNOS. So maybe that is involved. We don't know exactly when the MMPs are upregulated or when the aggrecanase activity is stimulated, but it does seem like persistent uncoupling of the chondrocytes from their surrounding HA turns those chondrocytes into more of a catabolic phenotype.

Q: *Vince Hascall*: If you remove the oligos, can you initiate a repair response?

A: *Cheryl Knudson*: We've done most of the work in the embryonic chick tibial system. When we culture the 12-day chick tibia and then wash out the HA oligos, those explants have a great capacity to recover and fill up their matrix. We haven't pursued similar studies on the adult articular cartilage explants.

Q: *Phil Osdoby*: There is a growing observation about changes in active vs. decoy receptors as a function of physiology. You mentioned that there is some variability in the expression of the short tail CD44. Have you had an opportunity to begin to dissect that relative to age or other parameters?

A: *Cheryl Knudson*: We've looked at 31 human articular cartilage samples. We see variability between about 5% of the short tail message being expressed up to about 35% in some of those donors. But, we don't have enough cases yet to really answer your question. In other cell types, you don't really see very much of the short-tailed message expressed. So it may be more relevant to chondrocytes and their capacity to either respond to growth factors or assemble matrix or whatever.

Discussion

Q: *Matt Warman*: This afternoon I was looking at the posters and saw that there was a CD44 knockout mouse in which the cartilage looks normal. So, what do you make of that?

A: *Cheryl Knudson*: That is a very surprising finding to me. The work that the doctors Tammi have done with Vince Hascall showed that CD44 is very important in the turnover of HA in the epidermis. Work by Stamenkovic targeted a knockout of CD44 to the epidermis where keratin is expressed. The skin is swollen and fragile presumably because the turnover of the HA is impaired. In the complete knockout mouse there appears to be something else that is compensating for all of these functions of CD44 because the skin in a knockout mouse appears normal. These mice are however, more susceptible to an insult such as a lung injury.

Skeletal turnover in health and arthritis

Introduction

Robin Poole

Shriners Hospital for Children, McGill University, 1529 Cedar Avenue, Montreal, Quebec H3G 1A6, Canada

Osteoarthritis involves multiple changes in different joint tissues. Most obvious are those in the articular cartilage and subchondral bone. Degeneration of cartilage is accompanied by a variety of changes in the turnover and structure of subchondral bone [1]. These may either preceed, accompany or be secondary to those in cartilage. Clearly the interface between the cartilage and bone is an area of great importance where changes in the dynamic strains and stresses of one tissue impact upon the other. Ted Oegema provides some novel insights into this important and poorly studied interface.

Articular cartilage plays a crucial biomechanical function in enabling joint articulation. The individual molecules of the extensive extracellular matrix and their interactions provide unique properties which enable this tissue to dissipate load and yet retain sufficient strength to resist destruction, in spite of the constant articulation involving articular cartilages. In osteoarthritis, damage to these matrix molecules leads to a loss of function, which is described by Alice Maroudas. These studies provide fascinating insights into the roles of different molecules such as those of the collagen fibrillar network and the large proteoglycan aggrecan and how these change in osteoarthritis.

Damage to these matrix molecules in cartilage has been the subject of intense study stretching back to the 1960s. Much recent progress has been made in identifying proteases involved in cartilage resorption. These are reviewed by Fred Woessner who also examines the mechanisms that may be involved in controlling their activities.

It is a special pleasure to have these speakers participate in this session since they have contributed enormously to our present knowledge of cartilage degeneration in arthritis.

References

1 Poole AR, Howell DS (2001) Etiopathogenesis of osteoarthritis. In: RW Moskowitz, DS Howell, VM Goldberg, HJ Mankin (eds): *Osteoarthritis: diagnosis, and management*, 3rd ed. WB Saunders Co., 29–47

Metalloproteinases and osteoarthritis

J. Frederick Woessner

Department of Biochemistry and Molecular Biology, University of Miami School of Medicine, P.O. Box 016960, Miami, FL 33101, USA

Metalloproteinases and cartilage

There are currently more than 250 known proteases dependent on zinc or other metal atoms for their catalytic activity. If we restrict these to endopeptidases (proteinases) found in humans and other mammals, we can reduce this number to about 75. Metalloproteinases of interest in cartilage metabolism and degradation fall mostly within the Clan MA – zinc proteases with two ligands in the sequence HEXXH and a third ligand that is either Glu or His. The members of this clan dependent on a third His residue constitute the metzincins (10 families) of which Family 10 – the collagenase family – and Family 12 – the astacin family – are of major interest. Family 10 contains the matrixin subfamily, which includes some 25 different matrix metalloproteinases (MMPs), which can digest almost all known components of the extracellular matrix (ECM). Family 12 has two subfamilies: an astacin subfamily, which includes procollagen C-peptidase, and the reprolysin or adamalysin subfamily, which includes the ADAMs (a disintegrin and metalloprotease domain) and the ADAMTSs (thrombospondin-like domains). There are 30 ADAMS, about half of which possess protease activity. There are currently nine ADAMTSs, of which –4 and –5 are aggrecanases. Information about these clans, families and enzymes can be found on the web at www.merops.co.uk.

The MMPs [1] have the basic structure: propeptide-catalytic domain-hemopexin-like domain. Additional domains are inserted in gelatinases (fibronectin type II repeats), membrane-type MMPs (transmembrane domain), etc. The enzymes are often activated by mercurials, which disrupt the link between propeptide Cys and active site Zn. There are hundreds of papers concerning these enzymes in cartilage in relation to rheumatoid and ostoearthritis [2]. More recently, it has been found that the major cartilage proteoglycan, aggrecan, is probably broken down by aggrecanases (ADAMTS-4,-5) [3]. These enzymes have some similarities to MMPs, but are not closely related evolutionarily. They have multiple domains including a proenzyme, metalloprotease, disintegrin-like, thrombospondin-like, cysteine-rich, and spacer domain. ADAMTS-2 is procollagen N-peptidase, needed to produce mature collagen. Finally, in cartilage ADAM 17 (tumor necrosis factor α converting enzyme or TACE), ADAM 10 (or MADM) and ADAM 15 (metargidin) are of interest [4]. The ADAMs have the domain structure: propeptide, metalloprotease, disintegrin-like, EGF-like, cysteine-rich, transmembrane, and cytoplasmic domain. They are typically activated intracellularly by furins [5]. Cartilage degradation is not mediated exclusively by metalloproteases, there are also serine proteases such as the plasmin system and lysosomal cathepsins such as cathepsin B that are involved. But this chapter will focus on metalloproteases, and specifically on MMPs.

Binding of MMPs and TIMPs to tissues

Cartilage is composed of a variety of ECM proteins and proteoglycans. An interesting feature of these components is that they do not diffuse readily into the synovial fluid, but rather tend to stay in position due to an interlocking network of connections between collagen fibers and other large molecules. Aggrecan, although already a large molecule, aggregates on hyaluronic acid strands, assisted by link protein, to form giant complexes that are trapped within the matrix. It should not be surprising that the MMPs also tend to interact with matrix components or cells so that they too do not readily diffuse from the tissue. In early studies of cartilage collagenase, it was found that little enzyme could be extracted from the tissue by homogenizing in buffer. Rather, strenuous means had to be used such as heating to 60°C in the presence of calcium, to remove the enzyme activity. This method was worked out already in 1976 to extract collagenase from the involuting uterus [6]. In recent years, I have focused my research on the reasons for this difficulty in extraction and what ECM components the MMPs might be binding to.

Recently, our work on matrilysin binding to the uterus has shown that the enzyme appears to interact specifically with heparan sulfate chains on proteoglycans [7]. If homogenates of uterus are treated with heparin at levels of 3–4 mg/ml, most of the MMP-7 can be extracted as compared to extraction directly in SDS. Other sulfated compounds can also be used (chondroitin sulfate, suramin, etc., but these are less effective). Heparanase digestion will release the enzyme from tissue sections. Confocal microscopy indicates co-localization of heparan sulfate and MMP-7 on epithelial cell surfaces and underlying basement membrane. In particular, CD44v3, a cell-surface hyaluronan-binding protein that also contains a chain of heparan sulfate, has been shown to bind MMP-7 to the apical surface of uterine epithelial cells (unpublished observation). In these studies, we obtained evidence that MMP-13, MMP-2 and MMP-9 also had some binding to the matrix in descending order relative to MMP-7. While MMP-7 is not prominent in cartilage, the other enzymes are present and are difficult to extract, suggesting binding in the tissue, perhaps to glycosaminoglycan (GAG) chains.

Table 1 - Why are MMPs bound

To cells?	1. Activation of other MMPs
	2. Positioned for activation by other proteases
	3. Shedding of receptors and growth factors
	4. Directed proteolysis and cell movement
To ECM?	1. Avoid loss from tissue by diffusion
	2. Reservoir of latent enzyme
	3. Release of bound growth factors
	4. Regulation of enzyme synthesis

Similarly, TIMP-3 has long been known to be bound to the ECM. We have found a similar story in the uterus – TIMP-3 is extractable with heparan sulfate and related compounds. Most of the binding is to heparan sulfate, but chondroitin sulfate probably plays a role as well [8]. In both cases, MMP-7 and TIMP-3, sodium dodecyl sulfate is an excellent extractant. It is not clear if this is due to the denaturation of the protein or to competition with sulfate binding sites, perhaps both.

If we consider the family of MMPs, we see that at least six members (membrane-type MMPs) are bound to cell membranes through insertion in the membrane. MMP-2 binds to the cell surface by interaction with MMP-14 and with integrins [1]. MMP-13 appears to have specific cell receptors [9]. We have evidence that MMP-1, -2, -7, -9 and -13 may bind to cell surface GAG chains as well as to similar GAG chains in the basement membrane and stroma [7]. Finally, there is evidence that MMP-1, -2 -9 and -13 may bind to collagen. Table 1 offers some suggestions as to why MMPs may bind to cells and matrix.

MMPs and TIMPs in osteoarthritic cartilage

Over the last 30 years, my research has had as one focus the degradation of ECM in osteoarthritic cartilage. This interest started with an exploration of cathepsin D. This enzyme is elevated in osteoarthritis (OA), but it is now considered that the lysosomes may not play a major role in OA. This was followed by studies of collagenase in human OA [10] and then of stromelysin [11]. A major finding at that time was that the increases in MMPs in OA were not offset by similar increases in TIMP activity, leading to an imbalance favoring degradative processes [11]. In recent years this line of study has been extended to the gelatinases and a more detailed look at TIMPs.

It is generally assumed that gelatinases are readily extractable from tissues, most workers merely prepare a homogenate in buffer and study the extract. However, in

Table 2 - Gelatinase activity in osteoarthritic patellae

Enzyme	Control	Osteoarthritis	p value
Total gelatinase A	5.9 ± 1.1	11.2 ± 1.8	$p<0.02$
Latent gelatinase A	0.7 ± 0.2	2.0 ± 0.5	$p<0.03$

Activity = µg MMP-2/g wet weight ± SEM, N = 9. p value from Student's t-test.

light of our previous experience with stromelysin, which required 1 M guanidine HCl; and collagenase, which required heat and calcium for extraction, we first examined the extractibility of gelatinase activity from human cartilage. Extraction with 1 M GuHCl was followed by subsequent extraction with 4 M GuHCl or 2% SDS. Either second step released further activity. Direct extraction with SDS (sodium dodecyl sulfate) proved optimal; 1 M GuHCl extracted only about 25% of the activity recovered by SDS. The major activity recovered was MMP-2 or gelatinase A. This enzyme is presumably binding to something – perhaps to collagen through its fibronectin-like domains. Extraction with guanidine HCl was not ideal because a great deal of proteoglycan then had to be separated from the enzyme.

A brief outline of methods follows. Human cartilage was minced and homogenized in Tris buffer, following which SDS was added directly to the homogenate to 2% final concentration. After 30 min this was centrifuged to provide an extract, and the pellet was re-extracted with SDS. The two extracts were combined and subjected to SDS/PAGE electrophoresis on gels containing 0.1 mg gelatin/ml to obtain zymograms by the usual means [12]. The zymograms were scanned and quantified relative to standards of MMP-2 and MMP-9. A wide range of sample sizes had to be applied to permit measurement in the linear range for both active and latent enzyme [12]. In this work, it was not possible to visualize any MMP-9 except at high overloading for MMP-2. We estimate that MMP-9 is less than 1% the amount found for MMP-2. The results of measurements on nine normal and nine OA (frank ulceration on the patellar cartilage surface) are presented in Table 2. It is seen that there is a doubling in total MMP-2 and almost a tripling in the activated form.

TIMPs were measured in the first SDS extract prepared above. In this case SDS/PAGE electrophoresis was used to separate protein bands, and standard TIMP-1 and mol. wt. markers were used on each plate. Part of the gel was cut off and stained with Coomassie blue, the remainder was then cut into slices with a razor in such a way that three peaks were distinguished (TIMP-1, -3 and -2) in decreasing M_r. These slices were transferred to assay tubes with Azocoll substrate and MMP-7 as the enzyme to be inhibited [13]. In each case, the slice of gel with standard TIMP-1 levels was included as a control. The tubes were incubated overnight and Azocoll

Table 3 - TIMP activity in osteoarthritic patellae

TIMP type	Control	Osteoarthritis
TIMP-1	3.5 ± 0.6	4.0 ± 0.3
TIMP-2	2.4 ± 0.5	3.2 ± 0.5
TIMP-3	3.2 ± 0.8	3.6 ± 0.7
Total TIMP	9.1 ± 1.6	10.7 ± 1.6

Activity = µg/g wet weight ± SEM, N = 9

digestion was measured in relation to digestion by MMP-7 standard. The results of this study are presented in Table 3. The individual TIMPs all were present at about the same level and these levels did not increase in OA.

Conclusions

Cartilage matrix in OA is degraded by a series of different proteases from different classes (serine, cysteine and metallo-). The focus in aggrecan degradation has been the metalloproteases MMP-3 and ADAMTS-4,-5. Both stromelysin and aggrecanase 1 are inhibited by similar inhibitors – hydroxamate-containing synthetic inhibitors [14] and by all four of the natural inhibitor TIMPs, especially TIMP-3 [15]. Collagenase, stromelysin and gelatinase A are all elevated in OA, gelatinase B is present at negligible levels. These increased MMPs activities are not offset by increases in TIMP, leading to an overall degradative balance.

References

1   Nagase H, Woessner JF (1999) Matrix metalloproteinases. *J Biol Chem* 274: 21491–21494
2   Elliott S, Cawston T (2001) The clinical potential of matrix metalloproteinase inhibitors in the rheumatic disorders. *Drugs Aging* 18: 87–99
3   Caterson B, Flannery CR, Hughes CE, Little CB (2000) Mechanisms involved in cartilage proteoglycan catabolism. *Matrix Biol* 19: 333–344
4   Flannery CR, Little CB, Caterson B, Hughes CE (1999) Effects of culture conditions and exposure to catabolic stimulators (IL-1 and retinoic acid) on the expression of matrix metalloproteinases (MMPs) and disintegrin metalloproteinases (ADAMs) by articular cartilage chondrocytes. *Matrix Biol* 18: 225–237

5   Black RA, White JM (1998) ADAMS - focus on the protease domain. *Curr Opin Cell Biol* 10: 654–659
6   Weeks JG, Halme J, Woessner JFJr (1976) Extraction of collagenase from the involuting rat uterus. *Biochim Biophys Acta* 445: 205–214
7   Yu WH, Woessner JF (2000) Heparan sulfate proteoglycans as extracellular docking molecules for matrilysin (matrix metalloproteinase 7). *J Biol Chem* 275: 4183–4191
8   Yu WH, Yu SSC, Meng Q, Brew K, Woessner JF (2000) TIMP-3 binds to sulfated glycosaminoglycans of the extracellular matrix. *J Biol Chem* 275: 31226–31232
9   Omura TH, Noguchi A, Johanns CA, Jeffrey JJ, Partridge NC (1994) Identification of a specific receptor for interstitial collagenase on osteoblastic cells. *J Biol Chem* 269: 24994–24998
10 Pelletier J-P, Martel-Pelletier J, Howell DS, Ghandur-Mnaymneh L, Enis JE, Woessner JFJr (1983) Collagenase and collagenolytic activity in human osteoarthritic cartilage. *Arthritis Rheum* 26: 63–68
11 Dean DD, Martel-Pelletier J, Pelletier J-P, Howell DS, Woessner JFJr (1989) Evidence for metalloproteinase and metalloproteinase inhibitor imbalance in human osteoarthritic cartilage. *J Clin Invest* 84: 678–685
12 Woessner JFJr (1995) Quantification of matrix metalloproteinases in tissue samples. Methods Enzymol 248: 510-528
13 Woessner JF Jr (1995) Matrilysin. *Methods Enzymol* 248: 485–495
14 Arner EC, Pratta MA, Trzaskos JM, Decicco CP, Tortorella MD (1999) Generation and characterization of aggrecanase. A soluble, cartilage-derived aggrecan-degrading activity. *J Biol Chem* 274: 6594–6601
15 Hashimoto G, Aoki T, Nakamura H, Tanzawa K, Okada Y (2001) Inhibition of ADAMTS4 (aggrecanase-1) by tissue inhibitors of metalloproteinases (TIMP-1, 2, 3 and 4). *FEBS Lett* 494: 192–195

Age-dependent changes in some physico-chemical properties of human articular cartilage

Haya ben-Zaken[1], Rosa Schneiderman[1], Hannah Kaufmann[2] and Alice Maroudas[1]

[1]Department of Biomedical Engineering, Technion – Israel Institute of Technology, Haifa 32000, Israel; [2]Department of Orthopaedics, Rambam Hospital, Haifa 32000, Israel

Introduction

It is well known that osteoarthritis (OA) is associated with advancing age. However, a large number of elderly people do not develop OA and a survey of aged cartilage, in the absence of degeneration, reveals unchanged physico-chemical characteristics. The object of this article is to describe some of the physical properties of femoral head and femoral condyle cartilage as a function of age and in OA. The properties with which we shall deal are the osmotic pressure of proteoglycan (PG) extracted from cartilage and the stiffness of the collagen network.

Chemical composition and osmotic pressure of extracted PG

The changes in the structure and composition of aggrecan take place primarily during maturation and with increasing age up to approximately 30–40 years. During this period the amount of keratan sulphate (KS) increases [1–3] due to the decrease in the size of the monomers (formation of truncated species) [4–6]. There are also changes in the type and amounts of non-collagenous, non-PG proteins, but these changes have not been quantified.

In view of the above phenomena, we thought that there might be alterations with age in the osmotic pressure of the aggrecan and other guanidinium chloride extractable macromolecules. In particular, there are reasons to expect the keratan to chondroitin sulphate (CS) ratio of the aggrecan to affect the electrostatic component of osmotic pressure because these two glycosaminoglycans have different inter-charge distances.

Using calibrated polyethylene glycol (PEG) solutions [7], we measured the osmotic pressure of guanidinium chloride extracts of cartilage which had been

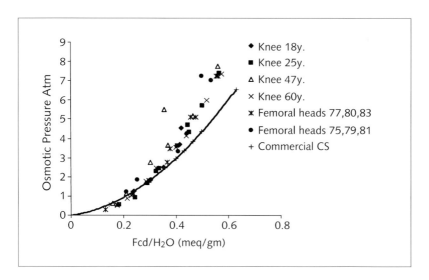

Figure 1
Variation of osmotic pressure of the GuHCl extracts (π_{PG}) vs. FCD in relation to age.

ultrafiltered and concentrated by passage through 1 kDa membranes. The cartilage ranged in age from 18 to 83 years. The results, plotted vs. fixed charge density (FCD) (see Fig. 1), show no age-related changes. This result, though at first sight surprising, can probably be explained by the calculations based on the Poisson-Boltzmann theory (Parker, Winlove, Ehrlich and Maroudas, unpublished calculations). According to these calculations, the effect of the variations in the CS:KS ratio that one finds in human cartilage with aging [8] on the electrostatic component of osmotic pressure is very small (~0.1 atm) and not detectable by our methodology. Our finding also means that the non-electrostatic component of osmotic pressure is little affected by the structural changes in the PG, which accompany aging.

Thus, the structural and compositional changes in PG that occur with aging do not affect *per se* their osmotic pressure. However, osmotic pressure is a function of FCD. FCD in cartilage increases with age, both when calculated per wet tissue weight [3] and on the basis of extrafibrillar water [9]. This increase is due to the increase in the absolute amount of KS and to a decrease in the water content.

Tension in the collagen network

The physiological role of the collagen network is to resist swelling and stretching of the cartilage. Equilibrium hydration of the tissue is achieved when the hydrostatic

Table 1 - Comparison between collagen tension in young and aged specimens

No. of samples	Age of specimens	Max slope of tension vs. vol.	Maximum tension
5	23, 24, 30	3.35 ± 0.47	2.13 ± 0.65
10	76, 78, 82, 110	3.67 ± 0.85	2.68 ± 0.72

pressure caused by the stress, P_c, exerted by the collagen network, combined with the externally applied stress (in our case osmotic stress, π_{PEG}), both of which tend to squeeze fluid out of the cartilage, are balanced by the osmotic pressure of the cartilage PG (π_{PG}).

Thus:

$$P_c = \pi_{PG} - \pi_{PEG} \qquad (1)$$

In order to determine P_c experimentally, using PEG to compress cartilage specimens, π_{PG} must be known. In our treatment π_{PG} is based on the curve given in Figure 1, the FCD being determined by experimental measurements on the given cartilage specimens and the extrafibrillar water is calculated as in Basser et al. [10].

It can be seen from Equation (1) that, in the absence of external pressure

$$P_c = \pi_{PG} \qquad (2)$$

Clearly, the maximum value of the collagen tension is equal to the osmotic pressure exerted by the PG in the extrafibrillar space.

Typical curves of P_c vs. normalized volume are shown in Figure 2, for both normal and osteoarthritic cartilage specimens. While it is difficult to see at a glance the influence of aging, the curves representing OA specimens are very clearly distinguishable from the normal.

A more quantitative assessment of aging on P_c was carried out on two groups of specimens: young (age range 20–30) and aged (age range 75–110). The results are presented in Table 1 which shows both the maximum value of P_c vs. normalized volume and the slope at this maximum point. It can be seen that, in the older age group, both these variables are higher, indicating a stiffer network, but the difference is not very great. The modest stiffening of the collagen network with age is likely to be in part due to the non-enzymatic glycation products formed during the aging process [11].

The flat appearance of the P_c curves (Fig. 2) for OA cartilage shows the weakening and disruption of the network. At times this is not accompanied by an

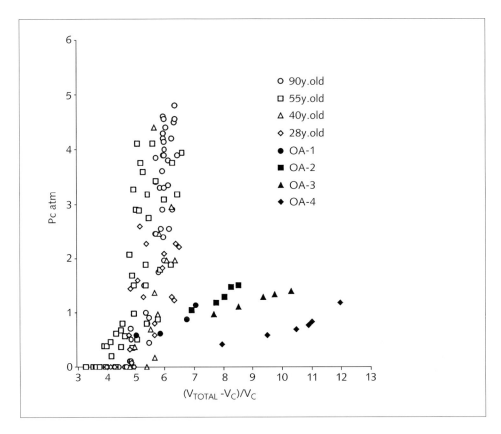

Figure 2
Tensile stiffness of the collagen network P_c calculated from the "balance of forces" equation. $(V_{TOTAL} - V_c)/V_c$ = normalized volume; where V_{TOTAL} = total volume per gm wet weight of cartilage; V_c = dry volume of collagen per gm wet weight of cartilage

absolute loss of PG, but simply by the fact that P_c is lower and hence less able to resist π_{PG}. Hence the cartilage swells and the concentration of PG based on wet tissue weight is lower than in normal cartilage [8].

Instantaneous deformation of normal and OA cartilage

It has been shown that the instantaneous deformation (ID) of cartilage in unconfined compression, i.e., change in shape, rather than change in volume, is primarily determined by the collagen network stiffness [12]. We have measured ID in normal specimens over a limited age range and could not find an obvious age-related trend.

Table 2 - *Zonal variation of instantaneous deformation (ID) parallel to the prick line of normal and osteoarthritic (OA) human cartilage under an unconfined compression of 0.56 MN/m²*

Zone	ID* of normal cartilage	ID* of OA cartilage	OA: normal cartilage ratio
Surface	1.2 ± 0.8	3.5 ± 0.9	2.9
Middle	2.5 ±1.2	4.9 ± 1.4	2.0
Deep	3.3 ± 1.4	5.0 ± 1.9	1.5

Values are the mean ± SD, expressed as the percentage of change in the diameter of the surface parallel to the prick lines. The data come from Bank et al. [13] by permission of the publishers.

On the other hand, the difference between the normal and the osteoarthritic samples is very significant (see Tab. 2) [13]. The increase in instantaneous deformation in OA cartilage is most pronounced in the superficial zone in the direction parallel to the collagen fibrils (as detected by the orientation of the prick-lines), and least pronounced in the deep zone. The above zonal variations are consistent with the observations of Hollander et al. [14] who have shown that damage to type II collagen in OA starts at the articular surface.

Conclusions

1. In spite of chemical modifications of the PG with age, their osmotic pressure vs. FCD relationship does not appear to vary with age.
2. The stiffness of the collagen network as obtained by the swelling pressure technique described in [10] shows a slight but significant increase with age, but this is not evident from instantaneous deformation experiments.

There is a large drop in the collagen stiffness in OA cartilage, whichever of the two methods is used.

It should be noted that for a given quantity of PG, the actual PG concentration and hence the ability of cartilage to resist compressive stresses depends on the hydration which, in turn, is a function of the stiffness of the collagen network.

Acknowledgments
We gratefully acknowledge the support of the Arthritis Foundation (USA) and the Israel Ministry of Health.

References

1. Stockwell RA (1970) Changes in the acid glycosaminoglycan content of the matrix of human cartilage. *Ann Rheum Dis* 29: 509–515
2. Hjertquist SO, Lemperg RC (1967) Identification and concentration of the glycosaminoglycans of human articular cartilage in relation to age and osteoarthritis. *Calcified Tissue Research* 10: 223–227
3. Venn MF (1978) Variation of chemical composition with age in human femoral head cartilage. *Ann Rheum Dis* 37: 168–174
4. Bayliss MT (1986) Proteoglycan structure in normal and osteoarthrotic human cartilage. In: K Kuettner et al (eds): *Articular cartilage biochemistry*. Raven Press, New York, 295–310
5. Bayliss MT (1990) Extraction and purification of proteoglycan and hyaluronan from human articular cartilage. In: A Maroudas, K Kuettner (eds): *Methods in cartilage research*. Academic Press, London, 220–222
6. Maroudas A, Bayliss MT, Uchitel-Kaushansky N, Schneiderman R, Gilav E (1998) Aggrecan turnover in human articular cartilage: Use of aspartic acid racemization as a marker of molecular age. *Arch Biochem Biophys* 350(1): 61–71
7. Urban JPG, Maroudas A, Bayliss MT, Dillon J (1979) Swelling pressures of proteoglycans at the concentrations found in cartilaginous tissues. *Biorheology* 16: 447–464
8. Maroudas A, Bayliss MT, Venn MF (1980) Further studies on the composition of human femoral head cartilage. *Ann Rheum Dis* 39: 514–523
9. Grushko G, Schneiderman R, Maroudas A (1989) Some biochemical and biophysical parameters for the study of the pathogenesis of osteoarthritis: A comparison between the processes of ageing and degeneration in human hip cartilage. *Conn Tiss Res* 19: 149–176
10. Basser PJ, Schneiderman R, Bank RA, Wachtel E, Maroudas A (1998) Mechanical properties of the collagen network in human articular cartilage as measured by osmotic stress technique. *Arch Biochem Biophys* 351(2): 207–219
11. Bank RA, Bayliss MT, Lafeber FPJG, Maroudas A, Tekoppele JM (1998) Ageing and zonal variation in posttranslational modification of collagen in normal human articular cartilage. *Biochem J* 330: 345–351
12. Mizrahi J, Maroudas A, Lanir Y, Ziv I, Weber TJ (1986) The (instantaneous) deformation of cartilage: Effects of collagen fiber orientation and osmotic stress. *Biorheology* 23: 311–330
13. Bank RA, Soudry M, Maroudas A, Mizrahi J, TeKoppele JM (2000) The increased swelling and instantaneous deformation of osteoarthritic cartilage is highly correlated with collagen degradation *Arthritis Rheum* 43(10): 2202–2210
14. Hollander AP, Pidoux I, Reiner A, Rorabeck C, Bourne R, Poole AR (1995) damage to type II collagen in ageing and osteoarthritis: starts at the articular surface, originates around chondrocytes, and extends into the cartilage with progressive degeneration. *J Clin Invest* 96: 2859–2869

The role of the bone/cartilage interface in osteoarthritis

Theodore R. Oegema, Jr.

Departments of Orthopaedic Surgery and Biochemistry, University of Minnesota, 420 Delaware St., SE, Minneapolis, MN 55455, USA; present address: Departments of Biochemistry and Orthopedic Surgery, Rush Medical College, 1653 West Congress Parkway, Chicago, IL 60612, USA

Introduction

In osteoarthritis (OA), changes to both the bone and the articular cartilage have been described. Exactly how bony changes relate to cartilage changes is still a matter of discussion. Bony changes are important to the pathology. For example, in end-stage disease in knee OA in patients with knee pain, only six out of 10 had relief of pain with intra-articular injection of local anesthesia [1], meaning bone could be the source of pain. In end-stage knee osteoarthritis, the pain may come from bony lesions detected by MRI. Bony lesions occurred in 72% of the group with pain as compared to 30% with knee OA, but no pain [2]. The bony lesions detected by MRI maybe caused by bone marrow edema, but why it is painful is unclear.

At least three types of bony changes occur. The first is a change in the subchondral bone, which can occur in human OA and animal models of OA. These data have been extensively reviewed [3]. As exemplified by the canine anterior cruciate transection model of OA, there is an initial loss of subchondral cancellous bone that is followed by a late increase in subchondral plate thickness [4]. Both changes may affect the mechanics of the joint. Osteophyte formation, which often starts at the attachment site of the synovium to bone and leads to major changes in joint mechanics and biology, is the second type of bony involvement [3, 5, 6]. Thirdly, the calcified bone/cartilage interface may change, and it is critical to the function of articular cartilage since it provides the mechanical interlock to bone and has intermediate mechanical properties that help transfer forces from cartilage to bone. By acting as a barrier, it influences nutrition of the articular cartilage and may change in OA. During growth, the deeper zone of the articular cartilage is part of a secondary growth plate. At maturity, the primary growth plate is obliterated, but this region of articular cartilage forms the zone of calcified cartilage. The cells just above the tidemark (or mineralizing front) continue to express some element of the early hypertrophic phenotype. The rate of movement of tidemark into the uncalcified articular cartilage dramatically slows at maturation. Thickness of the zone of calcified cartilage remains relatively constant, so mineralization at the tidemark is normally followed by resorption of the calcified cartilage and replacement by bone [7].

Central to this discussion is whether the cartilage-bone interface is involved in the pathology of OA since movement of the tidemark would alter the cartilage thickness and mechanics [7]. Two aspects of this will be addressed here. The first is a summary of evidence that there is remodeling of the zone of calcified cartilage in OA and the second is how the cells in the mature articular cartilage continue to function as a growth plate.

Chondrocytes in the growth plate undergo a carefully orchestrated series of steps that include proliferation, matrix formation, maturation, hypertrophy and then replacement by bone. There is considerable interest in determining the nature of the cells that are the precursors to the hypertrophic chondrocyte, the controls that determine progression, the different stages of commitment and the reversibility of the process at the different stages.

There are a number of cartilages that are considered permanent cartilages, which normally do not mineralize (i.e., tracheal, nasal, ear, costal, sternal and articular). So there is the question of how the chondrocytes in permanent cartilage are related to those in the hypertrophic pathway and become involved in the pathology of OA. Are chondrocytes with the "permanent" phenotype arrested in the hypertrophic phenotype before final commitment to hypertrophy, or are these cells in a branch that normally does not result in hypertrophy [8]? Cells from "permanent" cartilage can differentiate into hypertrophic chondrocytes if there is forced expression of the transcription factor, RUNX2 (Cbfa1) [9, 10]. Expression of chicken ERG (early growth response 1) gene [8] will similarly promote the hypertrophic phenotype. Over-expression of an alternatively spliced form of cERG (C-1-1) or extracellular exposure to TGFβ, FGF-2, PTHrP, HGF or GPF-5 will inhibit progression [8, 11]. How these regulatory mechanisms relate to adult articular cartilage in OA remains to be established.

What is the evidence that progression of the zone of calcified cartilage occurs *in vivo*, and what are the possible regulatory signals in adult articular cartilage that may regulate progression? In the adult, the zone of calcified cartilage is well organized with a relatively smooth tidemark or transition between calcified and uncalcified cartilage. The thickness of the zone of calcified cartilage is carefully regulated and is scaled to the joint [7, 12]. In normal adult cartilage, the cells immediately above the tidemark express at least some of the early hypertrophic markers such as alkaline phosphatase, but not type X collagen [13]. Cells deep in the zone of calcified cartilage can remain viable [7] and express type X collagen [13]. Where the zone of calcified cartilage abuts bone, the zone of calcified cartilage is highly irregular (Fig. 1). The subchondral bone contains blind capillaries that are surrounded with lamellar bone (Fig. 2). Occasional open vessels penetrate all the way to the tidemark [7]. The number of open-ended capillaries decreases in middle age, but increases in later life. Just like the growth plate, whose appearance is columnar and

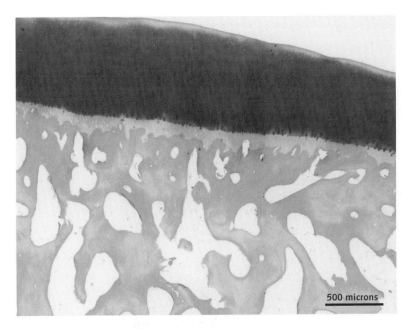

Figure 1
Photomicrograph of mature canine patella. The patella were fixed in formaldehyde with cetylpyridinium chloride, decalcified in formic acid, formaldehyde, paraffin-embedded, sectioned at 5 microns and stained with Safranin O, fast green [7]. The section illustrates blood vessels arising from the marrow space surrounded by bone in calcified cartilage, and the irregular bone-calcified cartilage interface, and the relative smooth tidemark.

where angiogenesis is carefully controlled, the deep cells are believed to actively inhibit blood vessel invasion whereas those in the mineralized region may promote the invasion of osteoclasts and endothelial cells [7].

One of the hallmarks of OA is the presence of multiple tidemarks. This has been interpreted as progression of the tidemark [7]. Alkaline phosphatase, type X collagen [13] and osteonectin [14] are also elevated just above the tidemark suggesting that the endochondral process is activated in OA. Using high definition macroradiography, patterns suggestive of an incremental advance of the zone of calcified cartilage have been found on the convex side of all the joints of osteoarthritic human hands. These changes were apparently unrelated to use since they do not depend on which hand was dominant [15].

In animal models, intraarticular fluorochromes will label the tidemarks and can be used to follow tidemark progression. In the patellae of rabbits, the tidemark moves as much as 8 microns/week at 4 months of age, but slows down to as little

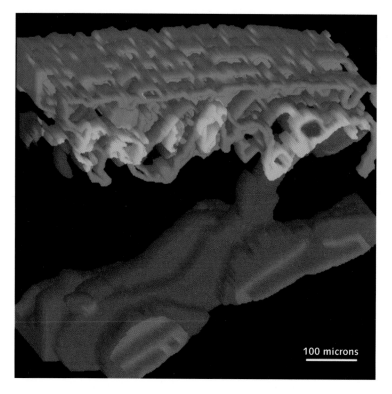

Figure 2
Three-dimensional reconstruction of canine patellae illustrating the zone of calcified cartilage (wire frame) and marrow space with blood vessel (dark filled volume). The transparent gray is bone. The vessel penetrates the calcified cartilage and remains covered with lamellar bone.

as 1 micron/week in 7-month-old rabbits and less in skeletally mature rabbits [7]. In older animals, the movement is difficult to measure since the label is lost over time if it remains at the interface too long. In OA induced by the Hulth model in immature rabbits, there was an increase in movement of the tidemark on the tibial plateau [12]. Movement was not seen in the same model in older animals.

In OA, there is a definite increase in apoptosis and the accumulation of matrix vesicles capable of mineralization in deeper layers, and changes are found in other layers of articular cartilage or around clones [16, 17]. There also may be an increase in accumulation of calcium pyrophosphate rather than apatite crystals [18]. This suggests that control of the progression of the tidemark seen in the young animal is no longer as well regulated, and progression may not even be controlled by the same signaling pathways.

One important question is what components of the regulatory pathways present in the rapidly growing growth plate are also present in mature articular cartilage. This has been partially evaluated using an *in vitro* system with bovine articular chondrocytes in suspension culture [19]. Mature full depth bovine articular chondrocytes were released with sequential pronase-collagenase digestion and plated in low-adhesion culture plates (Costar, ultra-low attachment cluster plates) at $1–10^6$ cell/ml in Ham's F12 containing 5% fetal bovine serum. After 7 days, the cells were evaluated for percent of alkaline phosphatase-positive cells and total cell number. There was a wave of alkaline phosphatase expression since the percentage of cells that stained for alkaline phosphatase after 6–7 days in culture was greater than after 3 or 10 days in culture. T_3 alone induced a 1.5 fold increase in alkaline phosphatase-positive cells. This was decreased by 50% with actinomycin D. When cultures were labeled with 5-bromodeoxyuridine (BUDR), 50% of the cells that were alkaline phosphatase-positive were also BUDR-positive, indicating the cells had divided. Exposure to FGF-2, PTHrP and TGFβ [11, 20] significantly inhibited the expression of alkaline phosphatase, but 1,25-dihydroxyvitamin D, as has been found by others, caused an increase [21]. When isolated deeper or surface cells were cultured alone, only the deep cells responded with a T_3-dependent increase in ALP. In contrast to human osteoarthritic surface chondrocytes [22], isolated bovine surface chondrocytes did not express ALP in suspension culture.

The interesting finding was that there was an interaction between deep and surface cells when corrected for cell number. The sum of the alkaline phosphatase-positive cells from cultures that contained only deep or surface cell chondrocytes was greater than what was found when the surface and deep cells were mixed together for the whole culture time. This suggested that the surface cells, analogous to the less mature cells in the growth plate [23], inhibit the deep cells from expressing ALP. Since surface chondrocytes are known to produce TGFβ and TGFβ alone inhibited progression, anti-TGFβ antiserum was tested in mixed surface and deep cell cultures. The anti-TGFβ antiserum inhibited not only the increase seen in response to exposure to T_3, but also the increase seen at 7 days with only 5% serum. This suggests that in these culture conditions, TGFβ is a major inhibitor of progression as measured by alkaline phosphatase expression. Since surface cells are lost in OA, the loss of the TGFβ they produce could relieve this inhibition, and this could lead to an increase in hypertrophic phenotype of deep cells.

In the primary growth plate, cells that express alkaline phosphatase have proceeded past the first of two checkpoints in the hypertrophic pathway. Progression past the checkpoint is not irreversible and can be reversed by exposure to PTHrp or PTH [8, 24]. So the question is, does the upregulation of the hypertrophic phenotype in osteoarthritic articular cartilage make any difference to progress. If it does, it may be a good place for intervention. Further understanding of the pathway may help design cell-based replacements and therapies where the control of where the bone-cartilage interface ends up is important to implant function.

Acknowledgements

This work was supported by NIH AR41975. The help of Andrea Chatfield for the preparation of the manuscript, Roby C. Thompson, Jr. and Jack L. Lewis for active collaboration, Laurel Deloria, Tim Nelson, Jill Vannelli for the experimental work, and Jerry Sedgewick from the Biomedical Imaging & Processing Laboratory for the three-dimensional reconstruction is gratefully acknowledged.

References

1. Creamer P, Hunt M, Dieppe P (1996) Pain mechanisms in osteoarthritis of the knee: effect of intraarticular anesthetic. *J Rheumatol* 23: 1031–1036
2. Felson DT, Chaisson CE, Hill CL, Totterman SMS, Gale E, Skinner KM, Kazis L, Gale DR (2001) The association of bone marrow lesions with pain in knee osteoarthritis. *Ann Int Med* 134: 541–549
3. Burr DB (1998) Subchondral bone. In: K Brandt, M Doherty, S Lohmander (eds): *Textbook on osteoarthritis*. Oxford University Press, 144–156
4. Boyd SK, Matyas JR, Wohl GR, Kantzas A, Zernicke RF (2000) Early regional adaptation of periarticular bone mineral density after anterior cruciate ligament injury. *J Appl Physiol* 89: 2359–2364
5. Williams JA, Thonar EJ (1989) Early osteophyte formation after chemically induced articular cartilage injury. *Am J Sports Med* 17: 7–15
6. van den Berg WB (1999) The role of cytokines and growth factors in cartilage destruction in osteoarthritis and rheumatoid arthritis. *Z Rheumatol* 58: 136–141
7. Oegema TR Jr, Carpenter RJ, Hofmeister F, Thompson RC Jr (1997) The interaction of the zone of calcified cartilage and subchondral bone in osteoarthritis. *Micros Res & Tech* 37: 324–332
8. Pacifici M, Koyama E, Iwamoto M, Gentili C (2000) Development of articular cartilage: What do we know about it and how may it occur. *Conn Tiss Res* 41(3): 175–184
9. Takeda S, Bonnamy JP, Owen MJ, Ducy P, Karsenty G (2001) Continuous expression of Cbfa1 in nonhypertrophic chondrocytes uncovers its ability to induce hypertrophic chondrocyte differentiation and partially rescues Cbfa1-deficient mice. *Genes Dev* 15: 467–481
10. Leboy P, Grasso-Knight G, D'Angelo M, Volk SW, Lian JV, Drissi H, Stein GS, Adams SL (2001) Smad-Runx interactions during chondrocyte maturation. *J Bone Joint Surg* (Am) 83-A (S): S15–22
11. Terkeltaub R, Lotz M, Johnson K, Deng D, Hashimoto S, Goldring MB, Burton D, Deftos LJ (1998) Parathyroid hormone-related protein is abundant in osteoarthritic cartilage, and the parathyroid hormone-related protein 1–173 isoform is selectively induced by transforming growth factor beta in articular chondrocytes and suppresses generation of extracellular inorganic pyrophosphate. *Arthritis Rheum* 41: 2152–2164

12 Oegema Jr TR, Johnson SJ, Meglitsch T, Carpenter RJ (1996) Prostaglandins and the zone of calcified cartilage in osteoarthritis. *Am J Therap* 3: 139–149
13 Boos N, Nerlich AG, Wiest I, von der Mark K, Ganz R, Aebi M (1999) Immunohistochemical analysis of type X collagen expression in osteoarthritis of the hip joint. *J Orthop Res* 17: 495–502
14 Nanba Y, Nishida K, Yoshikawa T, Sato T, Inoue H, Kuboki Y (1997) Expression of osteonectin in articular cartilage of osteoarthritic knees. *Acta Med Okayama* 51: 239–243
15 Buckland-Wright C, Patel N (2000) Pattern of advancement in the zone of calcified cartilage detected in hand osteoarthritis. *Osteoarthritis Cartilage* 8(SA): S41–44
16 Kirsch T, Swoboda B, Nah H (2000) Activation of annexin II and V expression, terminal differentiation, mineralization and apoptosis in human osteoarthritic cartilage. *Osteoarthritis Cartilage* 8: 294–302
17 Horton WE Jr, Feng L, Adams C (1998) Chondrocyte apoptosis in development, aging and disease. *Matrix Biol* 17: 107–15
18 Karpouzas GA, Terkeltaub RA (1999) New developments in the pathogenesis of articular cartilage calcification. *Curr Rheumatol Rep* 1: 121–127
19 Oegema TR Jr, Deloria LB, Nelson T, Thompson RC Jr (2000) Surface cells may regulate the stability of the zone of calcified cartilage in mature articular cartilage *via* TGF-β. 46th Ann. Mtg. of the Orthop. Res. Soc., March 12–15, 2000
20 Fergeson CM, Schwarz EM, Reynolds PR, Puzas JE, Rosier RN, O'Keefe RJ (2000) Smad2 and 3 mediate transforming growth factor-beta 1-induced inhibition of chondrocyte maturation. *Endocrinol* 141: 4728–4735
21 Boyan BD, Sylvia VL, Dean DD, Pedrozo H, Del Toro F, Nemere I, Posner GH, Schwartz Z (1999) 1,25-(OH)2D3 modulates growth plate chondrocytes *via* membrane receptor-mediated protein kinase C by a mechanism that involves changes in phospholipid metabolism and the action of arachidonic acid and PGE2. *Steroids* 64: 129–136
22 Stephens M, Kwan AP, Bayliss MT, Archer CW (1992) Human articular surface chondrocytes initiate alkaline phosphatase and type X collagen synthesis in suspension culture. *J Cell Sci* 103: 1111–1116
23 Pateder DB, Rosier RN, Schwarz EM, Reynolds PR, Puzas JE, D'Souza M, O'Keefe RJ (2000) PTHrP expression in chondrocytes, regulation by TGF-beta, and interactions between epiphyseal and growth plate chondrocytes. *Exp Cell Res* 256: 555–562
24 Zerega B, Cermelli S, Bianco P, Cancedda R, Cancedda FD (1999) Parathyroid hormone [PTH(1-34)] and parathyroid hormone-related protein [PTHrP(1-34)] promote reversion of hypertrophic chondrocytes to a prehypertrophic proliferating phenotype and prevent terminal differentiation of osteoblast-like cells. *J Bone Miner Res* 14(8): 1281–1289

Dualistic role of TGFβ in osteoarthritis cartilage destruction and osteophyte formation

Wim B. van den Berg, Peter M. van der Kraan, Alwin Scharstuhl, Henk M. van Beuningen, Andrew Bakker, Peter L.E.M. van Lent, Fons A.J. van de Loo

Rheumatology Research and Advanced Therapeutics, University Medical Center Nijmegen, 189, Geert Grooteplein 26–28, 6500 HB, Nijmegen, The Netherlands

Introduction

Osteoarthritis is a condition of focal cartilage lesions, combined with alterations in the subchondral bone and new bone formation (osteophytes) at the joint margins.

OA joint pathology is often associated with initiating events of excessive tissue damage, leading to sustained, but unsuccessful attempts at repair. In the search for mediators involved in OA pathology, transforming growth factor-β (TGFβ) seems a likely candidate. This growth factor is activated under conditions of tissue damage, and high levels of active TGFβ have been found in OA synovial fluid. Early stages of experimental OA show enhanced chondrocyte proteoglycan synthetic activity, indicative for dominant involvement of anabolic growth factors.

TGFβ stimulation of cartilage proteoglycan synthesis

Chondrocytes in normal cartilage do not show enhanced proteoglycan synthesis on first stimulation with TGFβ. In contrast, osteoarthritic cartilage displays enhanced sensitivity to TGFβ stimulation, and a similar condition can be achieved after prolonged exposure of normal cartilage to TGFβ, *in vitro* and *in vivo* [1]. This shows two major points: TGFβ transforms normal chondrocytes to sensitive cells, probably linked to shifts in receptor expression or signaling [2]; and, the sensitivity of OA cartilage is indicative for previous TGFβ exposure in the OA process, suggesting that TGFβ is a pivotal mediator in OA.

On repeated injection of TGFβ in the knee joint of a normal mouse, cartilage proteoglycan synthesis is increased two-fold [1]. TGFβ mainly is present in a latent form, and addition of the active form has a major impact on local levels. Intraarticular injection of BMP-2 (bone morphogenetic protein-2) causes a great and immediate rise in chondrocyte proteoglycan synthesis [3]. Opposite to the prolonged stimulation seen after TGFβ exposure, the BMP-induced stimulation is short-lasting. IGF-1 (Insulin-like growth factor-1) injection induced only marginal stimulation, probably due to considerable levels of endogenous IGF-1.

TGFβ-induced osteophyte formation and its comparison with BMP-2

Intraarticular injection of TGFβ induced cartilagenous outgrowth and enhanced chondrophyte formation [1, 4]. This was mainly noted as transformation and outgrowth from periosteal lining. In contrast, BMP-2 also induces chondrophytes, but the location was strikingly different. This growth factor seems to have major affinity for chondrocytes at the edges of the growth plates, and sprouting of new cartilage was almost selectively seen at those sites [3]. TGFβ induced chondrophytes, which later on mature into real osteophytes, resemble the pattern of osteophyte formation seen in experimental and spontaneous murine OA.

It does not exclude that other growth factors, including BMP-2 are second mediators in TGFβ induced osteophytes. Immunostaining of BMP-2 is highly upregulated at periosteal sites in experimental OA. It probably implies that TGFβ has to induce the transformation, but once cartilagenous cells are generated, BMP-2 may have a role in additional stimulation and maturation.

Osteophyte formation and cartilage lesions after excessive TGFβ exposure

Two months after triple TGFβ injections in the knee joint, focal proteoglycan depletion was found in deep layers of the tibial plateau [5]. Eventually, lesions occurred at the level of the tide mark, exactly the site where cartilage is torn off in experimental and spontaneous murine OA. In addition, the formation of mature osteophytes at the joint margins was obvious.

To mimic more closely the sustained, enhanced levels of TGFβ in an OA joint, TGFβ was applied in the knee joint by local gene transfer, using an adenoviral vector [6]. Findings were roughly similar as compared to observations after repeated injections.

Role of synovial lining macrophages in osteophyte formation

Since synovial lining macrophages are a major source of cytokines and growth factors expressed in the synovial tissue, we examined their contribution to the process of osteophyte formation. Lining cells can be selectively depleted by a single injection of chlodronate liposomes into the knee joint. When such pretreatment was done, osteophyte formation was markedly reduced after both repeated TGFβ injection or adenoviral TGFβ overexpression. This suggests that as yet unidentified factors from these lining macrophages are generated after TGFβ exposure and are involved in

TGFβ induced periosteal activation, chondrophyte growth and/or osteophyte maturation. Preliminary *in vitro* studies with macrophages and chondrogenic precursor cells confirmed such a pathway of TGFβ induced macrophage derived factors, other than TGFβ, in cartilagenous outgrowth.

TGFβ blocking reduces osteophytes

As a final proof of principle we have applied a TGFβ scavenger in a model of papain induced OA in the mouse. A low molecular weight (25 kDa), truncated form of the TGFβ type II receptor was cloned, expressed and supplied in Alzet minipumps, prior to induction of OA by local injection of papain. Intriguingly, blocking of endogenous TGFβ with the scavenging soluble receptor made it clear that chondrophyte formation, as a feature of osteoarthritic processes in the joint, is a TGFβ dependent phenomenon (Fig. 1).

Role of endogenous TGFβ in cartilage repair/pathology

As mentioned above, a truncated form of the soluble type II TGFβ receptor was used, which has access to the articular cartilage and may block TGFβ both in synovial tissue as well as in the cartilage itself. After soluble receptor treatment, proteoglycan depletion in the papain model was enhanced at day 7 at all cartilage layers. In addition, significant extension of cartilage damage was seen in the tibial plateau at day 14. This is compatible with a protective role of TGFβ in attempted cartilage repair in this model. Further studies are ongoing in mild and severe forms of this and other murine OA models, to substantiate the role of TGFβ in OA cartilage pathology.

Given the fact that the soluble receptor used is potent in neutralizing TGFβ1 and 3, but does not block TGFβ2 adequately, it cannot be excluded at present that the role of endogenous TGFβ was underestimated. On the other hand, preliminary immunostaining studies of OA joints identified mainly TGFβ1 and TGFβ3, and scant TGFβ2 expression.

TGFβ as a pathogenic or repair factor in OA

The above studies of TGFβ blocking make it clear that TGFβ is crucial in osteophyte formation, in line with the capacity of TGFβ to induce this phenomenon. The role of TGFβ in cartilage pathology or repair is less clear.

There is no doubt that TGFβ has two-sided effects, a pathologic and a protective one. The down side is identified in other organ disorders such as kidney and liver

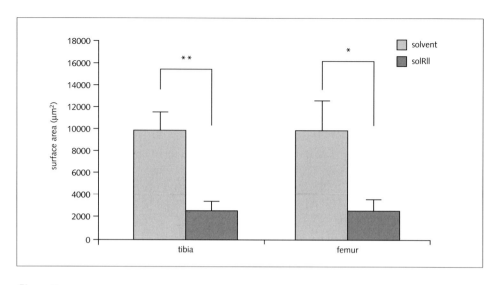

Figure 1
Suppression of chondrophyte formation in papain induced OA after treatment with TGFβ type II soluble receptor.

fibrosis and skin scarring, and this process was called the dark side of excessive tissue repair [7]. Apart from upregulation of enzyme inhibitors, which is considered as a protective effect, TGFβ stimulation of MMP-13 release from activated chondrocytes may underline its destructive or remodeling potential [8].

When joint injury occurs as an initiating early stage of OA, TGFβ will be activated and has a function in repair. This role is illustrated in the protective effect of added TGFβ in cartilage proteoglycan loss in mild forms of joint inflammation [9]. Enhanced TGFβ activity in cartilage of early stages of OA may mainly reflect attempted repair, with ultimate failure and collapse of the tissue in late OA.

When high concentrations of TGFβ are generated, under conditions of excessive damage and repair, pathogenic elements probably dominate, with excessive protease activity and induction of direct cartilage lesions. High levels of TGFβ also initiate pronounced cartilage formation in ligament structures, with concomitant stiffening of the joint. Occurrence of deep cartilage lesions in the tibial plateau may then be linked to inproper load bearing, apart from direct TGFβ mediated induction of protease activity in chondrocytes [5].

A further illustration that either too much or too little TGFβ signaling may cause pathologic changes, is provided by the occurrence of spontaneous OA in a transgenic mouse, carrying a truncated, nonfunctional TGFβ receptor [10]. Remarkably, these mice do develop osteophytes in the absence of TGFβ signaling, suggestive of redundancy under knockout conditions.

References

1 Van Beuningen HM, van der Kraan PM, Arntz OJ, van den Berg WB (1994) Transforming growth factor-β1 stimulates articular chondrocyte proteoglycan synthesis and induces osteophyte formation in the murine knee joint. *Lab Invest* 71: 279–290

2 Glansbeek HL, van der Kraan PM, Vitters EL, van den Berg WB (1993) Correlation of the size of type II transforming growth factor β (TGF-β) receptor with TGF-β responses of isolated bovine articular chondrocytes. *Ann Rheum Dis* 52: 812–816

3 Van Beuningen HM, Glansbeek HL, van der Kraan PM, van den Berg WB (1998) Differential effects of local application of BMP-2 or TGF-β1 on both articular cartilage composition and osteophyte formation. *Osteoarthritis Cartilage* 6: 306–317

4 Van Beuningen HM, Glansbeek HL, van der Kraan PM, van den Berg WB (1998) Differential effects of local application of BMP-2 or TGF-β1 on both articular cartilage composition and osteophyte formation. *Osteoarthritis Cartilage* 6: 306–317

5 Van Beuningen HM, Glansbeek HL, van der Kraan PM, van den Berg WB (2000) Osteoarthritis-like changes in the murine knee joint resulting from intra-articular TGFβ injections. *Osteoarthritis Cartilage* 8: 25–33

6 Bakker AC, van de Loo FAJ, van Beuningen HM, Sime P, van Lent PLEM, van der Kraan PM, Richards CD, van den Berg WB (2001) Overexpression of active TGFβ in the murine knee joint: evidence for synovial layer dependent chondro-osteophyte formation. *Osteoarthritis Cartilage* 9: 128–136

7 Border WA, Ruoslati E (1992) TGF-β in disease: The dark side of tissue repair. *J Clin Invest* 90: 1–7

8 Moldovan F, Pelletier JP, Hambor J, Cloutier JM, Martel-Pelletier J (1997) Collagenase-3 (MMP-13) is preferentially localized in the deep layer of human arthritic cartilage *in situ*. *In vitro* mimicking effect by TGFβ. *Arthritis Rheum* 40: 1653–1661

9 Glansbeek HL, van Beuningen HM, Vitters EL, van der Kraan PM, van den Berg WB (1998) Stimulation of articular cartilage repair in established arthritis by local administration of TGFβ into murine knee joints. *Lab Invest* 78: 133–142

10 Serra R, Johnson M, Filvaroff EH, La Borde J, Sheehan DM, Derynck R, Moses HL (1997) Expression of a truncated kinase-defcetive TGFβ tpe II receptor in mouse skeletal tissue promotes terminal chondrocyte differentiation and osteoarthritis. *J Cell Biol* 139: 541–552

Discussion

Metalloproteinases and osteoarthritis

Q: *Linda Sandell*: Does gelatin zymography permit you to detect different MMPs in OA cartilage?

A: *Fred Woessner*: You should be able to detect different MMPs, but the loading is critical, and it is very hard to detect collagenases. It requires 25 times more protein to see collagenases than it does to see gelatinases.

Q: *Linda Sandell*: What's your view, then, on the relative activities of different active MMPs in OA, MMP-7 versus MMP-13 for instance?

A: *Fred Woessner*: We haven't looked at MMP-7. I am guessing that it is probably mostly MMP-13. Everybody likes to use Robin Poole's method, which tells you there is a lot of collagenase, but it doesn't tell you which one it is. There is certainly not much gelatinase there. Jim Quigley reported that gelatinase A (MMP-2) can digest collagen. But others disagree. I don't think there is enough to do a lot of damage to collagen. So we're looking more at the involvement of MMP-1 and MMP-13. Of course Klaus's group has shown that there is some MMP-8. So these collagenases, gelatinase A and stromelysin (MMP-3), in my opinion, are the major players. But the last half of the list of 25 has not been studied in cartilage.

Q: *Dick Heinegard*: When proteins bind to one another, they often change conformation. So what about MMPs? For example, some MMPs bind to heparin. What would be the chance that this could activate MMPs?

A: *Fred Woessner*: Good question. My student has a marvelous theory about that but no supporting data. Pro MMP-7 has something like 14 more arginines and lysines on its surface compared to active MMP-7. They almost completely surround the molecule. So it's attractive to think that maybe the heparin chains sort of wrap around the molecule. The force of the chains pulling on it might make it more susceptible to activation. Also we have a recent paper in Analytical Biochemistry that shows that heparin enhances the activity of these proteases in the zymogram. That, too, may have some connection to your idea about heparin changing the conformation and improving the activity of the protein. At this time it's only speculation, however.

Q: *Ada Cole*: Klaus' group has shown MMP-1, -8, -13 and -14 in normal ankle cartilages. There are about equal amounts of inactive and active forms, which really surprised us.

A: *Fred Woessner*: Okay, we will put you in the next edition of some book or other.

Age-dependent changes in some physico-chemical properties of human articular cartilage

Q: *Robin Poole*: In view of your comments, do you think collagen or proteoglycans should be a particular target for arrest of degradation in arthritis?

A: *Alice Maroudas*: I believe that collagen is the prime target. Clearly, if you have no proteoglycans, you lose much of the function related to swelling of these molecules. However, if the collagen is defective, it doesn't matter how much proteoglycan you produce – it won't be of any use.

Q: *Richard Loeser*: Of the age-related changes, what proportion of these are cell mediated, and what proportion would occur without the cells even being there? Do agents released by the cells, like reactive oxygen species play a role in any of this?

A: *Alice Maroudas*: The changes are small. As far as the increase in keratan sulfate is concerned, this is due to the increase in the smaller degraded and KS-rich aggrecan hyaluronan-binding region. For the reduction of water content, it's the stiffness of the collagen network – the stiffer the network the more it will push on the tissue and reduce water content. So one doesn't need to invoke the cell in these particular processes.

Q: *Robert Sah*: Could you clarify the aging studies; were these all of the femoral head?

A: *Alice Maroudas*: Yes. However, I should have mentioned that in the knee, if we used the same methodology and the same student, we got the opposite result. Here, I believe that stiffness slightly decreases with age. Is that right, Bob?

Q: *Bob Sah*: I think so. Do your results relate to the tensile studies of Geoffrey Kempson and Jill Urban of femoral head cartilages? Their results suggested little change in aging in the tensile network in the same joints and not much change in the tensile properties.

A: *Alice Maroudas*: Kempson found a decrease in the tensile properties with age. We are not measuring the tensile properties directly. We have used, in addition to the osmotic compression method, instantaneous deformation as well – where you actually squeeze cartilage like a pancake and look at the increase in the surface area. This is more or less related to both shear and tensile properties. Kempson found a decrease in tensile properties with age. We find an increase. In studies of instantaneous deformation as a function of three different ages, the oldest specimen is definitely the stiffest. It would be fair to say that the changes with aging are small.

Q: *Chris Handley*: Can you comment on the approaches we should be thinking about in terms of looking at the involvement of noncollagenous proteins in the micro and macro biomechanics of cartilage? Are we going to have to rely on knock-out mouse experiments for this? Or can we do some direct measurements?

A: *Alice Maroudas*: I think that the proteins play a part in linking the collagen and making up the collagen network. So one would have to have really selective enzymes that attack, shall we say, the collagen but not the proteoglycans. It is very difficult. We have had a go at it, and we have given up at the moment. The knockout animal is an option of course, but then you have very little cartilage to work with.

Q: *Thomas Andriacchi*: The difference between your tensile properties with the aged specimen and the OA specimen, suggest that some disruptive event takes place in OA, and from a mechanical viewpoint we think of failure or fracture. So it leads us back to looking at the fracture properties of the collagen. Stiffening tends to be linked to the brittleness of a material. Perhaps the collagen network is more predisposed to fracture in aged specimens. Have you looked at that?

A: *Alice Maroudas*: We have a fatigue apparatus standing on our bench, but nobody at the moment to use it. It is possible that with aging and with biochemical and mechanical insults, the cartilage becomes stiffer and thereby more resistant to compression. It may also become more brittle. I believe that the best test will be a good controlled fatigue testing protocol.

The role of the bone/cartilage interface in osteoarthritis

Q: Dick Heinegard: In the calcifying cartilage, is there a difference in pyrophosphate levels?

A: *Ted Oegema*: In normal calcified cartilage only hydroxyapatite crystals are present. In different disease states, pyrophosphate crystals start to form, but they don't form in the calcifying zone. They form around the cells just above the zone.

Q: *Ernst Hunziker*: Mineralization is a cell-controlled process. How does the tidemark move up? Is it a process where chondrocytes in the deep zone continuously suppress mineralization whereas under pathological conditions, when the tidemark moves, these chondrocytes are activated and start to induce mineralization?

A: *Ted Oegema*: The regulated movement of the tidemark might occur prior to when we see the pathology. So far we have noted that it can move under normal circumstances when you change the mechanical loading –even in our arthritic model. Chondrocytes normally produce a lot of pyrophosphate, which inhibits mineralization, but angiogenesis is absent. I doubt that these nice tidemark progressions have anything to do with the disease process. I think they happened earlier.

Q: *David Howell*: Scintigraphic studies by Dieppe and coworkers of patients with early finger OA have revealed that enhanced bone turnover is correlated with the subsequent development of OA. I wonder about these studies since some patients had finger OA adjacent to fingers with no OA. It looks as if some trauma may have occurred. Could trauma initiate the advancement of the tidemarks?

A: *Ted Oegema*: Tidemarks move without any trauma. But in OA patients it gets very disregulated.

Dualistic role of TGFβ in OA cartilage destruction and osteophyte formation

Q: *Srinivasan Chandrasekhar*: There is a knockout mouse of the type II receptor. They also develop osteophyte formation. Is TGFβ involved here?

A: *Wim van den Berg*: The OA pathology in these mice reflect the protective role of TGFβ, implying that in the absence of TGFβ signaling cartilage homeostasis is disturbed. The fact that these animals still develop osteophytes suggests that other growth factors substitute under these knockout conditions. Our data indicate that TGFβ is a dominant growth factor in osteophyte formation in normal mice.

Q: *Robin Poole*: Regarding the synovial cell depletion can you be absolutely sure that you know this is truly selective for synovial cells?

A: *Wim van den Berg*: Using labeled liposomes we have no evidence that the liposomes reach the periosteal cells in significant quantities. Moreover, we do not see any cell loss at the periosteal layer.

From biomarkers to surrogate outcome measures in osteoarthritis

Introduction

Stefan Lohmander

Department of Orthopedics, University Hospital, 22185 Lund, Sweden

Identification of individuals at risk for disease progression and the monitoring of outcome in clinical trials and treatment of osteoarthritis (OA) presents a challenge. Lack of readily usable methods for these purposes remains a great impediment for progress in the development of effective treatment of joint diseases. The chosen patient group, intervention target and outcome measure provide an envelope for clinical trial design. The relationship between outcome measures that monitor cartilage, joint and patient in OA urgently needs to be clarified in order to define surrogate outcome measures for use in studies of OA natural history or treatment.

Many interventions for OA now being discussed concentrate on agents that inhibit proteolytic degradation of cartilage matrix, or stimulate its regeneration. A considerable literature exists on molecular markers in human arthritis and OA. It supports a relationship between markers in joint fluid, serum or urine, and joint cartilage turnover, and provides face validity for these markers to monitor dynamic changes. Other aspects of validity are less well supported, but a few examples can be given to suggest that markers may indeed become useful.

Increased serum concentrations of hyaluronan, CRP and COMP appear to predict future OA progression. Such information could be used to select high risk individuals in early trials. Information on within- and between-patient variability for select molecular markers in joint fluid, serum and urine is available in OA cohorts. These data suggest (a) that variability differs between different markers in the same fluid compartment, (b) variability is lower within than between patients, and (c) that markers are responsive to change. Using variability data for one joint fluid marker, it was calculated that some 30 patients per treatment arm would be needed to show a change by treatment of 0.5 SD with 80% power. Such information is needed for other patient cohorts, biomarkers and body fluid compartments, to facilitate the planning of clinical trials. Results of other investigations provide new evidence for the molecular mechanisms and dynamics of cartilage matrix turnover, in particular in early-stage OA. These and other examples serve to show that research on molecular markers to support selection and monitoring of patients in early stage trials of joint disease and repair is now poised to yield useful results. However, the final answer to the question of the utility of these surrogate measures will have to await the availability of interventions that can change the disease progression. The greatest challenge will not be to identify new targets or tentative biomarkers at the "discovery level", but to validate them by the use of well characterized patient cohorts.

This session of the workshop "The Many Faces of Osteoarthritis" brings together three contributions on biomarkers in OA with two contributions that provide evidence that OA indeed has many faces: new technologies are being used to explore expression and regulation of known and unknown genes in OA. Finally, a close examination of human OA tissues provides additional evidence that OA is not always OA is not always OA. Indeed, this disease formerly regarded as monotonously degenerative is now beginning to show considerable and exciting variety on both the genotypic and phenotypic levels.

Collagen cross-links as markers of bone and cartilage degradation

David R. Eyre, Lynne M. Atley and Jiann-Jiu Wu

Department of Orthopædics and Sports Medicine, Orthopædic Research Laboratories, University of Washington, Box 356500, Seattle, WA 98195-6500, USA

Introduction

An early study showed that desmosine and isodesmosine from degraded elastin are not metabolized, but are quantitatively excreted in urine [1]. The structurally related pyridinoline cross-links of collagen were later also found in urine [2]. Robins and colleagues introduced methods for measuring urinary pyridinoline and deoxypyridinoline residues as markers of systemic collagen degradation, and in particular of bone resorption [3]. Finding that most of the urinary pyridinolines were in a narrow size-range of small peptides, we identified the amino acid sequences containing the cross-linking residues for immunoassay on the premise that this would provide greater specificity to their originating tissues and collagen types [4]. First, we targeted cross-linked N-telopeptide fragments of type I collagen, most of which must have originated from bone based on their HP:LP ratio [5]. Commercial versions of this assay (NTx) for urine and serum have seen widespread use in clinical studies, for example of anti-osteoporotic drugs [6, 7]. Immunoassays for the cross-linked C-telopeptides of type I collagen (CTx), also identified in urine [8], have been introduced as comparable bone resorption markers [9].

Bone resorption

These various products (free and total pyridinoline and deoxypyridinoline, cross-linked N- and C-telopeptides of type I collagen) might be expected to respond in unison in clinical studies if all follow the same pathway and originate mostly from resorbed bone. While true in general, in practice consistent differences emerge indicating that their metabolic origins and fates are not entirely the same. For example, in monitoring patients in whom bone resorption is inhibited by a potent bisphosphonate, the telopeptide markers in urine show the greatest decreases. Total deoxypyridinoline comes next followed by total pyridinoline. But the free pools of deoxypyridinoline and pyridinoline in urine can show little or no response. These and other consistent differences can be explained if tissues other than bone contribute and the fate of bone collagen fragments at the level of the osteoclast and subsequent liver and kidney metabolism are taken into account. The antigen for the NTx assay is the eight-residue amino terminus of the α2(I) collagen chain, JYDGXGVG (where J is pyroglutamic acid and X is the cross-link), linked to a second telopeptide fragment (from α1(I), or a second α2(I)). The antibody, 1H11, that forms the basis of the original NTx assay, requires the C-terminal glycine to be free, i.e., a proteolytic neoepitope. The key protease that makes this cleavage (at a G/L bond) is cathepsin K. Cathepsin K is restricted to osteoclasts and responsible for the cell's ability to degrade bone collagen after mineral removal in the resorption lacuna. Studies *in vitro* showed that recombinant human cathepsin K was able to solubilize demineralized bone collagen and release the cross-linked N-telopeptide domain as immunoreactive NTx with faster kinetics than even bacterial collagenase [10], which can also generate the neoepitope.

This finding explains the observation that osteoclasts resorbing bone particles or dentin slices in culture rapidly release immunoreactive NTx into the culture medium in a molar yield equivalent to the amount of collagen solubilized [11]. Free pyridinolines were not generated by osteoclast action, only the peptide-bound cross-links, suggesting that peptide-bound pyridinolines are degraded to the free cross-links in the liver and kidneys. This begins to explain why the various cross-link assays do not respond in unison clinically.

Particularly instructive is the behavior of these "resorption" markers in pycnodysostosis, a rare, recessive genetic disorder in which the gene for cathepsin K is effectively null because of mutations in affected homozygous or compound heterozygous patients [12]. Very little of the telopeptide markers (NTx and CTx) is excreted in the urine of pycnodysostosis patients [13]. In contrast, another C-telopeptide marker, ICTP, in which the antibody recognizes an epitope located in a proximally longer C-telopeptide sequence measurable in serum, not urine, was markedly increased. Finally, total pyridinolines, measured in urine after acid hydrolysis, were in the normal range, and their ratio was typical of bone collagen. These seemingly paradoxical findings can be explained if one takes into account the likely proteolytic origins and metabolic fates of the collagen degradation products. Pathways from mineralized bone to urine can be proposed that explain the findings in normal physiology and in pycnodysostosis. Normally, demineralized bone matrix is degraded to peptides by cathepsin K secreted by the osteoclast into the resorption lacuna. All or part of this digest enters the osteoclast, which subjects the protein fragments to further lysosomal degradation before their release to the circulation. The cross-linked telopeptide fragments produced from bone matrix by cathepsin K *in vitro* [10] are small and similar in molecular size to those found in urine (2 kDa). We speculate, therefore, that short cross-linked telopeptides with little or no triple-helical sequence remaining, escape metabolism in the liver and are excreted by the kidneys after reabsorption and removal of any amino acids that can be readily scavenged (the urinary NTx peptides lack a few residues compared with the cathepsin K

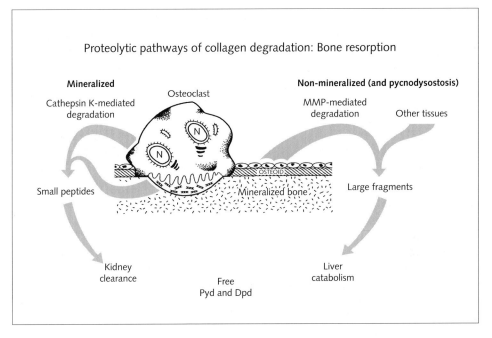

Figure 1
Metabolic origins of free and telopeptide-bound collagen cross-links in urine. Different catabolic pathways for collagen fragments from cathepsin K-initiated versus MMP-initiated degradation mechanisms are proposed.

digestion products). In pycnodysostosis, mineral is effectively removed by the osteoclast but, without cathepsin K, the demineralized collagen is inefficiently degraded. The high ICTP level in serum can be explained if larger collagen fragments are released to the blood, perhaps through an alternative pathway of proteolysis subsequent to osteoclast demineralization. In the cathepsin K-null mouse, demineralized collagen is seen to persist in the resorption pits by electronmicroscopy [14]. Cathepsin K cleaves within and distal to the ICTP epitope and so can be expected to destroy it, explaining why serum ICTP is a poor bone resorption marker. The normal levels of pyridinoline and deoxypyridinoline in pycnodysostosis urines can be explained if the cross-linked domains of degraded bone collagen are in a form (i.e., larger fragments) that are picked up by the liver and subjected to a more thorough catabolism than can occur by the action of osteoclasts alone, so generating free pyridinoline and deoxypyridinoline, which the kidneys then excrete.

The basic concepts are shown in Figure 1. Though pycnodysostosis is an uncommon and extreme form of disturbed bone collagen catabolism, it illustrates how in more general physiology, the different cross-link forms may behave disparately.

Even the products of the two telopeptide sites (N-telo and C-telo) show diversity. For example, in Paget's disease of bone, NTx follows what can be a massive increase in bone resorption (trackable even by urinary hydroxyproline measurements), but the CTx response is much lower in comparison, particularly of the AspGly β-isomer [15]. Serum CTx also can show very large circadian increases at night that are related to fasting [16]. This implies a secondary effect on the generation of the CTx antigen (which also is a proteolytic neoepitope), perhaps at the level of the osteoclast or the liver. Cathepsin K alone is a poor generator of the CTx epitope (a free C-terminal arginine), which may be a basis for explaining the low values for CTx in Paget's disease and increases on fasting compared with NTx.

Cartilage collagen cross-links

As an extension of this concept, we have pursued comparable peptide fragments containing cross-linking residues from other collagen types. In particular, type II collagen is an attractive candidate marker of cartilage degradation. The goal would be to develop an assay capable of tracking collagen breakdown from articular cartilage to joint fluid to blood and perhaps urine comparable to the bone telopeptide marker.

First, however, it is important to understand the polymeric form of cartilage collagen. In adult articular cartilage the fibrils are covalently interlinked heteropolymers of collagens II, III, IX and XI. Collagen II molecules account for > 90% of this. Pyridinoline (HP) is the predominant cross-linking residue (> 95% HP, < 5% deoxypyridinoline (LP)) in type II collagen. Collagen IX is linked to collagen II fibrils by pyridinolines (lower HP:LP ratio than in collagen II) and borohydride-reducible divalent cross-links [17, 18]. Collagen XI molecules, though also copolymerized in the same fibrils as collagens II and IX [19], are cross-linked to each other in a head-to-tail fashion mostly by borohydride-reducible divalent cross-links [20].

Figure 2 proposes a best-fit model for how collagen IX molecules interact with the surface of the thin type II collagen fibrils forming the meshwork of the basket (chondron) around chondrocytes. In this model [21], the folded back COL1 domain accommodates the collagen IX-to-IX covalent bonds and all the sites of collagen IX-to-II bonds that have been identified ([18, 22] and unpublished observations). In growth cartilage, all fibrils are thin (< 20 nm) and coated with collagen IX molecules, whereas in adult articular cartilage, collagen IX is concentrated close to the cells in similarly thin fibrils, with little on the thicker, banded fibrils of the interterritorial matrix. In considering how the chondrocyte assembles and maintains this pericellular basket of thin collagen fibrils and orchestrates their maturation into thicker fibrils farther from the cell, selective proteolysis that results in removal of surface collagen IX molecules, lateral fusion of thin fibrils and addition of new collagen monomers to existing fibrils all seem likely processes.

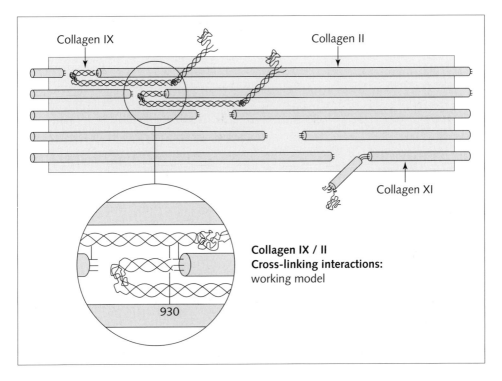

Figure 2
Proposed mode of interaction of collagen IX molecules with the surface of the collagen II fibril that accounts for all the identified IX-to-II and IX-to-IX intermolecular cross-linking sites.

These questions and concepts are clearly central to an understanding of the changes that occur within the collagen framework in osteoarthritis. Both the early swelling of the collagen network and the later overt fibrillation and degradation are believed to be caused in part by the action of proteases produced by the chondrocytes in response to molecular and mechanical signals. If clinically useful degradation markers of cartilage collagen are to be developed, the pathways of breakdown within the tissue, both for normal maintenance and in pathology, need to be defined. The lessons from bone resorption here are clear.

Degradation pathways of articular cartilage collagen

Current dogma holds that fibrillar collagens (types I, II and III) are degraded by collagenases (MMP1, 8 and/or 13), which initiate the process by cleaving at the 3/4-length locus in the triple-helical domains. Further cleavages then occur by the same

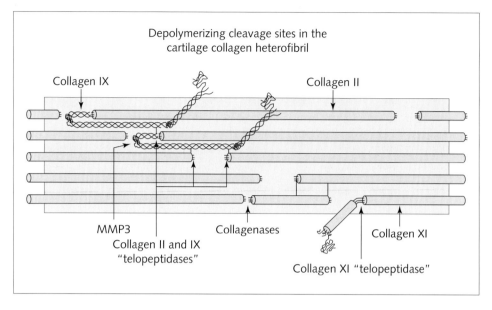

Figure 3
Speculative mechanisms of collagen heterofibril remodeling and depolymerization by "telopeptidases". A regulatory role for telopeptidase cleavages is proposed.

enzymes or other MMP family members after the 3/4- and 1/4-length fragments unwind (denature) at body temperature. Such a concept rests largely on experimental evidence from the degradation of soluble collagen monomers *in vitro*. Cross-linked fibrils are degraded slowly and incompletely by collagenases alone [23]. We propose a concept that cleavages in the telopeptide domains near the cross-links are the initial, controlling step for fibril degradation and remodeling. Collagenases are brought into action in situations requiring rapid collagen degradation and the removal of a high flux of released collagen monomers. The findings of Krane and colleagues with the mouse engineered to lack the collagenase cleavage site in the α1(I) chain are consistent with this [24]. No phenotype was apparent during development; only later did thickened skin and post-partum fibrotic uteri emerge. Presumably in animals lacking the collagenase-cleavable 3/4-length locus in type I collagen, most tissues could manage the collagen remodeling necessary for growth.

Among the wide range of MMP, ADAM and ADAMTS proteases potentially involved in matrix and matricellular protein processing, the candidacy for collagen telopeptidases is wide open. The native substrates for most of these zinc proteases, except perhaps for the collagenases, is essentially unknown. Molecules cleaved in the test-tube may not be the target in tissues. In studying how chondrocytes manage the growth and remodeling of their extracellular collagen, it will be important to study cross-linked, polymeric substrates as well as isolated monomers in seeking the responsible proteases. Some years ago we showed that MMP3 (stromelysin-1) cleaved at two sites in the N-telopeptide of type II collagen between the cross-link and the triple-helix [25], i.e., MMP3 was a telopeptidase theoretically capable of partial depolymerization of collagen II fibrils. MMP3 also rapidly cleaved through all three chains of collagen IX at a single locus in the NC2 domain and removed NC4 from the amino terminus of $\alpha1$(IX). Whether these cleavages have physiological relevance in the biology of cartilage or the pathogenesis of osteoarthritis is unknown.

We speculate that the chondrocyte has proteases in its arsenal that have evolved to cleave within the telopeptide domains of collagens II and XI and in the NC domains of collagen IX (Fig. 3). Such activities can selectively remove collagen IX molecules from fibril surfaces, collagen XI N-propeptide extensions and allow a controlled depolymerization of the heteromeric fibrils. In this way, the cells are able to remodel the fine meshwork of collagen II/XI copolymeric fibrils coated with collagen IX molecules and control their growth into the meshwork of thicker, more typical collagen fibrils of mature cartilage that contain less collagen XI and little or no collagen IX farther from the cell. Whether additional proteases cause the collagen network swelling seen early in joint-injury models of osteoarthritis, and the overt collagen disruption and degradation seen at later stages of osteoarthritis, remains to be seen.

Markers of cartilage collagen degradation

Developing and validating markers of cartilage collagen breakdown will not be easy given the complexities in structure and poorly understood catabolic pathways. Immunoassays for cartilage collagen fragments, with the potential for monitoring joint disease, have mostly targeted type II collagen. Antibodies that recognize denatured triple-helical epitopes [26], collagenase-cleavage products from the 3/4-length locus [27, 28] and telopeptide domain epitopes [29] have been described. A collagen II degradation marker is attractive. It is the most abundant protein in the tissue; normally it turns over extremely slowly and its overt degradation is thought to mark a critical, perhaps irreversible, stage in joint destruction. The barriers include low concentrations in body fluids requiring sensitive assays, the potential that non-joint cartilage sources will dominate and complicate interpretation and that systemic levels (blood, urine) may not inform on single-joint disease activity.

In our efforts, we focused on pyridinoline-containing fragments of type II collagen, the logic being that such peptides might survive proteolysis further downstream from the originating tissue better than simple linear peptide fragments.

Based on our finding in the urine of growing children of a hexapeptide sequence EXGPDP in a prominent pyridinoline-containing peptide peak [30], we targeted this sequence for monoclonal antibody production. The mAb, 2B4, was selected because it recognized this sequence with high affinity when X was the cross-linked form of lysine and only when cleaved after the C-terminal proline and not in more extended sequences. Further studies with a range of recombinant MMPs showed that MMP7 (matrilysin) was the most active in cleaving the P-L bond that released this neoepitope from human cartilage collagen. MMP3 and MMP13 were also active but in lower yields. In inhibition ELISA format with 2B4, much higher levels of the peptide were present in children's than in adult urines. On average, patients with OA gave about two-fold higher levels than adult controls, and patients with multiple joints affected gave higher values than those with only a single knee [19, 31]. Measurements of the same epitope in synovial fluid showed higher levels in operated joints of dogs and rabbits in models of surgically induced OA and in human knees after injury (unpublished observations). The highest levels are within 2 weeks of injury suggesting that collagen breakdown is an early event. What form this breakdown takes and the identity of the enzyme(s) responsible for the epitope generation *in vivo* are not known.

We are currently targeting for immunoassay other peptide fragments of collagen II telopeptides with the goal of defining the primary depolymerizing cleavage sites in the tissue and the subsequent metabolic events going from the tissue to joint space and through the lymphatic system to blood and urine. In this way an assay that has both face validity and a sound molecular basis might be achievable.

Acknowledgements

Work described in the authors' laboratory was supported in part by the National Institute of Arthritis, Musculoskeletal and Skin Diseases, the Burgess Chair Endowment of the University of Washington, and Ostex International.

References

1. Starcher BC, Goldstein RA (1979) Studies on the absorption of desmosine and isodesmosine. *J Lab Clin Med* 94(6): 848–852
2. Gunja-Smith Z, Boucek RJ (1981) Collagen cross-linking compounds in human urine. *Biochem J* 197(3): 759–762
3. Robins SP (1982) An enzyme-linked immunoassay for the collagen cross-link pyridinoline. *Biochem J* 207(3): 617–620
4. Eyre DR, Ericsson LH, Simon LS, Krane SM (1988) Identification of urinary peptides derived from cross-linking sites in bone collagen in Paget's disease. *J Bone Miner Res* 3(Suppl. 1): S210

5   Hanson DA, Weis MA, Bollen AM, Maslan SL, Singer FR, Eyre DR (1992) A specific immunoassay for monitoring human bone resorption: Quantitation of type I collagen cross-linked N-telopeptides in urine. *J Bone Miner Res* 7(11): 1251–1258
6   Chesnut CH III, Bell NH, Clark GS, Drinkwater BL, English SC, Johnson CC, Jr., Notelovitz M, Rosen C, Cain DF, Flessland, KA, Mallinak NJ (1997) Hormone replacement therapy in postmenopausal women: urinary N-telopeptide of type I collagen monitors therapeutic effect and predicts response of bone mineral density. *Am J Med* 102(1): 29–37
7   Gertz BJ, Clemens JD, Holland SD, Yuan W, Greenspan S (1998) Application of a new serum assay for type I collagen cross-linked N-telopeptides: Assessment of diurnal changes in bone turnover with and without alendronate treatment. *Calcif Tissue Int* 63(2): 102–106
8   Bonde M, Qvist P, Fledelius C, Riis BJ, Christiansen C (1994) Immunoassay for quantifying type I collagen degradation products in urine evaluated. *Clin Chem* 40: 2022–2025
9   Greenspan SL, Rosen HN, Parker RA (2000) Early changes in serum N-telopeptide and C-telopeptide cross-linked collagen type 1 predict long-term response to alendronate therapy in elderly women. *J Clin Endocrinol Metab* 85(10): 3537–3540
10 Atley LM, Mort JS, Lalumiere M, Eyre DR (2000) Proteolysis of human bone collagen by cathepsin K: Characterization of the cleavage sites generating the cross-linked N-telopeptide neoepitope. *Bone* 26(3): 241–247
11 Apone S, Lee MY, Eyre DR (1997) Osteoclasts generate cross-linked collagen N-telopeptides (NTx) but not free pyridinolines when cultured on human bone. *Bone* 21(2): 129–136
12 Gelb BD, Shi GP, Chapman HA, Desnick RJ (1996) Pycnodysostosis, a lysosomal disease caused by cathepsin K deficiency. *Science* 273(5279): 1236–1238
13 Nishi Y, Atley L, Eyre DR, Edelson JG, Superti-Gurga A, Yasuda T, Desnick RJ, Gelb BD (1999) Determination of bone markers in pycnodysostosis: Effects of cathepsin K deficiency on bone matrix degradation. *J Bone Miner Res* 14(11): 1902–1908
14 Saftig P, Hunziker E, Wehmeyer O, Jones S, Boyde A, Rommerskirch W, Moritz J, Schu P, von Figura K (1998) Impaired osteoclastic bone resorption leads to osteopetrosis in cathepsin-K-deficient mice. *Proc Natl Acad Sci USA* 95:13453–13458
15 Delmas PD (1999) Biochemical markers of bone turnover in Paget's disease of bone. *J Bone Miner Res* 14(S2): 66–69
16 Wichers M, Schmidt E, Bidlingmaier F, Klingmuller D (1999) Diurnal rhythm of crosslaps in human serum. *Clin Chem* 45(10): 1858–1860
17 Wu JJ, Eyre DR (1984) Cartilage type IX collagen is cross-linked by hydroxypyridinium residues. *Biochem Biophys Res Comm* 123: 1033–1039
18 Wu JJ, Lark MW, Chun LE, Eyre DR (1991) Sites of stromelysin cleavage in collagen types II, IX, X and XI of cartilage. *J Biol Chem* 266: 5625–5628
19 Mendler M, Eich-Bender SG, Vaughan L, Winterhalter KH, Bruckner P (1989) Cartilage contains mixed fibrils of collagen types II, IX, and XI. *J Cell Biol* 108: 191–197

20 Wu J-J, Eyre DR (1995) Structural analysis of cross-linking domains in cartilage type XI collagen: Insights on polymeric assembly. *J Biol Chem* 270(32): 18865–18870
21 Miles CA, Knott L, Sumner IG, Bailey AJ (1988) Differences between the thermal stabilities of the three triple-helical domains of type IX collagen. *J Mol Biol* 277: 135–144
22 Diab M, Wu J-J, Eyre DR (1996) Collagen type IX from human cartilage: A structural profile of intermolecular cross-linking sites. *Biochem J* 314: 327–332
23 Vater CA, Harris ED, Jr., Siegel RC (1979) Native cross-links in collagen fibrils induce resistance to human synovial collagenase. *Biochem J* 181(3): 639–645
24 Liu X, Wu H, Byrne M, Jeffrey J, Krane S, Jaenisch R (1995) A targeted mutation at the known collagenase cleavage site in mouse type I collagen impairs tissue remodeling. *J Cell Biol* 130(1): 227–237
25 Wu J-J, Lark MW, Chun LE, Eyre DR (1991) Sites of stromelysin cleavage in collagen types II, IX, X, and XI of cartilage. *J Biol Chem* 266(9): 5625–5628
26 Hollander AP, Heathfield TF, Webber C, Iwata Y, Bourne R, Rorabeck C, Poole AR (1994) Increased damage to type II collagen in osteoarthritic articular cartilage detected by a new immunoassay. *J Clin Invest* 93(4): 1722–1732
27 Billinghurst RC, Dahlberg L, Ionescu M, Reiner A, Bourne R, Rorabeck C, Mitchell P, Hambor J, Diekmann O, Tschesche H, Chen J, Van Wart H, Poole AR (1997) Enhanced cleavage of type II collagen by collagenases in osteoarthritic articular cartilage. *J Clin Invest* 99(7): 1534–1545
28 Otterness IG, Downs JT, Lane C, Bliven ML, Stukenbrok H, Scampoli DN, Milici AJ, Mezes PS (1999) Detection of collagenase-induced damage of collagen by 9A4, a monoclonal C-terminal neoepitope antibody. *Matrix Biol* 18(4): 331–341
29 Atley LM, Shao P, Ochs V, Shaffer K, Eyre DR (1998) Matrix metalloproteinase-mediated release of immunoreactive telopeptides from cartilage type II collagen. *Trans Orthop Res Soc*, New Orleans 23(2): 850
30 Eyre DR, Shao P, Vosberg-Smith K, Weis M, Shaffer K, Yoshihara P (1996) Cross-linked telopeptides from collagen types I, II and III in human urine. *J Bone Miner Res* 11(S1): S413
31 Atley LM, Sharma L, Clemens JD, Shaffer K, Pietka TA, Riggins JA, Eyre DR (2000) The collagen II CTx degradation marker is generated by collagenase 3 and in urine reflects disease burden in knee OA patients. *Trans Orthop Res Soc*, Orlando, FL 25: 0168

Markers of joint tissue turnover in osteoarthritis

Tore Saxne, Dick Heinegård and Bengt Månsson

Department of Rheumatology and Department of Cell and Molecular Biology, Section for Connective Tissue Biology, Lund University, S-221 85 Lund, Sweden

Introduction to the molecular marker concept in rheumatology

The tissues in the joint, i.e. cartilage and bone, continuously remodel in a finely tuned balance between matrix synthesis and degradation. In common diseases like osteoarthritis (OA) and rheumatoid arthritis (RA), the normal balance is disturbed and shifted towards degradation in established disease. This eventually leads to disruption of the structural and functional integrity of the joint. As a consequence of the disturbed matrix turnover, increased amounts of macromolecules or fragments thereof are released into synovial fluid and may subsequently reach the blood stream. Some fragments may also be found in the urine. This sequence of events forms the rationale for efforts to identify alterations in the tissue by quantifying matrix macromolecules, "molecular markers", in body fluids with the purpose to define non-invasive methods to monitor pathological tissue processes [1].

Analyses of such molecular markers are currently explored for diagnostic and prognostic purposes and for monitoring effects of therapy on joint tissues [1]. At present, most studies in OA and RA focus on markers of cartilage and this paper will concentrate on this tissue. Increased release of cartilage markers indicates altered tissue turnover, which, if persistent, eventually will lead to permanent joint damage. Thus, increased synovial fluid or serum levels of cartilage macromolecules or increasing such levels over time represent potential markers that are prognostic for future cartilage destruction. Another approach, which has proven feasible in studies of the pathophysiology of tissue destruction in arthritis, is the use of concomitant quantification of a number of tissue markers in the same fluid sample as a "fingerprint" to distinguish and characterize the process in the tissue and to assess the damage at the molecular level. Delineating the fragment pattern will also aid in identifying the enzymes responsible for the tissue breakdown. The molecular marker technology will thus facilitate the identification of processes in the tissue and in a long-term perspective allow the design of agents that block destructive pathways in joint diseases.

Table 1 - Examples of applications of molecular molecular markers in human and experimental joint disease

Elucidate pathophysiological mechanisms of tissue destruction, e.g.
- Define cleavage patterns, reveal enzymatic mechanisms
- Aid in understanding how mediators, e.g. cytokines drive the process

Diagnostic test
Monitor and quantify tissue damage
Predict progression of tissue damage
Predict or measure response to therapy/provide proof of concept in drug trials

In this communication, some of the basic principles of the molecular marker technology for assessment of pathological processes in cartilage will be highlighted focusing on its potentials both for elucidation of pathophysiological mechanisms for tissue destruction and for clinical purposes (Tab. 1). Cartilage oligomeric matrix protein (COMP) [2, 3] is probably the most explored cartilage marker to date, and we will use this molecule as a prototype for a tissue marker. Rather than detailing all pros and cons regarding the marker technology, which have been dealt with elsewhere [1], we will highlight a few principal issues by reviewing key findings pertaining to the utility of the marker technology.

What is COMP?

COMP belongs to the thrombospondin family of proteins. It consists of five identical subunits, each with a Mw of 83 kDa, held together in a coiled-coil domain close to the N-terminal end and further stabilised by disulphide bridges. Each subunit has a globular domain in its C-terminal end. In this way a bouquet-like structure is formed. Each subunit is composed of 8 Ca-binding domains and 4 epidermal growth factor (EGF)-like domains. COMP is located pericellularly and territorially in developing cartilage, but mainly interterritorially in mature cartilage. Its true function is not fully elucidated, but it has been shown to interact with fibril-forming collagens type I and type II at very specific sites on the collagen molecule suggesting a regulating role in fibril formation. Its interterritorial localisation in mature cartilage indicates a structural role in the tissue. The different glycosylation and localisation in foetal cartilage suggests yet another function. Point mutations in COMP result in severe chondrodysplasias or epiphyseal dysplasias indicating a critical role in cartilage formation and/or function [4, 5]. COMP is most prominent in cartilage, but the protein is also present in other pressure loaded tissues, e.g. tendon

Table 2 - Some key issues of relevance for interpretations of tissue marker analyses using COMP as a marker prototype

- Which assay is used?
- Is age a possible confounder?
- What about diurnal variation or variation over time?
- Does inflammation affect serum COMP levels?
- Do changes in serum COMP reflect therapeutic effects on cartilage?

and meniscus. More recently, the protein has also been found in synovial membrane [5]. The relative abundance of the protein differs considerably, being a hundred-fold higher in cartilage as compared to synovial membrane (Heinegård D, Saxne T, Månsson B, unpublished observations).

Although COMP is present in tissues other than cartilage, the evidence strongly indicates that circulating COMP is mainly derived from this tissue. Thus there is a clear-cut correlation between pathological events in the cartilage and changes in serum COMP, e.g. in well-controlled experimental situations [6, 7]. Furthermore, upregulation of COMP in cartilage is seen in early OA, supporting that the increasing serum COMP levels seen in developing OA reflect increased cartilage turnover [8, 9].

Some key issues of relevance for interpretations of tissue marker analyses

The utility of the applications of molecular marker measurements detailed in Table 1 relies on several assumptions regarding the normal turnover of a cartilage protein. If we restrict the discussion to COMP and in particular serum analyses, there are several key issues that need to be addressed (Tab. 2). First, different assays of COMP using different antibodies may not be comparable. Most publications to date have used the original inhibition-ELISA based on a polyclonal antiserum [3], but in animal models other antibodies are used [6, 7]. Furthermore, monoclonal antibodies that recognize select epitopes are now available ([10] and Heinegård, unpublished observations). Somewhat surprisingly perhaps, studies using these more specific antibodies confirm and extend results obtained using the original assay, although in subgroups of patients the face-to-face correlations may not be impressive, thus corroborating different specificity of the antibodies. The conclusion so far is that it matters which test system is applied and that results obtained using one COMP assay may not hold true using another system.

The other issues highlighted in Table 2 are practical, clinical ones, but with pathophysiological implications, and in the following we will highlight these questions by reviewing some recently obtained and some published results.

Is age a possible confounder?

In children, as for many other cartilage and bone markers, serum COMP shows high levels, which typically decrease to adult levels around age 15 years (Saxne T, Månsson B, Heinegård D, unpublished observations). This is most likely due to the higher cartilage turnover during growth. In blood donors, the serum concentrations of COMP do not vary with age in individuals between age 20–50 years. However, in blood donors between 50–65 years of age about 20% show serum concentrations above the upper normal range for the blood donors in the younger age group [11]. In a recent publication by Clark and coworkers, increasing serum levels of COMP in similar age groups, i.e. after age 50 years both in controls and patients with OA, were found using another COMP assay [12]. Furthermore this study also corroborated the findings of increased serum COMP levels in patients with early radiographic changes compatible with osteoarthritis in the knee as previously described [13]. It is tempting to speculate that the blood donors in this age group with increased serum COMP may have early osteoarthritis. No information regarding features of osteoarthritis in our blood donors is available, and it is likely that they represent stages prior to those where currently used clinical diagnostic procedures are of value. Therefore this remains an attractive, but unproven hypothesis. In support, in patients with hemophilia with repeated joint bleedings and increased risk of developing OA, increasing COMP levels are seen already at the age of 35 years (Månsson et al., unpublished observations). From a practical viewpoint, these observations emphasize the difficulty of identifying "normal controls," especially in the age groups prone to osteoarthritis.

What about diurnal variation of COMP levels or variation over time?

When applying molecular marker measurements for monitoring tissue effects of therapy, it is important to know how serum concentrations vary over time irrespective of treatment. If serum levels are stable in normal individuals or in patients with benign disease, changes in, for example, serum COMP, observed after intervention or during phases of rapid disease progression, are likely to relate to the tissue process. If, on the other hand, serum levels vary in an unpredictable manner in healthy individuals or in patients with a stable disease, the feasibility of serum measurements of these markers for monitoring the disease can be questioned.

In studies of healthy hospital workers between 30–50 years of age, serum levels of COMP differ between individuals, but in each individual the level remains remarkably stable over an 8-year period (Saxne et al., unpublished observation). In a subgroup of 10 individuals from the Spenshult cohort of patients with knee joint pain monitored longitudinally to elucidate early signs of OA [13], we analysed serum COMP by a novel sandwich-ELISA based on two monoclonal antibodies (Anamar Medical, Lund, Sweden) in samples obtained at regular intervals during 24 h with defined physical activity. Two important findings emerged. First, no significant changes in the COMP levels were observed during day-time between 8 a.m. and 9 p.m., i.e., during the period when most routine blood sampling is likely to take place. Second, a significant decrease in serum COMP was apparent during bedrest at night, reaching the lowest levels around 5 a.m., indicating a significant turnover of COMP in the circulation (Petersson et al., unpublished observation). Taken together, these observations indicate that measurement of serum COMP should be useful for identifying pathological processes in the cartilage and for monitoring interventions aimed at modifying such processes.

Does inflammation affect serum COMP levels?

As indicated, COMP in the circulation is primarily thought to reflect the turnover of cartilage, which will be further discussed in relation to therapeutic interventions in the next section. However, since COMP is also present in the synovial membrane, although in small quantities compared to cartilage and synovial fluid, it could be speculated that some of the circulating COMP is derived from this source and that serum COMP thus might reflect an inflammatory component of the disease. In RA, several studies convincingly show that serum COMP does not correlate to variables of inflammation, i.e., ESR, CRP, and not to the recently described YKL-40 protein (1, Saxne et al, unpublished observation). In OA, COMP does not correlate to serum levels of YKL-40 (Saxne et al., unpublished observation). Is there a correlation to CRP in OA? This question is relevant, since a few studies using sensitive assays for CRP show increased levels in established OA [14, 15]. We have developed a sensitive ELISA for serum CRP and examined whether serum CRP distinguishes individuals with knee pain with or without radiographic OA at baseline and after 3 years, and whether it correlates to serum COMP in the Spenshult cohort [16]. We could show that CRP was increased already at baseline and remained elevated at the 3-year follow up in the individuals showing radiographic knee OA at follow up. In contrast, the levels were low and did not differ from levels in age-matched blood donors in the individuals with only knee pain. As already mentioned, serum COMP changed in a different manner, and importantly, there was no significant correlation between serum CRP and serum COMP in any of the groups. This strongly argues that CRP and COMP reflect different features characterizing OA, the inflammation and the cartilage process, respectively. Furthermore, the increased serum CRP in early phases of OA suggests the presence of low-grade inflammation, which could play a pathophysiologic role in OA [17].

Do changes in serum COMP reflect therapeutic effects on cartilage?

An ideal tissue marker would be one that is exclusively synthesized and laid down only in a certain tissue, e.g. cartilage. A joint cartilage specific marker would be even better for assessing joint processes, and a protein confined to a certain joint cartilage compartment would be even more desirable. Tissue-specificity would be ideal, but the lack of complete tissue-specificity does not by itself disqualify a matrix protein as a marker for processes in a certain tissue. This view is best illustrated by the current status of COMP as a marker for cartilage processes in arthritis. Originally isolated from cartilage, this protein has now, as already mentioned, been detected in or produced by cells from meniscus, synovial membrane and tendon. Therefore, at first sight one could question the role of COMP as a cartilage marker. However, the crucial experiments would be those that show that levels of synovial fluid or circulating COMP change in a manner which can be correlated to events in the cartilage, and that neither events in other tissues, e.g. the synovial membrane or the tendon (synovitis or tendinitis), nor generalized inflammation influence the COMP levels. In experimental arthritides, it is possible to examine the cartilage at different timepoints, and thus in these experiments serum concentrations can be correlated to cartilage histopathology. Such experiments have been performed and prove the validity of serum COMP as a marker for processes in the cartilage (see e.g. [6, 7]). Unfortunately, as yet there are no data on the effect on serum COMP of therapeutic interventions in OA or in experimental models of OA, but experimental arthritis such as collagen-induced arthritis show that treatment that retards the cartilage damage normalises serum concentrations regardless of whether the treatment affects clinical symptoms. A very convincing example is the comparison of TNFα blockade and IL-1α/β blockade in a murine collagen-induced arthritis model. Here, both treatment modalities ameliorated inflammation as determined by joint score and serum IL-6, whereas only IL-1 blockade ameliorated the cartilage damage as shown by histopathological examination. The COMP levels in serum were normalized during IL-1 blockade but remained increased during TNF-blockade, which correlated to the histopathological findings and supports the role of COMP as a cartilage marker [7].

Thus measurement of cartilage markers, e.g. COMP, shows promise for monitoring treatment effects on the tissue in trials of new, potentially disease modifying drugs. Already at this stage such monitoring is ongoing, and results are awaited. In RA, we have shown that glucocorticoid treatment alters the cartilage turnover as measured by serum COMP. Thus serum COMP decreases during such treatment, although in a fashion not closely linked to the ameliorated inflammatory component (Saxne et al., unpublished observation). The pathophysiological explanation for this is not fully elucidated, but in view of the experimental data above it is tempting to speculate that this is due to the IL-1 blocking effect of glucocorticoids.

Future perspectives

The molecular marker approach is a novel strategy for studying early tissue alterations in OA and other joint diseases. Already at present it has showed promise as a useful tool for monitoring the disease process in the clinic. Future work along these lines, importantly in close collaboration between basic connective tissue biology research and clinical research, will aid in understanding these complex diseases including the identification of novel targets for therapy. COMP is but one of a number of candidate markers, and in the future, we hope that a combination of markers sensitive to any cartilage process (global markers) with markers of processes affecting specific tissue structures can be used to delineate the destructive process in detail.

References

1   Saxne T, Månsson B (2000) Molecular markers for assessment of cartilage damage in rheumatoid arthritis. In: GS Firestein, GS Panayi, FA Wollheim (eds): *Rheumatoid arthritis: new frontiers in pathogenesis and treatment*. Oxford University Press, Oxford, 291–304
2   Hedbom E, Antonsson P, Hjerpe A, Aeschlimann D, Paulsson M, Rosa PE, Sommarin Y, Wendel M, Oldberg Å, Heinegård D (1992) Cartilage matrix proteins. An acidic oligomeric protein (COMP) detected only in cartilage. *J Biol Chem* 267: 6132–6136
3   Saxne T, Heinegård D (1992) Cartilage oligomeric matrix protein. A novel marker of cartilage turnover detectable in synovial fluid and blood. *Br J Rheumatol* 31: 583–591
4   Heinegård D, Lorenzo P, Saxne T (1999). Noncollagenous proteins: glycoproteins and related proteins. In: MJ Seibel, SP Robins, JP Bilezikian (eds): *Dynamics of bone and cartilage metabolism*. Academic Press, San Diego, 59–69
5   Heinegård D, Lorenzo P, Saxne T (2000) Matrix Glycoproteins and Proteoglycans in Cartilage. In: S Ruddy, ED Harris Jr, CB Sledge (eds): *Kelley's textbook of rheumatology*, 6th ed. WB Saunders Company, Philadelphia, 41–53
6   Vingsbo-Lundgren C, Saxne T, Olsson H, Holmdahl R (1998) Increased serum levels of cartilage oligomeric matrix protein in chronic erosive arthritis in rats. *Arthritis Rheum* 41: 544–550
7   Joosten LAB, Helsen MA, Saxne T, Van de Loo FAJ, Heinegård D, van den Berg WB (1999) IL-1α,β blockade prevents cartilage and bone destruction in murine type II col-

lagen-induced arthritis, whereas TNFα blockade only ameliorates joint inflammation. *J Immunol* 163: 5049–5055

8   Salminen H, Perälä M, Lorenzo P, Saxne T, Heinegård D, Säämänen AM, Vuorio E (2000) Up-regulation of cartilage oligomeric matrix protein at the onset of articular cartilage degeneration in a transgenic mouse model of osteoarthritis. *Arthritis Rheum* 43: 1742–1748

9   Heinegård D, Bayliss MT, Lorenzo P (1998) Biochemistry and metabolism of normal and osteoarthritic cartilage. In: KD Brandt, M Doherty, LS Lohmander (eds): *Osteoarthritis*. Oxford University Press, 74–78

10 Vilim V, Lenz ME, Vytasek R, Masuda K, Pavelka K, Kuettner KE, Thonar EJ (1997) Characterization of monoclonal antibodies recognizing different fragments of cartilage oligomeric matrix protein in human body fluids. *Arch Biochem Biophys* 341: 8–16

11 Månsson B, Heinegård D, Saxne T (2000) Diagnosis of osteoarthritis in relation to molecular processes in cartilage: comment on the article by Clark et al. *Arthritis Rheum* 43: 1425–1426

12 Clark AG, Jordan JM, Vilim V, Renner JB, Dragomir AD, Luta G, Kraus VB (1999) Serum cartilage oligomeric matrix protein reflects osteoarthritis presence and severity: the Johnston County Osteoarthritis Project. *Arthritis Rheum* 42: 2356–2364

13 Petersson IF, Boegård T, Svensson B, Heinegård D, Saxne T (1998) Changes in cartilage and bone metabolism identified by serum markers in early osteoarthritis of the knee joint. *Br J Rheumatol* 37: 46–50

14 Spector TD, Hart DJ, Nandra D, Doyle DV, Mackillop N, Gallimore JR, Pepys MB (1997) Low-level increases in serum C-reactive protein are present in early osteoarthritis of the knee and predict progressive disease. *Arthritis Rheum* 40: 723–727

15 Sharif M, Shepstone L, Elson CJ, Dieppe PA, Kirwan JR (2000) Increased serum C reactive protein may reflect events that precede radiographic progression in osteoarthritis of the knee. *Ann Rheum Dis* 59: 71–74

16 Lindell M, Månsson B, Petersson IF, Saxne T (2000) Serum C-reactive protein and serum COMP in knee joint osteoarthritis (abstract). *Arthritis Rheum* 43 (Suppl): 339

17 Pelletier JP, Martel-Pelletier J, Abramson SB (2001) Osteoarthritis, an inflammatory disease: potential implication for the selection of new therapeutic targets. *Arthritis Rheum* 44: 1237–1247

Gene expression profiling by the cDNA array technology: Molecular portraying of chondrocytes

Thomas Aigner[1], Pia M. Gebhard[1] and Alexander Zien[2]

[1]Cartilage Research, Department of Pathology, University of Erlangen-Nürnberg, Krankenhausstr. 8–10, D-91054 Erlangen, Germany; [2]GMD (German National Research Center for Information Technology), Schloss Birlinghoven, D-53754 Sankt Augustin, Germany

Introduction

Many studies have investigated morphological and biochemical changes during osteoarthritic cartilage degeneration. These studies have never allowed to paint a picture of an expression profile in the chondrocytes during the disease process, based on a broad number of genes simultaneously investigated using the same material. cDNA-array technology is a new, very powerful tool for mRNA expression profiling of large numbers of genes at the same time. The major limitation in using this technology for gene expression profiling in diseased *versus* normal tissue is the inherent heterogeneity of cell populations present in most tissues. Thus, one is not able to distinguish expression signals derived from endothelial cells, neural tissue, inflammatory cell populations, fibroblasts and other cells physiologically or pathologically present in the tissues. In this respect, adult human articular cartilage offers the great advantage that it contains – both in the normal and in the diseased state – only one cell population, the chondrocytes. Therefore, the detected gene expression levels can be fully attributed to the chondrocytes.

After an initial trial with an own "in-house" cDNA array containing overall 20 spots, we used the human cancer 1.2 cDNA array from Clontech, which was spotted with 1185 genes. Our main focus was to understand matrix turnover processes and expression levels of gene products (matrix components and matrix degrading proteases) involved in this process.

Material and methods

For the study, cartilage from human femoral condyles of the knee joints were used. Normal articular cartilage (9) and early degenerated cartilage (6) were obtained from donors at autopsy. Late stage OA cartilage was obtained from total knee replacement surgery (n). The cartilage was frozen in liquid nitrogen immediately after removal and stored at −80°C. RNA was isolated as described previously [1].

The Clontech Human Cancer 1.2 cDNA array was probed with ^{32}P-dATP labeled cDNA probes. Membranes were hybridized for 18 h at 68°C, washed and exposed to a phosphor plate for 72 h [2]. Images were captured on a Molecular Dynamics Storm phosphor imager, using Image Quant software (Molecular Dynamics). For primary spot quantification the software from Clontech (ATLASIMAGE 1.01) was used. The biomathematics including a specifically developed normalization procedure is described elsewhere [3].

Collagen expression pattern

Expression of cartilage fibrillar collagens types II, IX, and XI were not detectable in normal and strongly upregulated in late stage osteoarthritic cartilage. The same was true for collagen type III and pericellular collagen type VI.

Proteoglycans

For aggrecan we were not able to show a significant upregulation in diseased tissue either by cDNA hybridization or qPCR. For decorin, a strong expression signal was found in all cartilage samples without significant differences in between the investigated specimen groups. In contrast, another small proteoglycan biglycan appeared to be upregulated in late stage osteoarthritic chondrocytes. Versican was barely detectable in any of the specimens in line with previous PCR data.

Non-collagenous non-proteoglycaneous cartilage matrix proteins

One very prominent gene expression product of articular cartilage is fibronectin. It is known biochemically to show increased synthesis in osteoarthritic chondrocytes, results also seen in our arrays on the mRNA level. Of note, an increase of fibronectin synthesis was already found in the early degeneration. This is particularly interesting, since fibronectin fragments were shown to enhance catabolic activity of articular chondrocytes. Thus, fibronectin could well be a crucial regulator of matrix turnover activity of chondrocytes during early disease development [4]. Another molecule upregulated in particular in late stage osteoarthritic cartilage was osteonectin. Osteonectin is usually a bone associated protein, but is also expressed by hypertrophic chondrocytes in fetal growth plate cartilage. Its expression by osteoarthritic chondrocytes fits to the increased expression of other gene products of hypertrophic chondrocytes in a portion of the osteoarthritic cartilage cells such as type X collagen and osteopontin. These processes are primarily observed in the lower zones of late stage osteoarthritic cartilage and support the concept of a re-initiation of the fetal hypertrophic differentiation program leading, for example, to the extension of the calcified cartilage zone, a characteristic feature of osteoarthritic cartilage.

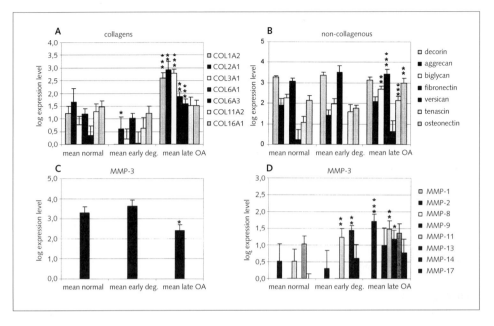

Figure 1
cDNA-array analysis of mRNA expression levels in chondrocytes in normal (n=9), early degenerative (n=6) and late stage osteoarthritic articular cartilage (n=6) for matrix components ((A): collagens; (B): proteoglycans and other matrix proteins) and matrix metalloproteinases (C+D). ($^*p<0.05$; $^{**}p<0.01$; $^{***}p<0.001$).

Matrix-degrading enzymes – matrix metalloproteinases

The dominantly expressed form in articular cartilage appears to be MMP-3 (stromelysin 1), which is well known as a crucial enzyme of matrix turnover also in normal articular cartilage. MMP-3 was down-regulated in the late stage specimens. This indicates that in late stages of cartilage degeneration other degradation pathways might be more important involving other enzymes such as MMP-2 and MMP-11, both of which were upregulated in late stage disease. MMP-11, an endopeptidase mainly so far thought to be involved in placentation and tumor invasion was upregulated also in the early degenerative samples.

Collagen degrading enzymes – collagenases

Most interestingly, neither MMP-1 (collagenase 1) nor MMP-8 (collagenase 2) were detectable in any sample. Instead, in late stage OA cartilage, a significant increase of MMP-13 expression was seen suggesting that MMP-13 is involved in terminal breakdown of collagen fibres in osteoarthritic cartilage degeneration. In contrast, no detectable expression levels of MMP-13 were found in normal articular chondrocytes in line with the virtually non-existent collagen turnover in normal articular cartilage. Also early stages appear to be more characterized by degradation of other matrix components such as aggrecan and other non-collagenous molecules. MMP-2 (gelatinase A), an enzyme known to be able to degrade denatured collagens (gelatins), is also significantly upregulated in late stage OA cartilage.

The use of the cDNA array technology allows one to identify differential gene expression patterns in early degenerative and late stage osteoarthritic compared to normal articular cartilage. Overall, cDNA technology offers cartilage research a powerful tool for investigation of gene expression patterns of a high number of different genes simultaneously. Despite inherent draw-backs of this technique, such as limitation in sensitivity and quantification, the incapacity to detect splice variants and to account for posttranscriptional or posttranslational regulation of protein synthesis, this technique provides a highly powerful tool to get an overview on gene expression pattern, not available using other techniques: it allows one to analyze biological systems rather than single aspects of it.

Besides adding to previous studies, which provided interesting data on single or a few genes, this investigation was intended to contribute to an overall picture of genes involved in the osteoarthritic disease process. "Molecular portraits" of these processes will stimulate the testing of new disease markers, which are desperately needed for the diagnosis and monitoring of the disease, in particular in earlier disease phases, and will allow the identification of new pathways. They will also provide new targets for studying disease mechanisms and outline new targets for therapeutic intervention.

References

1 McKenna LA, Gehrsitz A, Soeder S, Eger W, Kirchner T, Aigner T (2000) Effective isolation of high quality total RNA from human adult articular cartilage. *Anal Biochem* 286(1): 80–85
2 Aigner T, Zien A, Gehrsitz A, Gebhard PM, McKenna LA (2001) Anabolic and catabolic gene expression pattern analysis in normal *versus* osteoarthritic chondrocytes using the cDNA-array technology. *Arthritis Rheum* 44: 2777–2789
3 Zien A, Aigner T, Zimmer R, Lengauer T (2001) Centralization: a new paradigm for the normalization of gene expression data. *Bioinformatics* 17: S323–S331
4 Homandberg GA, Davis G, Maniglia C, Shirikhande A (1997) Cartilage chondrolysis by fibronectin fragments cause cleavage of aggrecan at the same site as found in osteoarthritic cartilage. *Osteoarthritis Cartilage* 5: 450–453

Clinical evaluation of markers for osteoarthritis

Ivan G. Otterness[1] and Dolores Vázquez-Abad[2]

[1]241 Monument St. #5, Groton, CT 06320, USA; [2]Pfizer Global Research and Development, New London, CT 06320, USA

Introduction

The ideal uses of molecular markers for osteoarthritis (OA) – diagnosis, prognosis and monitoring therapeutic responses – remain elusive. The application of markers to other diseases, i.e., insulin and blood sugar in diabetes, LDH and cholesterol in heart disease, temperature in bacterial infections – highlight the importance of markers in both staging of disease and monitoring treatment. Clinical management of OA could be similarly improved by the discovery of a suitable marker. In addition, markers could identify activity in the different pathologic processes that mediate pain, inflammation or cartilage degradation, and clinical management targeted appropriately. The current status of markers for OA has been reviewed recently [1–3], and the methodology for validating the use of markers for OA has been well presented elsewhere [4, 5]. This discussion is focused on criteria for evaluation of the utility of markers.

Requirements for marker assessment

The distribution of marker values

Marker concentrations in the OA population must first be compared to those in the control population to determine if there is a difference that can be relied on for clinical decisions. For this, the distribution of marker values in the control and OA populations needs to be known. Standard statistical tests such as the Kolmogorov-Smirnov test or the Shapiro-Wilk W-test [6] determine the fit of the data to a normal distribution. Routinely, both the control and the OA patient marker values fail to conform to a normal distribution. One remedy is to convert the marker values to their logarithms. For all of the 23 markers we studied, the $\log_{10}$ values conformed to a normal distribution. This makes it possible to use standard statistical tests based on a normal distributions. Otherwise, less powerful non-parametric tests must be used.

Marker practicality

To estimate the ease of determining an improvement in marker values in the OA population, we advocate using a simplified power calculation expressed in terms of the number of patients (N*) required to detect a 50% improvement in marker values with a $p < 0.05$. N* equals 50 × the square of the ratio of the standard deviation of the OA group divided by the difference in means between the OA and the control groups. A lower N* implies better discrimination. For different markers, N* ranges from lows of 5–10 patients to highs of hundreds of patients. N* gives a quantitative measure of the practicality of different markers.

Study populations

Definition of the control population

It is often assumed that an individual without joint space narrowing does not have OA and should have normal marker values. However, the radiographic and symptomatic occurrence of OA in most individuals who are 70 years of age suggests that early disease is present but undetected by radiography or clinical exam in the majority of 50–60 year olds. These individuals are likely to have elevated marker levels. Hence we suggest that normal marker values should be found in individuals in the 30–40 year-old range – when the exuberant turnover of cartilage components seen during growth is over, and marker values should be more normal than in the following decades when deterioration of cartilage becomes much more common.

In addition, for many markers there is a small number of individuals in the control group who have high values. Currently, it is unclear whether these individuals are normal with atypically high marker values or whether these high values are indicative of early, undiagnosed disease. In the first case, our control values are representative; in the second case, our control values have been biased upward. The status of these individuals must be followed over time to determine the correct interpretation.

Definition of the OA population

In many marker studies, a categorical definition of OA has been used, i.e., by some measure, for example, the ACR criteria [7], the individuals either have or have not OA. Using categorical criteria, a number of markers have been found to discriminate OA. Such data give only a poor indication of the utility of the marker. It is not possible to define the sensitivity of OA detection or determine which disease parameters are associated with the marker. Moreover, as OA cases in a secondary or tertiary care facility are likely to be more severe, this potentially gives an overly optimistic assessment of the ability of the marker to diagnose OA.

Since cartilage damage appears irreversible, a marker should detect early disease so that treatment can be applied before OA becomes too advanced. Thus it is important that a range of OA disease severity be studied including early, even pre-clinical disease so that the limits of detection of disease are defined.

Most published clinical studies follow a signal joint. This is fraught with problems when the object is to correlate clinical outcome with marker values. The degree of elevation of a marker depends not only on the release of the marker from the signal joint, but additionally on the contributions of all joints. Recording only a signal joint may substantially distort the interpretation of marker results.

Defining the OA diseased population for marker studies is a further problem. The clinical definition of OA of the knee, for example, relies on joint pain, osteophytes, radiographic joint space narrowing (JSN) and clinical history [7]. Numerous studies have declared a lack of correlation of pain and JSN. Osteophytes have been suggested to correlate better with pain than with JSN. JSN as a measure of cartilage loss is compromised by meniscal extrusion in about half the cases [8]. Clinical history gives a method of integrating all the findings with medical knowledge of the stage and course of the disease, but it is only an indirect reflection of what is transpiring in cartilage. This means that the ability to separate OA from controls needs to be independently analyzed against each of the parameters used to define OA as well as against a global OA assessment. Only then can the meaning of a marker be teased out. We have previously presented methodology for associating the various clinical measures with marker values [9].

Study population size

Most published clinical studies have patient numbers of 20–40 per arm. That may be sufficient to find the range of normal values. But to address the sensitivity of early detection, it will be necessary to stratify patients according to severity, and perhaps other clinical features (age, gender, disease duration, body mass index, etc.), so that larger numbers will be required in the OA group than estimated from our practicality index.

Diagnosis

$N^*/4$ gives the 95% confidence limits for discriminating an OA patient from a control. Thus $N^*/4$ should be a good measure of a marker's diagnostic capabilities. However, with the question of early detection unaddressed, the diagnostic capability of markers remains unresolved. If one focuses on individuals with more severe disease, many markers are quite strongly elevated. Hence there is a subpopulation that can be readily diagnosed because of the substantial abnormality in their marker values. These are the very patients that are easily pinpointed by standard clinical measurements and for whom there is no question of their diagnosis. Such a subpopulation stratagem does not allow the detection of early disease where therapeutic intervention is believed to be most effective and where clinical definition of OA is most difficult.

Use of multiple markers

We have found that markers can be classified into independent groups of dependent markers [10]. The dependent markers can be used to strengthen associations (akin to replication); the independent markers can be used to enhance discriminatory power, for example, between OA and controls. Combined use of two or more markers could potentially be used to overcome the weakness of a single marker for clinical diagnosis or prognosis.

Assessment of OA progression

Two general methods have been used to assess OA progression: MRI and radiographs, measuring cartilage loss [11] and joint space narrowing [12], respectively, over a 6–12 month period in single joints. Clinical instruments such as the WOMAC [13] provide information on pain and disability of a signal joint, but are at best an imprecise measure of the underlying cartilage changes. By contrast, a marker should give an instantaneous measure of the rate of overall disease progression. Similarly, a marker should give a rapid indication of any change in progression rate in response to treatment. Regretfully, no marker has been validated to make such judgments.

Conclusions

Validation that a marker measures progression of OA requires (i) an effective therapeutic agent to demonstrate change in progression measured by standard techniques that is associated with change in marker value, or (ii) a long-term epidemiological study in which marker values can be linked to outcome. In either case, this means examining the marker with each of the measures of disease progression over a sufficiently protracted time to establish whether or not there is a real correlation of marker with progression. In either case, a repository of archival samples linked to patient clinical progression is needed to enable rapid validation of markers. Development of such archival sets is in progress, and this means that the conditions to validate markers are at last being established.

References

1 Lohmander LS (1997) What is the current status of biochemical markers in the diagnosis, prognosis and monitoring of osteoarthritis? *Baillieres Clin Rheumatol* 11: 711–726
2 Otterness IG, Saltarelli MJ (2001) Using molecular markers to monitor osteoarthritis. In: GC Tsokos, LW Moreland, GM Kammer, J-P Pelletier, J Martel-Pelletier, S Gay (eds): *Modern therapeutics in rheumatic disease*. Humana Press, Totowa, NJ, 215–236
3 Poole AR (2000) NIH White Paper: Biomarkers, the osteoarthritis initiative. A basis for discussion. http://www.nih.gov/niams/news/oisg/oabiomarwhipap.htm
4 Lohmander LS, Felson DT (1997) Defining the role of molecular markers to monitor disease, intervention, and cartilage breakdown in osteoarthritis. *J Rheumatol* 24: 782–785
5 Lohmander LS, Felson DT (1998) Defining and validating the clinical role of molecular markers in osteoarthritis. In: KD Brandt, M Doherty, LS Lohmander (eds): *Osteoarthritis*. Oxford University Press, Oxford, 519–539
6 Shapiro SS, Wilk MB, Chen HJ (1968) A comparative study of various tests of normality. *J Am Stat Assoc* 63: 1343–1372
7 Altman R et al. (1986) Development of criteria for the classification and reporting of osteoarthritis: classification of osteoarthritis of the knee. *Arthritis Rheum* 29: 1039–1049
8 Adams JG, McAlindon T, Dimasi M, Carey J, Eustace S (1999) Contribution of meniscal extrusion and cartilage loss to joint space narrowing in osteoarthritis. *Clinical Radiology* 54: 502–506
9 Otterness IG, Weiner E, Swindell AC, Zimmerer RO, Ionescu M, Poole AR (2001) An analysis of 14 molecular markers for monitoring osteoarthritis. Relationship of the markers to clinical endpoints. *Osteoarthritis Cart* 9: 224–231
10 Otterness IG, Swindell AC, Zimmerer RO, Poole AR, Ionescu M, Weiner E (2000) An analysis of 14 molecular markers for monitoring osteoarthritis: segregation of the markers into clusters and distinguishing osteoarthritis at baseline. *Osteoarthritis Cart* 8: 180–185
11 Eckstein F, Reiser M, Englmeier K-H, Putz R (2001) *In vivo* morphometry and functional analysis of human articular cartilage with quantitative magnetic resonance imaging-from image to data, from data to theory. *Anat Embryol* 203: 147–173
12 Buckland-Wright JC, Wolfe F, Ward RJ, Flowers N, Hayne CT (1999) Substantial superiority of semiflexed (MTP) views in knee osteoarthritis: A comparative radiographic

study, without fluoroscopy, of standing extended, semiflexed (MTP), and Schuss view. *J Rheumatol* 26: 2664–2674

13 Bellamy N, Buchanan WW, Goldsmith CH, Campbell J, Stitt LW (1988) Validation study of WOMAC: a health status instrument for measuring clinically important patient relevant outcomes to antirheumatic drug therapy in patients with osteoarthritis of the hip or knee. *J Rheumatol* 15: 1833–1840

Joint degradation in rapidly destructive and hypertrophic osteoarthritis of the hip

Wolfgang Eger[1], Stefan Söder[2], Dietmar Thomas[1], Thomas Aigner[2] and Günther Zeiler[1]

[1]Department of Orthopaedic Surgery at Rummelsberg Hospital, Rummelsberg 71, D-90592 Schwarzenbruck, Germany; [2]Cartilage Research Group, Institute of Pathology, University of Erlangen-Nürnberg, Krankenhausstrasse 8–10, D-91054 Erlangen, Germany

Introduction

The clinical and radiological faces of osteoarthritis of the hip are variable. Osteoarthritis (OA) of the hip is generally divided into primary and secondary varieties, the former being attributed to an intrinsic cartilage degradation process and the latter resulting from previous cartilage damage or major mechanical abnormality resulting in abnormal or incongruous loading of the joint for long periods [1]. Most common primary and secondary OA of the hip is accompanied by hypertrophic bone reaction with increased subchondral sclerosis, onset of osteophytes and development of subchondral cysts. Clinical symptoms like pain and dysfunction in hypertrophic hip OA (HH-OA) progress slowly over years.

Rapidly destructive hip OA (RDH-OA), introduced by Postel and Kerboull [2], usually occurs in a single normal hip and cumulates in severe joint destruction usually within 1 year. The arthropathy is apparently different from other disorders like rheumatoid arthritis, infection of the joint or aseptic necrosis of the femoral head. The radiographic signs include a cystic joint destruction with fast concentric loss of joint space, cortical erosion and bone atrophy without formation of osteopytes or subchondral sclerosis. This study addresses the morphology of joint degradation in RDH- and HH-OA in order to better understand the faces of the disease, distinguish between characteristic forms and develop different therapeutic strategies.

Material and methods

Patients undergoing total hip arthroplasty for either RDH- or HH-OA were screened for pain, disability and function. ESR and CRP were determined routinely. Patients were examined for OA of other joints as well as other diseases. Microbiological examinations of the synovial fluids were performed. Frontal slices from the central load bearing zone of the femoral head and synovial tissue from 12 patients with either HH- or RDH-OA were fixed in 4% paraformaldehyde. 3-mm-

Table 1 - Erythrocyte sedimentation rate (ESR) after 1 and 2 h and serum C-reactive protein (CRP) form 12 patients with RDH-OA and 12 patients with HH-OA 1 day before surgery.

	HH-OA	RDH-OA	T-test
ESR (mm/1 h)	8.3 ± 4.1	22.9 ± 9.4	$p < 0.0001$
ESR (mm/2 h)	19.8 ± 9.0	49.0 ± 18.0	$p < 0.0001$
CRP (mg/dL)	0.5 ± 0.1	1.2 ± 0.8	$p = 0.04$

thick slices were investigated by microradiography and decalcified with EDTA. After deparaffinization, samples were either stained with hematoxylin-eosin (HE) and toluidine blue or incubated with mono- and polyclonal antibodies recognizing collagen type I, II, VI, X and aggrecan.

Results

In RDH-OA, severe joint damage occurred within 11.3 ± 5.5 months. Ten of 12 patients showed a fixed contracture. In HH-OA, disease duration was 41 ± 21 months. Fixed contracture was detected only in one hip. ESR and CRP were significantly higher in RDH- than HH-OA (Tab. 1). In HH-OA, cortical bone appeared intact and hypertrophied, subchondral cysts lay beneath the surface, and cartilage showed deep fissuring and fibrillation with depletion of proteoglycan and collagen II. Expression of collagen VI was highly upregulated. Fibrosis of the capsule was the predominant feature of synovial morphology. Synovial membrane appeared weakly activated. RDH-OA was characterized by complete cortical bone erosion and deep crater-like cysts within the load-bearing zone (Fig. 1). Cystic lesions were infiltrated by fibrous connective tissue surrounding necrotic bone and cartilage fragments. The peripheral joint surface was covered by a thick layer of cellular fibrous connective tissue without inflammatory cells. Bone marrow fibrosis and chondrocytic metaplasia with cartilage tufts growing onto the surface were characteristic features (Fig. 2). Signals for collagen type II and aggrecan were increased at the roots of the tufts and within the metaplastic chondrocyte nests and correlated with intense Toluidine blue staining. Signals for collagen type VI could be observed throughout the whole tufts with homogeneous intensity while collagen type I was most concentrated within the metaplastic chondrocyte nests. Synovial membrane showed moderately activated micro- and macro-detritus synovitis containing fragments of necrotic bone and cartilage. Aggregates of lymphocytic infiltration were detected in six of 12 patients, and increased synovial cell proliferation and extensive villous hyperplasia was observed in all synovial membranes.

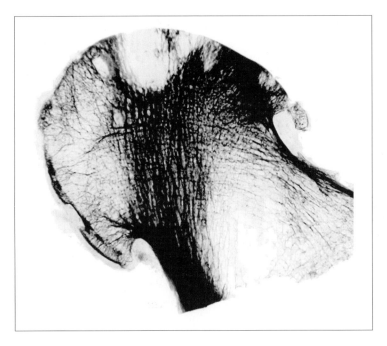

Figure 1
Microradiography from a frontal section of a femoral head form of RDH-OA shows cortical erosion and deep cystic bone defects.

Discussion

The mechanism of destruction in rapidly destructive hip arthropathy is still unknown.

Former investigations showed that RDH-OA is associated with inta-articular deposition of hydroxyapatite crystals that stimulate bone resorbing activity liberated from macrophages in the presence of prostaglandin E_2 [3–6]. However, basic calcium phosphate and calcium pyrophosphate dihydrate crystals are frequently found in synovial fluid from patients with established OA [7, 8], so basic calcium phosphates including apatites are suggested to be of doubtful pathogenic significance [9]. Chondrocalcinosis is accompanied by crystal deposition in hyaline cartilage forming macroscopically visible white and rough spots. In none of the investigated hips could changes like that be observed on the femoral or acetabular surfaces.

CRP was found to be significantly higher in rapidly destructive than in slowly progressive hip OA [10]. In the present study, synovial membranes in RDH-OA, but not in HH-OA, were characterized by a moderate activated detritus-synovitis with focal lymphocytic aggregates or follicles that may be reflected by an increase in

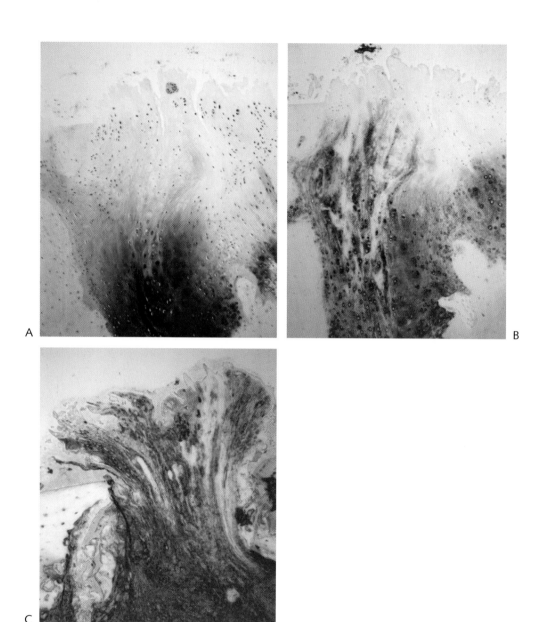

Figure 2
In RDH-OA cartilage, tufts grow out of metaplastic chondrocyte nests onto the surface. Proteoglyans, detected by Toluidine blue staining (A), as well as collagen type II (B) were found to be highest in the deep parts of the cartilage tufts. Signals for collagen type VI (C) could be observed throughout the fibrous-cartilage repair tissue. (Magnification × 80.)

serum CRP and ESR. Synovial membranes, however, showed less inflammation than in rheumatoid arthritis.

Access to bone marrow compartment due to destruction of cortical and subchondral bone stimulated proliferation of metaplastic chondrocytes forming fibrous cartilage tufts that grew towards the joint surface. This repair mechanism was frequently seen in all RDH-OA, but in none of the HH-OA.

In a previous study, gelatinolytic activity in synovial fluid was detected in RDH- but not HH-OA, and synthesis of proteolytic enzymes by synovial cells in the presence of IL-1 was increased in RDH-OA [11]. Masuhara et al. demonstrated MMP-3 and MMP-9 synthesis by fibroblasts from fibrous connective tissue within subchondral cystic defects in RDH- but not in HH-OA [12]. Articular cartilage completely disappeared in all RDH-OA joints, so cartilage MMP- and TIMP-expression could not be investigated. It is still unclear whether the rapid cartilage degradation is triggered by an increased synthesis of degrading enzymes overwhelming inhibitors such as tissue inhibitors of metalloproteases or inhibitors to block the action of bone resorbing factors.

The clinical and morphological differences of RDH-OA may be related to a process of rapid cartilage degradation with increased synovial inflammation that again contributes to cartilage damage and initiates bone resorption, and could explain the rapid progression of the disease.

References

1   Solomon L (1976) Patterns of osteoarthritis of the hip. *J Bone Joint Surg* 58-B: 176–183
2   Postel M, Kerboull M (1970) Total prosthetic replacement in rapidly degenerative arthrosis of the hip joint. *Clin Orthop* 72: 138–144
3   Dieppe PA, Doherty M, Macfarlane DG, Hutton CW, Bradfield JW, Watt I (1984) Apatite associated destructive arthritis. *Br J Rheumatol* 23: 84–91
4   Alwan WH, Dieppe PA, Elson CJ, Bradfield JW (1988) Bone resorbing activity in synovial fluids in destructive osteoarthritis and rheumatoid arthritis. *Ann Rheum Dis* 47: 198–205
5   Alwan WH, Dieppe PA, Elson CJ, Bradfield JW (1989) Hydroxyapatite and urate crystal induced cytokine release by macrophages. *Ann Rheum Dis* 48: 476–482
6   Ledingham J, Dawson S, Preston B, Milligan G, Doherty M (1992) Radiographic patterns and associations of osteoarthritis of the hip. *Ann Rheum Dis* 51: 1111–1116
7   Swan A, Chapman B, Heap P, Seward H, Dieppe P (1994) Submicroscopic crystals in osteoarthritic synovial fluids. *Ann Rheum Dis* 53: 467–470, 1994
8   van Linthout D, Beutler A, Clayburne G, Sieck M, Fernandes L, Schumacher HR Jr (1997) Morphometric studies on synovium in advanced osteoarthritis: Is there an association between apatite-like material and collagen deposits? *Clin Exp Rheumatol* 15: 493–497

9   Dieppe P, Swan A: Identification of crystals in synovial fluid (1999) *Ann Rheum Dis* 58: 261–263
10 Conroizer T, Chappuis-Cellier C, Richard M, Mathieu P, Richard S, Vignon E (1998) Increased serum C-reactive protein levels by immunonephelometry in patients with rapidly destructive hip osteoarthritis. *Rev Rhum Engl Ed* 65: 759–765
11 Komiya S, Inoue A, Sasaguri Y, Minamitani K, Morimatsu M (1992) Rapidly destructive arthropathy of the hip. *Clin Orthop* 284: 273–282
12 Masuhara K, Bak lee S, Nakai T, Sugano N, Ochi T, Sasaguri Y (2000) Matrix metalloproteinases in patients with osteoarthritis of the hip. *Intern Orthop* 24: 92–96

Differential gene trap: A new strategy for identifying genes regulated during cartilage differentiation

Tamayuki Shinomura[1], Kazuo Ito[2] and Magnus Höök[2]

[1]Department of Hard Tissue Engineering, Tokyo Medical and Dental University, Yushima, Bunkyo-ku, Tokyo 113-8549, Japan; [2]Institute of Biosciences and Technology, Texas A&M University, Houston, Texas 77030, USA

To gain a better understanding of cartilage tissue and its development, it is essential for us to know all genes expressed in chondrocytes. Our experiments described here are aimed toward identifying most of genes specifically activated in the tissue.

There are several methods for detecting genes that are specifically expressed in cartilage, such as subtractive hybridization or differential display. However, these methods show a strong bias toward high copy number of the message, potentially missing the identification of mRNA that exist in low numbers. To solve these problems, we sought to apply a gene trap method that is a powerful way to study gene-specific changes in transcription during cellular differentiation. The most important point is that the trapping event is, in principle, independent of the abundance of the message.

The integration of a trap vector adjacent to an actively transcribed gene places the selectable marker under the control of the endogenous transcription unit. Thus, we have developed a new strategy for identifying cartilage-specific genes by selective activation of a fusion gene for a dominant positive and negative selectable marker during chondrogenic differentiation (Fig. 1). To achieve this goal, we have constructed and tested a gene trap vector containing a fusion gene between the herpes simplex virus thymidine kinase (HSV-tk) and the hygromycin phosphotransferase (hyg) genes [1]. The HSV-tk serves as a negative selectable marker, and the hyg serves as a positive selectable marker.

We here describe experiments showing a strategy to identify genes activated during cartilage differentiation of cultured mouse ATDC5 cells [2]. When the cells, growing in insulin deficient medium, are placed into medium with insulin, they undergo chondrogenic differentiation. In this culture system, the transfer of our gene trap vector into undifferentiating ATDC5 cells brings about the circumstance that the activation of the fusion gene will be lethal in the presence of gancyclovir. On the contrary, the activation is required to survive in the presence of hygromycin after chondrogenic differentiation. Therefore, the sequential negative and positive selections coupled with the cellular differentiation allow us to screen many genes that are newly activated during the cartilage differentiation of ATDC5 cells. The

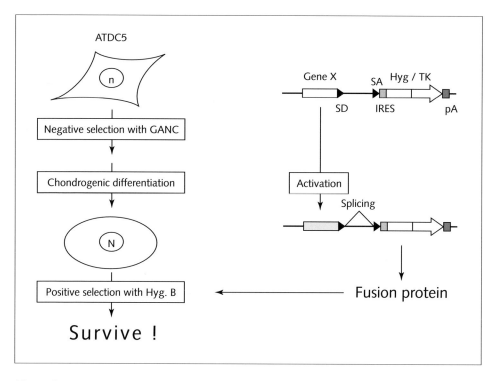

Figure 1
Strategy of differential gene trapping. When the trapping vector is inserted into an intron of a gene X, the successful splicing would produce a fusion transcript containing sequences of the trapped gene at the 5' end and Hyg and TK fusion sequences at the 3' end. To allow for the expression of Hyg/TK fusion protein, an internal ribosome entry site was placed upstream of the Hyg/TK fusion gene in the construct. See text for further explanation. SD, splice donor sequence, SA, Splice acceptor sequence; IRES, internal ribosome entry site; Hyg, hygromycin phosphotransferase gene; TK, Herpes simplex virus thymidine kinase gene; pA, poly A signal.

genes that we have identified in this manner were confirmed to be up-regulated during cartilage differentiation by northern blot analyses. Furthermore, the northern analyses indicated that the trapped genes were expressed in various temporal patterns during the cartilage differentiation of ATDC5 cells. In relation to this, our preliminary data suggested that the timing of hygromycin-selection would be particularly critical to trap genes that were activated at early stage or at late stage of the differentiation more effectively.

We believe that our gene trap approach will offer a powerful tool for identifying and cataloging the genes activated during chondrogenic differentiation. Once we have identified many of those genes, DNA chip technology will allow us to perform systematic studies of the expression of cartilage specific genes in normal and disease states. Conversely, the analysis of the alterations in gene expression in a cartilage can provide detailed information about its state.

References

1 Lupton SD, Brunton LL, Kalberg VA, Overell RW (1991) Dominant positive and negative selection using a hygromycin phosphotranferase-thymidine kinase fusion gene. *Mol Cell Biol* 11: 3374–3378
2 Shukunami C, Shigeno C, Atsumi T, Ishizeki K, Suzuki F, Hiraki Y (1996) Chondrogenic differentiation of clonal mouse embryonic cell line ATDC5 *in vitro*: Differentiation-dependent gene expression of parathyroid hormone (PTH)/PTH-related peptide receptor. *J Cell Biol* 133: 457–468

Discussion

Collagen cross-links as markers of bone and cartilage degradation

Q: *Linda Sandell*: How can you sort out the newly synthesized from older type II products in cartilage when you look at the telopeptide? Is there any way to subtract out N-propeptide or C-propeptide results?

A: *David Eyre*: We select antibodies that will only recognize or prefer to recognize those sequences in which the lysine residue is involved in the crosslink. If newly made collagen has not become cross-linked, then the lysine or the hydroxy-lysine will still be in the telopeptide. However, it is invisible to the antibodies that we use.

Q: *Linda Sandell*: How long does it take to get cross-linked?

A: *David Eyre*: If you put labeled lysine into a joint and then look at the maturation of C-14-lysine into crosslinks, the halftime to maximal level of pyridinoline formation is 2–3 weeks, a fairly short term.

Q: *Linda Sandell*: What I'm getting at is that Robin Poole's data show that collagen type II C propeptide changes in concentration in OA.

A: *David Eyre*: The devil is in the details. There could be greater synthesis of type II collagen, but if the fragment you want to measure, whether it is a propeptide or telopeptide, has been degraded very rapidly by a downstream process that is part of the catabolic route, then you would actually see a decrease in your formation marker. It gets very difficult to interpret.

Q: *Dick Heinegard*: I wasn't quite clear about the osteoclastic breakdown. Is the free pyridinoline derived primarily from osteoid breakdown while your crosslink assay is derived from the calcified bone underneath?

A: *David Eyre*: Osteoclasts *in vitro* in our hands did not generate any free pyridinolines. So, the free pyridinoline in normal children and normal adults I think has been generated downstream, which would probably mean the liver and the kidneys.

Q: *Dick Heinegard*: Then are the larger fragments that are not taken up by the kidney generated from the osteoid?

A: *David Eyre*: Are you thinking in terms of pycnodysostosis? In pycnodysostosis all of the fragments from bone, whether it was calcified collagen or not, end up being degraded primarily to the free pyridinolines. Whereas in normal circumstances, where cathepsin-K is the primary degrader of the collagen, the resulting pool of very small peptides escapes degradation in the liver and gets excreted in urine.

Q: *Stefan Lohmander*: My interpretation of your presentation is that it would be difficult, if at all possible, to develop assays for breakdown or turnover of the other and minor collagens in cartilage. Is that a fair assessment?

A: *David Eyre*: If you mean type IX collagen or type XI collagen, I think it could be done. However, the levels in blood even from type II collagen are very low compared to the levels of type I collagen that are derived from bone. So to go to the next level down and measure collagen IX breakdown products you would have to be at about 1% of the level of type II collagen. So it's not impossible, but I think it's a difficult task.

Q: *Robin Poole*: We have set up assays for the col 2 and NC4 domains of type IX collagen. In forming cartilage, like growth plate, the ratio of the col 2 domain of type II collagen to col 2 domains of type IX is 5 to 1. It fits very nicely with Karl Kadler's concept of the microfibril. But when we looked more recently at adult human articular cartilage, we found ratios between several hundred and several thousand. There seems to be a selective drop-off of type IX, at least in our hands. There thus seems to be a lot less (as David said) of the type IX. So, type II seems to offer better opportunities for measuring fibril turnover.

A: *David Eyre*: In fact there's a gradient, as you know, from around the cell to the matrix. The interterritorial matrix in adult articular cartilage has little or no type IX collagen. They're essentially bare fibrils that are clearly banded, and they're thick. My sense is that the fine fibril meshwork around the chondrocyte is set up in its organization by the collagen IX, and by the small proteoglycans. That sets the template for the organization that is farther out.

Q: *Alice Maroudas*: If the collagen network gets stiffer with age, then why should we have less type IX?

A: *David Eyre*: I don't think type IX collagen is responsible for the stiffness of the network of a mature hyaline cartilage. I really think that type IX may be more important in setting up the fibril organization of type II, which is more random than that of any other tissue or collagen type. The matrix will then mature and last as you have shown, Alice, with a half-life of maybe 50 to 60 years. Thus most of the fabric of the matrix was made long ago. But I think it can be retread, if that's the right word.

Markers of joint tissue turnover in osteoarthritis

Q: *Paul Dieppe*: Can I put it to you that markers such as serum COMP might actually be much more useful in rheumatoid arthritis than a disease like osteoarthritis? My reason for suggesting that is this: Rheumatoid arthritis is a polyarticular disease, where as far as we can tell all the joints are in the same phase of process at any given timepoint. If the disease is active, then lots of joints are active. In osteoarthritis, there are two complications. There is the possibility of a monoarthritis multiplex type phenomenon where one big joint is in one phase and another joint is in a completely different phase. Then there is the complication in older people who are developing spinal disease at the same time. There is a mass of synovial joints in the spine with a huge amount of disc material, and we have absolutely no idea what is going on because we have no way of measuring what's going on in the spine. It seems to be that these two factors make it very complicated to unravel a serum marker in osteoarthritis.

A: *Dick Heinegard*: I guess that one of the uses today would be to monitor effects of intervention. If the COMP levels changed, this would indicate that the process in the cartilage has changed. That would also hold for osteoarthritis. Another situation is when you have gone out skiing and the skis go one way and you go another way resulting in damage to some of your joints, which could also be a high risk for progressive joint damage. You could perhaps use COMP levels to find out how the cartilage is affected and what is going on. But I think we are only at the starting point with COMP. There are certainly proteins that are unique to the meniscus and others that are unique to articular cartilage that might be used to monitor joint damage. I think that there is a difference between different joints and how they are affected in RA and OA: It would be very interesting to find molecules that differ. I wouldn't be as pessimistic with regard to OA as you may be.

Q: *Joel Block*: Whenever we see marker data for OA, there's always a huge amount of splay, and we're always making conclusions based on population trends. I know that your group and others have looked at the relationships among several markers. Have you looked at individual relationships? For instance, in the data you presented among those individuals who had progressive OA and low COMP levels, were those the same people that had progressive OA and low CRP? Or did they have high CRP and low COMP levels?

A: *Dick Heinegard*: Of course it's interesting to combine the different markers and make ratios of various types, but I don't think that's been done with the CRP and COMP.

Q: *Joel Block*: But I'm specifically interested in the outliers. For instance, the trend of the population is that the COMP goes up, but there are always 2 or 3 outliers that are going down. I'm wondering whether those people who are going down but still have progressive disease, have another marker that also goes in the other direction.

A: *Dick Heinegard*: It's too early to tell.

Q: *Stefan Lohmander*: With collagen II markers in serum or urine we think we have a fairly good idea that it is coming from cartilage. Now what about the COMP? What's the source of COMP we find in serum?

A: *Dick Heinegard*: My feeling is that it is fairly selective for cartilage. We've looked at tendon, and cartilaginous, pressure loaded parts of tendon have fairly high contents of COMP. I don't know where those fragments would end up. In the synovium, there is a low expression of COMP. However, we can prevent synovial inflammation, but that will not change the COMP. We have to deal with the cartilage to change the COMP level. There's a much higher concentration of COMP in the synovial fluid than there is in the synovial capsule, about two to three orders of magnitude difference. So, if there is a production in the capsule, it is very low.

Q: *Stefan Lohmander*: You've looked at knee joints to a large extent and there is COMP in the meniscus.

A: *Dick Heinegard*: I don't think we can distinguish today between meniscus and articular cartilage in the knee. But we have very similar data from the hip joint, so I think it will hold up.

Gene expression profiling by the cDNA array technology: molecular portraying of chondrocytes

Q: *Vince Hascall*: I'm reminded of a quote of Martin Matthews quite some time ago. When asked to define cartilage, he said, "cartilage is like pornography... it's hard to define, but it's easily recognized by experts". The point is how do you define normal? When I think of a chondrocyte, I think that a chondrocyte is not a chondrocyte is not a chondrocyte. If you did gene expression rate analysis between the superficial chondrocytes, deep zone chondrocytes or calcified chondrocytes, I suspect you'd find quite different spectra. In the end when you look at the osteoarthritic cartilage, what is your real baseline of comparison?

A: *Thomas Aigner*: That's an important point. There is no answer to that. What you could do is make laser dissection experiments taking a few cells out of a certain very well defined zone and try to quantify the RNA. If you look at late stage OA cartilage, you have lost the surface zone, and that immediately shifts your expression patterns. For the late stage OA cartilage, we are not going for the most damaged areas, but are sticking to the peripheral areas, and these peripheral areas have a significant portion of the matrix left over.

Q: *Stefan Lohmander*: One thing that strikes me in listening to these presentations, is that it is tempting to look at the clear-cut cases of overt osteoarthritis *versus* normals. I think it is more interesting to look at the early stage disease because that's where things are happening. I can deal with the late stage pretty well as an orthopaedic surgeon, but the early stage is something where I think if we could identify those events that start to shift the slope of the line, that would be a very significant finding.

A: *Thomas Aigner*: Well, the problem is obviously to get the material from the early stages. The best we can do, which is excellently done in Chicago, is to take degenerative lesions from the autopsies of donors.

Clinical evaluation of markers for osteoarthritis

Stefan Lohmander: Thank you, Ivan. Lots of us are spending much effort and time on the discovery level of biomarkers, but the perspective you are providing us with here is as important for the utility of the markers.

Q: *John McPherson*: What about having patients serve as their own controls? Wouldn't it make more sense to do a paired *t*-test strategy where you look at the patient at time zero and 1 year later and determine whether or not you've changed the degradation rate of type II collagen within that patient?

A: *Ivan Otterness*: Yes, that's the way you would like to do that kind of trial. But, in fact, the measurement between the diseased and the control population gives you an idea of the window that you have to correct.

Q: *John McPherson*: I guess the problem is that even if you could show statistically significant reduction in the rate of degradation within a patient, you are faced with the problem of knowing whether that is clinically relevant in the long term.

A: *Ivan Otterness*: You have to do the clinical studies, and you have to have the imaging data to determine clinical relevance of the marker change.

Q: *Paul Dieppe*: One of the problems when one is thinking about using markers in therapy is that we don't have a positive control because we don't have an intervention that works in OA. There is one possibility I've thought about and that Stefan has thought about, but not got around to doing any work on, and that is osteotomy at the knee joint. Osteotomy at the knee joint, in a proportion of cases, leads to apparent healing. Healing is in terms of fibrocartilage rather than hyaline cartilage, but with reformation of a joint surface and effective remodeling of the bone. I wonder if you or Stefan or anyone else has actually got any data of markers using that as a positive control? That might really help us get a feel of whether we can apply them to therapeutic trials.

A: *Stefan Lohmander*: We have looked at osteotomy patients using each patient as his or her own control and comparing with preoperative data: You do see some quite significant response in biomarkers. They appear to be reasonable changes that would go along with a lowering of the degradative activity. But I would say that the data we have thus far are useful in formulating a hypothesis for new trials rather than for confirming our previous hypothesis.

Q: *Joel Block*: If OA were really a variety of different diseases that all look alike at the joint level, then it would make sense in theory that there would be a bunch of outliers for any individual marker. From a statistical point of view, wouldn't it make sense, then, to do an outlier analysis rather than a median or population analysis?

A: *Ivan Otterness*: We've followed markers over a year's time to measure stability, and we've measured other markers at the same time. The variation of marker samples between clinical visits is much higher than the variation in the particular measurement you make. So I would rather not treat individual outliers. I would rather treat the patient's results over a year of time.

Q: *David Eyre*: The units of TIINE in urine, are they expressed per milliliter or do you normalize to creatinine?

A: *Ivan Otterness*: No, in fact we don't normalize to creatinine. There is a different secretion pathway for peptides. If we normalize to creatinine our between-visit measurement error on our patients goes up from about 23% to around 45%.

Joint degradation in rapidly destructive and hypertrophic osteoarthritis of the hip

Q: *Paul Dieppe*: You mentioned some possibilities of the etiology, but there are a couple of others that have been quite seriously investigated in the literature that I wondered if you had data on. Early on Solomon and others actually described the atrophic form of hip OA as indomethacin hips, and there is quite a lot of literature implicating drugs as a cause of this rapid destruction in the hip. So one question is whether you investigated that and looked at the drug history. Then there is another area of investigation, which is the suggestion that this type of rapid destruction is driven by a synovial metaplasia reaction driven by apatite crystals perhaps derived from the bone.

A: *Wolfgang Eger*: There was no evidence that those patients with RDH-OH have received any other medicine differed from the other patients with hip sclerotic OA. We haven't looked at the synovial tissues in detail yet to answer if there is a higher crystal depositum in that tissue.

Q: *Paul Dieppe*: It is, of course, extremely difficult to get comprehensive drug histories on these patients, but I would encourage you to try and do that and to go back as far as possible.

A: *Stefan Lohmander*: I would also suggest that on the list of further investigations that you get their DNA histories.

Q: *Hans Häuselman*: Is there a different diagnosis of cell-negative arthropathy? Did you look for whether these patients have an undifferentiated sero-negative form of arthropathy, or an inflammatory bowel disease associated arthropathy, or another type of arthropathy?

A: *Wolfgang Eger*: We checked routine parameters, but we haven't looked at the extended parameters that the rheumatologists usually measure. From our clinical observations, the patients were healthy individuals, except for the disease in the hip.

Q: *Matt Warman*: Following up on Stefan's question that addresses the genetic component, what is the percentage of patients that have cystic *versus* sclerotic OA?

A: *Wolfgang Eger*: It's about 10–15% in our hospital.

Q: *Matt Warman*: And do you know anything about whether there is familial aggregation in this type?

A: *Wolfgang Eger*: No, we ask all those questions and there is no evidence for that.

Q: *Matt Warman*: If you did a 5-year follow up of a patient with this particular type of OA what is the likelihood of the contralateral joint (or other hip joint) also becoming affected?

A: *Wolfgang Eger*: We haven't checked that, but most patients that develop a cystic hip OA in one joint have a healthy hip on the other side. We have patients who have a cystic hip on one side and a sclerotic hip OA on the other.

Differential gene trap: A new strategy for identifying genes regulated during cartilage differentiation

Q: *Linda Sandell*: How many different genes do you have altogether that are activated?

A: *Tamayuki Shinomuro*: This time I just showed five different genes, but we have 50 genes already isolated.

Q: *Linda Sandell*: If you have about 50 that are newly activated, we'll be seeing this for a while!

Q: *Manas Majumdar*: What is the percentage of cells that differentiate after adding insulin?

A: *Tamayuki Shinomuro*: About 70% or 80% of the isolated colonies differentiate to chondrocytes.

Q: *Vince Hascall*: It's an extremely clever approach. I assume that in each case you are going to have to go back to the primary cultures and see whether that particular gene actually does behave in the way that you presumed it would from the assay. The question then is have you done that enough to know the level of fidelity of the system? How many false positives might you get?

A: *Tamayuki Shinomuro*: After the selection condition, if the gene is not activated and not detected in the cartilage, the gene is discarded. Now around 15% of the genes are newly activated in our system.

Assessment of joint damage in osteoarthritis

Introduction

Paul Dieppe

MRC HSRC, Department of Social Medicine, University of Bristol, Canynge Hall, Bristol BS8 2PR, UK

What is osteoarthritis: Joint damage, joint pain or both?

Osteoarthritis (OA) may be defined pathologically as a characteristic form of synovial joint damage, or clinically on the basis of pain (particularly use-related joint pain), gelling of joints after inactivity, crepitus and reduced mobility [1]. However, regional use-related joint pain is extremely common in the community, and when this occurs in joints such as the hip or knee (which have a high prevalence of OA) in older people (the risk group) clinicians pragmatically tend to call this 'OA' even when there is no evidence for joint damage.

But, as researchers working within the biomedical model of disease, we like to see evidence of pathology in association with symptoms. The pathology (joint damage) that is characteristic of OA features: focal areas of articular cartilage destruction, subchondral bone change, marginal osteophyte (or chondrophyte) formation, capsular thickening, and variable degrees of mild synovitis. When severe enough, these can be seen on the radiograph, which is the investigation most commonly used to diagnose OA and assess the degree of joint damage.

Major problems inherent in this research-based definition and in the scientific investigation of OA include:

1) The radiograph is insensitive to joint damage and cannot detect early changes or visualise cartilage or soft tissues.
2) The correlation between radiographic evidence of joint damage and the pain is poor [2]. Some people with severe joint damage never experience symptoms, whereas others may experience severe pain of the sort we associate with OA, in the absence of any evidence of joint damage.

What processes lead to the joint damage?

OA is a disease of the whole joint organ, involving linked pathological processes affecting cartilage, bone, synovium and capsule.

Table 1 - *Mutually adjusted relative risk and 95% confidence intervals for different risk factors for OA and their association with initiation or progression of x-ray changes (from Cooper et al., 2000).*

Risk factor	Initiation n = 30	Progression n = 49
Obesity	9 (2.6–32.2)	2.6 (1.0–6.8)
Pain	2.9 (1.2–7.2)	0.8 (0.4–1.7)
Heberden's nodes	2.7 (1.0–6.9)	0.7 (0.4–1.6)
Previous injury	4.8 (1.0–24.1)	1.2 (0.5–3.0)
Elite sport	3.2 (1.1–9.1)	0.7 (0.4–1.6)

Recent work suggests that subtle changes in the stability or shape of joints may initiate these processes [1, 3], which can be regarded as an attempt by the synovial joint to contain the change in loading by altering its shape and stability. Longitudinal studies suggest that this attempt at containment is often successful – radiographs often show no change in joint damage over a period of years or decades, even when there is quite advanced OA [4, 5]. However, in some patients the processes continue, increasing the degree of joint damage, often leading to complete joint "failure".

We and others have recently published evidence that suggest that the risk factors for the initiation and progression of joint damage are different [6]. A community based cohort of people with and without knee pain was examined and radiographed twice, with an interval of 5 years between examinations. Risk factor data, collected at the time of the first examination, was related to the initiation of x-ray changes, or to their progression. The data (Tab. 1) showed that the risk factors normally associated with OA and with trauma as a key factor in its pathogenesis, were associated with the initiation but not the progression of the x-ray changes.

What processes lead to the joint pain?

Many of the pathological processes involved in developing joint damage can be painful [7]. For example, the growth of osteophyte can result in pain, probably by periosteal elevation, and some of the subchondral bone changes cause increased intraosseous pressure, resulting in pain. Synovial and capsular changes, and meniscal extrusion are other potential causes of pain resulting directly from the pathological changes of OA.

However, local pain of the type that we associate with OA (use-related pain with inactivity stiffness) can also arise from periarticular problems and muscles, which

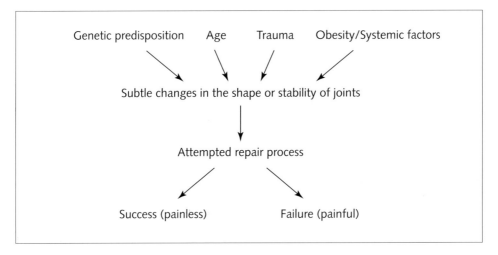

Figure 1
A diagrammatic representation of some of the pathways involved in the development of OA, which is seen as the attempt of the joint to adjust to mechanical abnormalities or damage by changing its shape and stability to accommodate them.

have little or no direct relationship to the pathological processes or joint damage of OA.

Recent data from the Baltimore Longitudinal study of ageing [8] indicate that the relationship between the severity of joint damage and severity of pain is poor, suggesting that there may be a threshold of joint damage that predisposes to pain (a threshold that may be different in different individuals), after which further joint damage has little effect on symptoms. In addition, it is clear that there are many other factors that affect pain and its severity in OA, including psychososcial factors such as anxiety, depression, social isolation and poverty [9].

Putting these concepts together leads us to a diagrammatic representation of the pathogenesis of OA (Fig. 1).

Radiographic changes of OA

The thesis outlined above suggests that OA can be regarded as the reaction of joints to mechanical insult, a reaction that is successful in many people, but one that leaves an imprint of radiographic change, and one that can be progressive in a minority of people. This would result in their being three different groups of people in the community, each of which have very similar radiographic changes of OA, but in whom these changes have quite different significance (Fig. 2). This might explain why

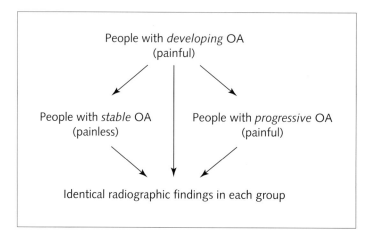

Figure 2
Longitudinal data suggest that in most cases the OA process becomes quiescent, perhaps when the process has successfully contained the abnormal biomechanics that were driving it. A minority of cases progress inexorably. However, radiographs, the commonly used tool to diagnose and assess OA, will be identical whether the process is active or not. Pain occurs when it is active, but joints may also be painful in the stable phase for different reasons.

attempts to sort out OA on the basis of cohorts of people recruited from the community by radiographic change have not been very successful.

Assessing the severity of joint damage in OA

There are four different methodologies available to us to assess joint damage in OA:

- Clinical
- Imaging
- Direct visualisation
- Pathology

There are both generic and specific problems with these methods.
The generic problem is ascertainment – as outlined above. In other words, which people are assessed, and which of their joints are looked at. Some of the specific problems relating to each method are discussed in the following papers in this section of our proceedings – in this paper two generic ascertainment problems will be highlighted – which types of people get assessed for the presence of OA, and which of their joints get examined.

Table 2 - *The frequency of osteophytes and eburnation of bone at selected joint sites in a population of 563 skeletons recovered from an English burial site (burial times ranged from 900 AD to 1850 AD). (From Shepstone et al., 2001.)*

Joint site	% with osteophytes	% with bony eburnation
Hip	15.4	2.3
Knee	11.0	3.9
Elbow	16.5	4.2

Which people get assessed for the presence or severity of OA?

Many people in the community have radiographic changes without pain, and many have pain of the sort associated with OA, without radiographic changes. In addition, many of those with pain, whether or not they have OA, never seek any help from the medical profession, and are never examined. Most of those who do seek help are dealt with by primary care physicians, and never reach a hospital-based department. And yet, the majority of in-depth studies of OA are undertaken on people ascertained from hospital-based departments of rheumatology or orthopaedics. Such studies will not provide us with a representative picture of the nature of OA.

Which joints get examined for OA severity?

OA is thought of as a disorder that mainly affects the knees, hips, small joints of the hands and feet, and the spinal joints. This is because it is these joints that cause pain and disability as a consequence of OA pathology. So, if a person comes in to see us with OA in the knee, for example, we are likely to look for changes in the spine, hips, hands and feet, but not, for example, the elbow. It is extremely unlikely that the elbow will be radiographed in the absence of pain in that joint. Paleopathological studies carried out by Juliet Rogers in the UK have provided a variety of new insights into the disease [10]. For example, her data suggest that OA pathology is very frequent at the elbow [11]. As shown in Table 2, both osteophytosis and eburnation of bone are seen more commonly at the elbow than at the hip or knee when skeletons are examined. This suggests that an understanding of OA which comes from the biased clinical and radiographic data alone will be an inadequate one. It also suggests that the elbow might be worthy of study in order to help understand why some joints with OA do or do not develop pain.

Conclusions

The term "OA" covers a heterogeneous group of disorders and disease processes which result in similar morphological changes in joints. These morphological changes sometimes lead to pain in some of the affected joints.

In order to understand the condition better we need to extend our research beyond the morphology to take account of the determinants of pain and disability. This is likely to require us to embrace a broader biopsychosocial model rather than the restricted biomedical one which dominates our present research activities.

References

1   Felson DT, Lawrence RC, Dieppe PA, Hirsch R, Helmick CG, Jordan JM, Kington RS, Lane NE, Nevitt MC, Zhang Y et al (2000) Osteoarthritis: New insights. Part 1: The disease and its risk factors. *Ann Intern Med* 133: 635–646
2   Lawrence JS, Bremner JM, Bier F (1966) Osteoarthrosis: prevalence in the population and relationship between symptoms and x-ray changes. *Ann Rheum Dis* 25: 1–24
3   Sharma L, Lou C, Felson DT, Dunlop DD, Kirwan-Mellis G, Hayes KW, Weinrach D, Buchanan TS (1999) Laxity in healthy and osteoarthritic knees. *Arthritis Rheum* 42: 861–870
4   Dieppe P, Cushnaghan J, Shepstone L (1997) The Bristol "OA500" study: progression over 3 years and the relationship between clinical and radiographic change at the knee joint. *Osteoarthritis and Cartilage* 5: 87–97
5   Dieppe P, Cushnaghan J, Tucker M, Browning S, Shepstone L. (2000) The Bristol "OA500" study: progression and impact of the disease after 8 years. *Osteoarthritis and Cartilage* 8: 63–68
6   Cooper C, Snow S, McAlindon T, Kellingray S, Stuart B, Cogon D, Dieppe P (2000) Risk factors for the incidence and progression of radiographic knee osteoarthritis. *Arthritis Rheum* 43: 995–1000
7   Dieppe P (1999) Osteoarthritis: time to change the paradigm. *BMJ* 318: 1299–1300
8   Creamer P, Lethbridge-Cejku M, Hochberg M (1999) Determinants of pain severity in knee osteoarthritis. *J Rheumatol* 26: 1785–1792
9   Creamer P, Lethbridge-Cejku M, Costa P, Tobin J, Herbst J, Hochberg M (1999) The relationship of anxiety and depression with self-reported knee pain in the community. *Arthritis Care and Research* 12: 3–7
10 Rogers J and Dieppe P (2001) Paleopathology. In: K Brandt, M Doherty, S Lohmander (eds): *Osteoarthritis*, 2nd ed. Oxford Medical Press; *in press*
11 Shepstone L, Rogers J, Browning S, Dieppe P (2001) A paleopathological study of osteoarthritis. *Osteoarthritis and Cartilage* 9 (Suppl B): S7 (abstract)

Imaging cartilage changes in osteoarthritis

Charles G. Peterfy

Synarc, Inc., 455 Market Street, San Francisco, CA 94105, USA

Introduction

Imaging articular cartilage has become one of the most dynamic areas of development in musculoskeletal radiology, and over the past several years numerous advances have been made in both *in vitro* and *in vivo* imaging of this tissue. Yet, integrating these techniques into epidemiological studies and clinical trials of osteoarthritis (OA) has been surprisingly difficult and slow, particularly in light of the fact that the imaging technologies upon which these methods are based have been an integral part of day-to-day clinical practice for decades. Moreover, the role that imaging is asked to play in epidemiological studies is the same role that it currently plays in clinical practice, namely patient selection (diagnosis and staging) and monitoring disease progression and treatment response. Why should such a disparity exist between clinical practice and clinical research? Why is radiology so late to this party?

At least part of the reason has been the lack of effective therapies for combating OA. In the absence of therapy, clinical practice has no need for methods of identifying patients most appropriate therapy or for determining how well therapy worked. Accordingly, the methods available in clinical practice (loosely-standardized assessments of radiographic joint-space width, subjective assessments of articular cartilage integrity) evolved to support only low-level structural information thus far. However, considerably more precise and validated methods are required for clinical research in OA and for gaining regulatory approval of any putative new therapies. Thus, the demand for these innovations arises first during the clinical testing of new therapies, and it is therefore, in a Darwinian sense, the priorities and unique regulatory and logistical constraints of the clinical trials process that shape the early evolution of these radiological techniques (Fig. 1). During this phase of technical/analytical evolution, mainstream clinical practice has little value for the enhanced performance that these methods provide, and it is only once the therapies are approved for clinical use and available to clinicians that a demand for patient-selection and efficacy-monitoring tools emerges in the mainstream.

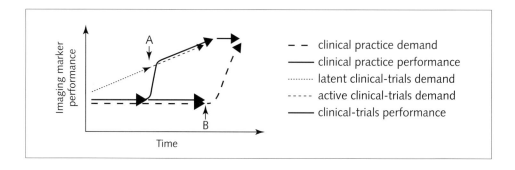

Figure 1
Therapy is a key driver of innovation in medical imaging. The thick broken line (- - - -) depicts the level of performance demanded by mainstream clinical practice for imaging markers of articular cartilage. Prior to the availability of cartilage-modifying therapy, this trajectory is relatively low and flat. The thick solid line (———) shows the actual performance of available imaging techniques and markers (radiographic joint-space width, subjective cartilage integrity with MRI), which tracks the clinical demand closely. The performance required for clinical trials of cartilage-modifying therapy (·········) is greater, particularly with respect to validity, multicenter stability and measurement precision, but this is only a latent demand until putative therapies begin to appear in the pipelines of pharmaceutical and biotechnology companies. Entry of these new therapies into clinical testing (A) triggers the development of new imaging techniques and markers (———) along performance criteria aligned with the priorities of clinical trials (- - - - - - -). Initially, the enhanced performance of these imaging endpoints is not valued by mainstream clinical practice. The technology demand in clinical practice, therefore, continues along its original, shallow trajectory. However, once the new therapy enters the market (B), a new demand emerges in clinical practice for methods of selecting patients appropriate for the new therapy and monitoring treatment effectiveness and safety. In this way, the tools and endpoints adapted to suit clinical trials research eventually find their way back into mainstream clinical practice. (Adapted from: Peterfy CG (2001) Magnetic resonance imaging of rheumatoid arthritis: The evolution of clinical applications through clinical trials. Semin Arthritis Rheum. 30(6): 375–396.)

In osteoporosis research, for example, it was the promise of bisphosphonate therapy that created the initial demand for more precise methods of non-invasively measuring bone mineral density. This demand fueled the evolution of dual-energy x-ray absorptiometry (DXA), a unique imaging technology that prior to the appearance of bisphosphonate therapy had no place in mainstream clinical practice, but that today is part of the routine clinical management of patients with osteoporosis. It was during the clinical trials of these therapies that DXA was refined and validated, and gained its initial foothold in the clinical infrastructure. Moreover, the availability of DXA and other tools, such as vertebral morphometry, for monitoring disease progression and treatment response in osteoporosis has stimulated further drug development in this area. So, therapeutic innovation and innovation in diagnostics are locked in a co-evolutionary process, in which advances in one stimulate advances in the other.

A similar pattern is seen in OA research, in which the recent emergence of structure-modifying therapies in the pipelines of biotechnology and pharmaceutical companies has spawned numerous innovations in imaging evaluation of this disease and its therapy. Tools that have been adapted for multi-center clinical trials are useful not only for global drug development but also large epidemiological studies, and extend more easily into clinical use than do some of the cutting-edge technologies typically used in single-site university research. Accordingly, innovation in clinical-trials radiology can advance understanding about the disease as well. This is exemplified by the recently launched Osteoarthritis Initiative, an historic collaborative between the National Institutes of Health (NIAMS, NIA) and the pharmaceutical industry to conduct a large, longitudinal study aimed at, among other things, creating new knowledge about the proper use of imaging and other biomarkers in clinical trials of osteoarthritis.

Therefore, in order to anticipate which of the numerous innovations in cartilage imaging that have been developed over the past several years will ultimately find their way into clinical use and what attributes they will have when they get there, one must understand the clinical trials process and the selective pressures at play in that environment. Familiarity with these dynamics is essential to contributing substantively to the development of these techniques.

One of the key challenges facing clinical trials and epidemiological research is dealing with multi-center data collection. This represents a fundamental departure from traditional university research, which typically operates on a single-site basis, but it is a necessity when hundreds or thousands of patients must be evaluated rapidly and efficiently in a single study. Radiology offers a unique advantage over clinical assessments in this regard, as it allows centralization of data generation, not just data management (Fig. 2).

In a typical multi-center clinical trial, hundreds or potentially thousands of patients must be evaluated at multiple sites throughout the country or around the world in a consistent and timely fashion. At each of these sites, clinical investigators using their medical expertise and judgement extract the relevant clinical information from their subset of patients and record the results on case-report forms that are then aggregated and databased by a central service, such as a contract research organization. The multiplicity of sources of data generation (clinical investigators) in this scenario is an inescapable source of variability in the clinical assessment data. Centralizing the data management is essential to any large multi-center study, but does not solve this fundamental problem.

Radiological assessment, however, is able to centralize the actual data generation and thereby contain this source of variability. In this scheme, the expert central radi-

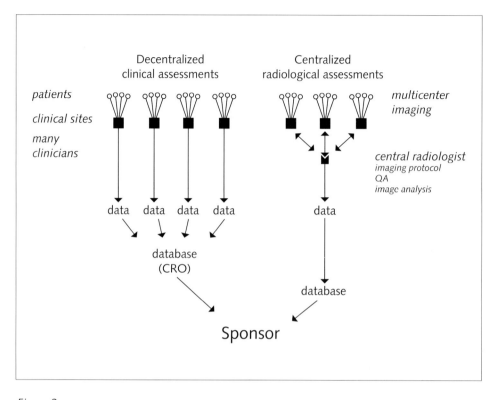

Figure 2
Clinical-trials radiology centralizes data generation, not just data management. The multiplicity of sources of data generation (clinical investigators) in the clinical assessments of multi-center trials is an inescapable source of variability. Centralized data management is essential to any large multi-center study, but does not solve this problem. Radiological assessment, however, is able to centralize data generation and thereby contain this source of variability. In this scheme, the expert central radiologist, based on the scientific and regulatory needs of the study, the capabilities of the sites selected, and an intimate understanding of how he/she will analyze the images, designs an imaging protocol that will generate the type of images needed consistently across all the patients, sites and equipment platforms, and throughout the duration of the study. The radiologist also provides any information or training necessary to the sites on an ongoing basis, and then checks the images to ensure protocol compliance and adequate image quality. Finally, using special expertise and sophisticated computer aids, the central radiologist extracts the relevant morphological, compositional and physiological information from the images and enters them directly in the central database. Centralizing data generation in this way not only reduces variability in radiological assessments, but facilitates integrating the critical components of this process, as image acquisition, QA, and image analysis/quantification must all come together seamlessly to work properly.

ologist, based on the scientific and regulatory needs of the study, the capabilities of the sites selected, and an intimate understanding of how he/she will analyze the images, designs an imaging protocol that will generate the type and quality of images needed, consistently across all the patients, sites and equipment platforms in the study, and for the entire duration of the study. This may include the use of specialized image-acquisition aids designed to improve the consistency of multi-center imaging (see below). The central radiologist also provides any necessary information or training to the sites on an ongoing basis, and checks the images to ensure protocol compliance and adequate image quality. Finally, using special expertise and computer algorithms, the central radiologist extracts the relevant morphological, compositional and process information from the images and enters the results directly into a central database. Centralizing data generation in this manner not only reduces variability in radiological assessments but facilitates integrating the essential components of this process.

In clinical-trials radiology, as in conventional clinical radiology, image acquisition, image analysis and quality assurance must all come together seamlessly into a single integrated process to work properly. It is difficult to dissociate these elements, even in routine clinical practice, but in clinical-trials radiology, proper integration is essential. Moreover, this integration must be bottom-up. That is, it is the image analysis method that dictates how the images must be acquired and what quality control considerations must be focused on. The imaging must furthermore be integrated with the clinical assessments and any molecular marker measurements or other tests included in the study. Integrating all of these elements properly in a clinical trial requires special expertise, experience and software.

The problem with conventional radiography in multi-center studies

Because radiography is unable to directly visualize the articular cartilage, cartilage loss in arthritis must be indirectly inferred from changes in the distance between opposing articular cortices, i.e., joint-space width (JSW). However, this measure is valid only where the two articular cartilage surfaces are in direct contact with each other. In many joints, including the knee, this necessitates loading to displace any intervening fluid. Joint loading is easily accomplished in the knee by acquiring the images while the subject is standing. Even then, however, only a small portion of the curved, incongruent articular surfaces of the femur and tibia are in direct contact in any one position. With increasing flexion of the knee, progressively posterior regions of the femoral cartilage articulate with the tibial surface. Thus, JSW as a measure of cartilage thickness is highly dependent on positioning in the knee (Fig. 3). Full extension of the knee is a more reproducible position than flexion, but articulates a region of the femoral cartilage that is less frequently thinned in osteoarthritis than is the central-posterior region that articulates when the knee is

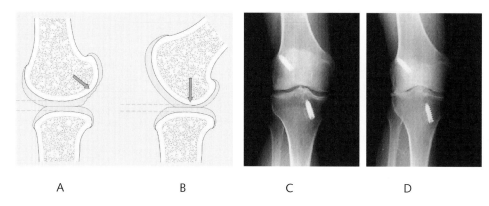

A   B   C   D

Figure 3
The importance of knee positioning in radiographic joint-space width measurement. The degree of flexion of the knee determines the region of femoral cartilage represented in JSW measurements. Mild flexion articulates the central-posterior femoral cartilage, which tends to show thinning earlier. Panel C shows a radiograph acquired in extension and shows ample lateral femorotibial joint space. Following mild flexion (D), this joint-space collapses to bone-on-bone contact.

mildly flexed. Slight flexion, as in the Schuss view [1], is therefore preferred over extension for evaluating JSW in osteoarthritis. However, positioning the knee in exactly the same degree of flexion on serial examinations is difficult, and even minor variations in positioning can produce large errors in JSW measurements over time.

Another factor complicating JSW measurement is the projectional viewing perspective of conventional radiography. Since radiography casts two-dimensional shadows of three-dimensional anatomy onto flat sheets of film, the images show varying degrees of morphological distortion, geometric magnification and superimposition of overlapping structures. These distortions can interfere with accurate dimensional measurements [2]. X-ray beam alignment, joint positioning and the distance between the joint space and the radiographic film are, therefore, critical determinants of the accuracy of serial JSW measurements.

Optimizing radiography for clinical trials

In order to control some of these factors, Buckland-Wright, et al. [3, 4] advocated the use of antero-posterior (AP) fluoroscopic alignment of the anterior and posterior margins of the tibial plateau against a horizontal x-ray beam during active flexion (Fig. 4). With this and the use of foot maps to ensure reproducible foot positioning and metallic reference markers to correct for magnification, Buckland-

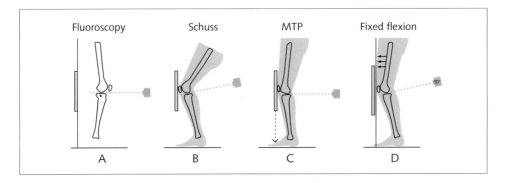

Figure 4
Comparison of techniques for radiography of the knee. The goal of serial knee radiography is optimization and fixation of the positions of the tibia and femur: 1. relative to each other (weight bearing, mild flexion); 2. relative to the film (knee-to-film distance as short as possible to minimize geometric magnification and blurring; ~10° external rotation to project tibial spines over femoral notch); 3. relative to the x-ray source (source-to-knee distance as large as possible to minimize magnification and scatter radiation); and 4. relative to the x-ray beam (beam angle in tangent to articular cortices and perpendicular to the film). This must be achieved with maximum precision and minimum cost in a multi-center setting. Four alternative techniques are depicted above. Reproducible foot positioning can be accomplished in each of these with foot maps or a fixation frame. A phantom can be applied to any of these methods to verify the side imaged (right or left) and beam angle used, and correct magnification and parallax errors. The main differences among these techniques are as follows: (A) The Buckland-Wright fluoroscopic method provides optimal alignment and good fixation of tibia, good fixation of the femur, but only minimal flexion of the knee. The knee-to-film distance is large and variable, and the source-to-knee distance is short and variable, leading to variable magnification, which must be corrected. The x-ray beam is perpendicular to the film, which is ideal. The technique is more expensive than the three alternatives, which cost the same as each other. Fluoroscopy carries greater radiation exposure to the subject and is difficult to perform in multi-center studies. (B) The Schuss view offers good alignment and fixation of the tibia, greater flexion of the knee (its main advantage), but poor fixation of the femur (its main limitation). The knee-to-film and source-to-knee distances are optimal and minimize magnification. Caudad beam angulation results in parallax error causing mild (3%) over-estimation of joint-space width, but this cancels out on serial examinations. (C) The "MTP" technique offers good alignment and fixation of the tibia, but only minimal flexion of the knee, and poor fixation of the femur (a serious limitation). The knee-to-film and source-to-knee distances are optimal, thus minimizing magnification, and the x-ray beam is perpendicular to the film. (D) The fixed-flexion technique provides good alignment and optimal fixation of the tibia, good flexion of the knee (a strength), and good fixation of the femur (its main strength). The knee-to-film and source-to-knee distances are optimal, minimizing any magnification. Caudad beam angulation results in mild parallax error, as for the Schuss view.

Wright and colleagues have successfully achieved reproducible measurements of the medial femorotibial JSW in both normal and osteoarthritic knees in single-site studies (SD = 0.19 mm). However, this technique necessitates the use of specialized equipment and technicians with special experience and expertise. These factors challenge the practicality of this technique in multicenter studies. In a recent investigation of osteoarthritis using this technique at five different clinical centers [5], JSW measurements showed poor reproducibility (within-unit SE = 0.32 mm, between-unit SE = 0.45 mm) despite careful training of the technicians. Only 64% of 348 radiographs showed proper flexion, while knee rotation (89%) and metallic reference marker placement (94%) were reproducibly done.

Aside from poor reproducibility, the degree of flexion achieved with this technique is minimal. Indeed, it is called a "semi-flexed" view. For many subjects the degree of knee flexion in their normal standing position is actually greater than that achieved in the "semi-flexed" position in that they must occasionally be instructed to extend their knees further in order to achieve alignment of the tibial rims. This semi-flexed position does not, therefore, engage the most commonly affected region of the femoral cartilage, as discussed above.

In a previous study [6], the original fluoroscopic technique used in the semi-flexed view was modified by positioning the subject with the feet in 10° external rotation and the toes touching a vertically positioned fluoroscopy table, and exposing the image posteroanterior (PA) rather than AP. This technique reduced both the magnitude and variability of magnification of the knee, and correctly projected the tibial spines between the femoral condyles. It did not, however, increase the degree of flexion of the knee or obviate the need for fluoroscopy and foot maps.

Flexing the knees until they also touched the vertical table, and pressing the thighs and pelvis against the table as well increased the degree of flexion of the knee (~20° flexion), and fixed the position of the femur without the need for fluoroscopy (Fig. 4). In this position, the angulation of the tibial plateau is determined primarily by the length of the tibia and the size of the foot, neither of which tend to change substantially over time in adults. Additionally, since the knee is positioned as close as possible (within a few cm) to the film, magnification is kept to a minimum and is unlikely to change substantially on serially acquired images, provided the same equipment is used. Also, while in the study cited above, the focus-to-film distance was kept the same for both techniques in order to allow a direct comparison, non-fluoroscopic radiography allows longer focus-to-film distances, and this further reduces geometric unsharpness and magnification.

With the knee in this fixed-flexion position, the x-ray beam must be angled caudal in order to be tangential to the floor of the medial tibial plateau. Using fluoroscopic guidance the optimal beam angulation for the patients used in the study above was 9° ± 3°. A 10° beam angulation is therefore recommended for clinical trials. The angulation of the x-ray beam relative to the x-ray film causes a minor projectional widening of the joint-space (parallax effect), but this amounts to only a 3%

Table 1 - *Intra-reader reproducibility errors for minimum medial JSW measurement using different radiographic techniques*

	FP normal	NFP normal	NFP osteoarthritic
manual	0.2 mm	0.3 mm	0.2 mm
automated	0.1 mm	0.1 mm	0.1 mm

Values are root-mean-square SD for replicate radiographs.
FP, fluoroscopic PA; NFP, non-fluoroscopic PA.

increase, and can be corrected for. Moreover, this parallax effect is propagated on serially acquired images and therefore cancels out during calculations of JSW change.

When compared with the optimized fluoroscopic acquisition technique, this simpler non-fluoroscopic method provided similar inter-reader reproducibility for manual JSW measurements in both normal and osteoarthritic knees (Tab. 1). As reported by others [4] automated JSW measurement showed greater reproducibility (root-mean-square SD = 0.1 mm) than did manual measurements (root-mean-squared SD = 0.2 mm – 0.3 mm).

This is similar to the non-fluoroscopic MTP (metatarsophalangeal) technique recently proposed by Buckland-Wright, which aligns the tip of the knee with the MTP joints of the foot, and exposes the knee PA with a horizontal x-ray beam (Fig. 4). Only a single-site study with the MTP technique has been reported [7] thus far, but this shows similar high precision for min-JSW measurement (SD = 0.1 mm). Advantages of the fixed-flexion technique over the MTP technique include greater knee flexion, no need for a space beneath the x-ray Bucky to accommodate the foot (this is not possible on some x-ray systems), and most importantly, fixation of the femur to ensure that the same region of femoral cartilage articulates with the tibia on serial examinations.

Further advances in knee radiography for multi-center research include the development of a positioning frame for reproducible foot positioning without the need for manually traced foot maps (Fig. 5). The frame also fixes the positions of both the tibia and femur to ensure that the exact same regions of articular cartilage are projected in the joint space on serial images. Additionally, the frame is equipped with a specially designed projection phantom that allows verification of the beam angle that was used during radiography and which knee (right/left) was imaged at each visit. Technicians occasionally mislabel knee side, and this can be very difficult to detect during image analysis or verify in retrospect. The phantom also corrects any magnification differences caused by equipment changes during the course of a study, and corrects parallax errors introduced by beam angulation.

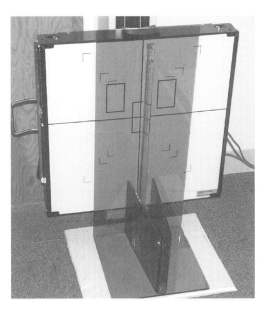

Figure 5
X-ray frame and phantom. The frame is shown positioned against a x-ray Bucky, as it would be used in imaging. The central vertical structure on the frame is a phantom that allows verification of knee side (right, left) and beam angle used, and correction of magnification and parallax errors. (Courtesy of Synarc, Inc., with permission.)

Regardless of the technique used to acquire the images, current state-of-the-art involves central digitization of the radiographs to allow safer and more economical archival, recovery and distribution of the images, as well as electronic submission to regulatory agencies. These electronic images are then analyzed with specialized reading software and monitors (Fig. 6). The software edge-enhances the digital images using unsharp masking to maximize the delineation of cortical and trabecular bone. It then strips the images of all identifying information to ensure reader blinding, and presents the images as serial sets for each subject in order to maximize the reader's ability to detect the smallest changes in semi-quantitative score (e.g., Kellgren-Lawrence or OARS(I)) and ensure anatomical consistency in serial JSW measurements. The software automatically delineates the articular surfaces of the joint and calculates the minimum JSW (Fig. 7). Electronic score sheets with reader sign-off and automatic databasing are used to record the results.

Coupling specialized image acquisition aids with computer-assisted analysis in this way makes clinical-trials radiology an integrated process and maximizes precision, speed and regulatory compliance. In a recent study of 457 osteoarthritic patients imaged with the fixed-flexion technique and the positioning frame at 58 dif-

Figure 6
Digital radiography viewing and analysis workstation. Serial radiographs of the knee from a clinical trial are projected on the two right monitors. The monitor on the left contains an electronic score sheet. (Courtesy of Synarc, Inc., with permission.)

ferent centers in the US, repeat films were obtained in a subset of 30 knees from 14 centers. The standard deviation of the medial min-JSW measurements was 0.25 mm (coefficient of variation = 4.1%) using the computer-assisted method. This was somewhat better computer-assisted precision than that reported (SE = 0.32 mm) in the five-center study using Buckland-Wright's fluoroscopic technique [5].

Imaging cartilage with MRI

MRI is well suited to imaging osteoarthritic joints. Not only does it provide direct multiplanar tomography with relatively high spatial resolution, it is unparalleled in its ability to depict soft-tissue detail. Thus, MRI is the only modality that can examine all components of the joint simultaneously and therefore the joint as a whole organ. Moreover, MRI is capable of probing not only the morphology of joint tissues but a variety of compositional and functional tissue parameters relevant to the arthritic process. Because MRI is noninvasive and generally well tolerated by patients, frequent serial examinations can be performed on even asymptomatic

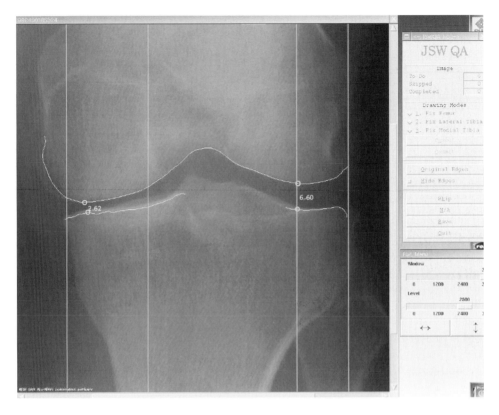

Figure 7
Computer-aided JSW measurement. Specialized software automatically traces the articular surfaces of the femorotibial joints and localizes and calculates min-JSW or joint-space area. An experienced reader verifies the results and can adjust the horizontal limits of the sampled joint space to ensure that only the weight bearing regions are included. Additionally, the reader can correct any tracing errors, such as those associated with projectional superimposition of tibial cortices. (Courtesy of Synarc, Inc., with permission.)

patients. With more than 9,000 MRI systems present world wide, lack of availability is less serious a limitation than in the past.

Morphological markers

Morphological markers of articular cartilage include both quantitative measures, such as thickness and volume, and semiquantitative measures that grade cartilage integrity by a variety of scoring methods. Intermediate-TE and long-TE fast spine echo (FSE) images are usually adequate for most of the current clinical applications

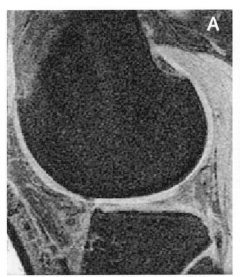 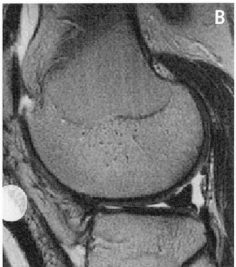

Figure 8
Fat-suppressed T1-weighted gradient-echo vs. T2-weighted fast spin-echo. (A) Sagittal T1-weighted, fat-suppressed 3D gradient-echo image of the knee shows a partial thickness defect in the femoral cartilage over the posterior horn of the lateral meniscus. (B) Sagittal T2-weighted 2D fast spin-echo image of the same knee was acquired in a fraction of the time required to generate image A, yet the same femoral cartilage defect is clearly delineated. (From Peterfy, CG (2000) Scratching the surface: articular cartilage disorders in the knee. MRI Clin N Am 8(2): 409–430.)

and in circumstances when lengthier, high-resolution techniques are not justified [8] (Fig. 8). However, thin-partitioned, fat-suppressed 3D spoiled-GRE images are preferable for delineating cartilage morphology. Advantages of this latter technique include greater contrast, higher resolution, wide availability, ease of use, stable performance, no chemical-shift artifact, and reasonable acquisition time (7 min – 10 min). Disadvantages include longer acquisition times than those required for FSE imaging and vulnerability to magnetic susceptibility and metallic artifacts. Failure of fat-suppression due to regional field heterogeneities is generally not a problem because of the cylindrical shape of the knee, but can arise if the knee is bent or if the patella protrudes excessively. Typically, however, failed fat-suppression in the region of the patella involves the marrow and superficial soft tissues and does not reach the articular cartilage.

Several studies have demonstrated the diagnostic accuracy of fat-suppressed 3D spoiled-GRE for identifying areas of cartilage loss in the knee [9, 10], and various scoring methods have been developed that could be used in clinical trials. However, the surrogate validity of these measurement schemes has not yet been thoroughly established. There is considerable face validity to the link between cartilage loss and the clinical manifestations of osteoarthritis, but exactly how much cartilage loss is clinically relevant is not yet known. The issue is complicated by the multi-factorial nature of joint failure and the inadequacy of viewing osteoarthritis as a disease of any single joint structure, even articular cartilage. Osteoarthritis is more appropriately modeled as a disease of organ failure, analogous to heart failure and renal failure, in which injury to one structure leads to damage of other structures and ultimately, the clinical manifestations of the disease. Nevertheless, cartilage loss is currently the most broadly accepted metric of structural progression in osteoarthritis and the focus of most therapies for the disease. Unresolved issues of surrogate validity not withstanding, semiquantitative scoring of cartilage loss can be relatively precise and able to resolve progression in 1 year. In a recent study of 29 patients with osteoarthritis in whom the articular cartilage was scored in 15 locations in the knee using a seven-point scale, the intra-class correlation coefficient (ICC) between two specially-trained radiologists was 0.99 [11]. A subsequent examination of 30 subjects from an ongoing cohort study of 3,075 elderly men and women imaged with a 15-min MRI protocol (T2-weighted FSE) found similar inter-reader precision for femorotibial cartilage using the same scoring method (ICC = 0.91) [12].

Aside from semiquantitative scoring, a number of quantitative markers of cartilage morphology have been developed, including cartilage volume [13–15]. This measurement can be derived from segmented images of the articular cartilage on fat-suppressed 3D spoiled GRE images using any of a variety of image analysis softwares (Fig. 9). A number of studies have validated the technical accuracy of these methods and found the precision error to range from 2% to 4% coefficient of variation (SD/mean volume) [13, 14]. Longitudinal data are limited, but in an investigation of 16 elderly women with osteoarthritis of the knee imaged at yearly intervals for 2 years, the mean annual rate of cartilage loss was determined to be $-6.7\% \pm 5.2\%$ for the femur, $-6.33 \pm 4.3\%$ for the tibia and $-3.4\% \pm 2.9\%$ for the patella, based on linear regression across the three time points [16].

Limitations of cartilage volume quantification include assumptions used to model cartilage volume change over time. For practical reasons, a linear model is usually the only feasible assumption, as most clinical trials and epidemiological studies involve four or fewer time points. More complicated models (quadratic, etc.) may turn out to be more accurate, but until careful natural history studies have refined these models, curve-fitting challenges limit their use in most studies. Regardless, measurement precision for cartilage volume change combines errors related both to the measurement technique and the cartilage-loss model used.

Another limitation of cartilage volume as a marker of disease severity and structural progression is its insensitivity to small local changes. Focal defects are more easily identified by semiquantitative scoring, or by regional cartilage volume map-

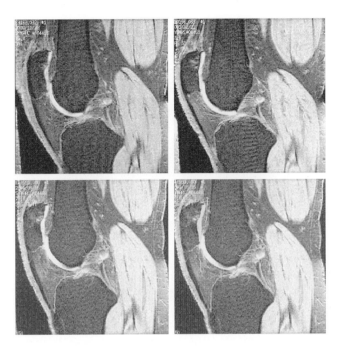

Figure 9
Articular cartilage segmentation. Lower right window of MRI analysis workstation specialized for clinical-trials radiology shows sagittal image of the knee with the articular cartilage of the patella (cyan) segmented using a seed-growing algorithm. A segmentation barrier (yellow line) is positioned where the patellar cartilage contacts the femoral cartilage to prevent the algorithm from crossing the interface between the two cartilage plates. (Courtesy of Synarc, Inc., with permission.)

ping [15]. Measurement precision and statistical power, however, decrease as the subdivisions get smaller. Accordingly, the trade off between sensitivity and measurement precision must be carefully balanced. One highly refined method of depicting regional variations in cartilage quantity is thickness-mapping [17, 18] (Fig. 10). As intuitive as cartilage thickness may seem, however, questions remain as to whether the minimum, maximum or average thickness is the most relevant, how to deal with multiple lesions, and to what extent the location of a lesion (e.g., weight bearing, non weight bearing) is important.

Perhaps the greatest limitation of all markers of cartilage morphology, however, is their fundamentally irreversible nature and relatively slow responsiveness. Regardless of how precisely change in cartilage morphology can be measured, its rate of change cannot be driven any faster than the disease process itself. For faster

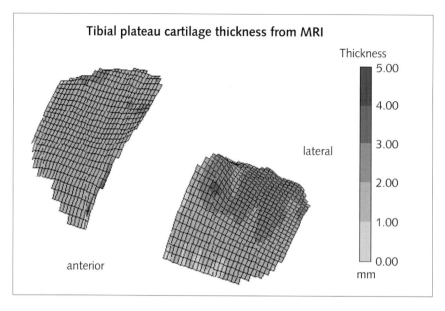

Figure 10
Cartilage thickness mapping. B-spline geometric model of the tibial cartilage of the knee was generated by manual segmentation and parametric bicubic B-spline representation. Regional cartilage thickness (perpendicular to the cartilage-bone interface) are depicted at intervals of 0.8 mm (From: Peterfy CG (1998) Magnetic resonance imaging. In: Brandt, Doherty, Lohmander (eds). Osteoarthritis. Oxford University Press, New York, 473–494; with permission).

responsiveness, one must look upstream to earlier stages in the disease process of cartilage degeneration. Accordingly, there has been a great deal of interest in MRI markers of cartilage matrix damage.

Compositional markers

MRI in markers of cartilage composition relate principally to the collagen matrix or proteoglycans. The most promising markers of collagen matrix integrity include T2 relaxation and magnetization transfer coefficient. Markers of proteoglycan integrity include water diffusion, Gd-DTPA^{2-} uptake and ^{23}Na concentration.

The principal determinant of T2 relaxation in articular cartilage is collagen [19, 20]. Accordingly, disruption of the fibrillar organization of collagen or actual decrease in collagen content reduces T2 relaxation and increases signal intensity on T2-weighted images. Areas of elevated signal in otherwise low signal-intensity car-

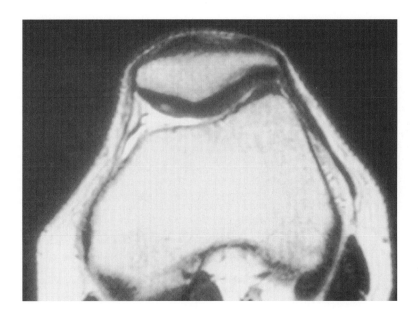

Figure 11
T2 lesions in articular cartilage. Axial T2-weghted fast spin-echo image of the knee shows a small, blistering T2 lesion in the cartilage over the medial facet of the patella. (From Peterfy, CG (2000) Scratching the surface: articular cartilage disorders in the knee. MRI Clin N Am 8(2): 409–430.)

tilage on long-TE MR images, therefore represent foci of chondromalacia (Fig. 11). Although careful prospective validation has is still needed, there is evidence to suggest that abnormal cartilage T2 relaxation is predictive of subsequent cartilage loss (Fig. 12). In one study [21], five (33%) of 15 meniscal surgery patients followed over three years post surgery developed a total of six T2-lesions in otherwise normal appearing articular cartilage. Two of these lesions progressed to focal cartilage defects during the study, while three persisted and one regressed. Interestingly, the four lesions that did not progress were in patients who had undergone meniscal repair, while the lesions that progressed were in patients who had meniscal resection. Accordingly, abnormal T2 may identify cartilage at risk of future loss and thereby identify patients in need of aggressive therapy - hopefully before the point of no return. Particularly intriguing is the potential utility of this marker in clinical trials of matrix metalloproteinase inhibitors, which target the collagen matrix directly. In addition to subjective evaluations of focal signal abnormalities in articular cartilage, regional changes in T2 relaxation can be quantified and monitored

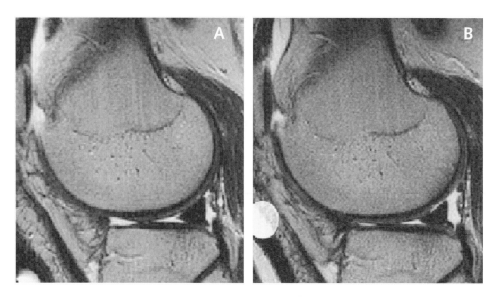

Figure 12
Progression of T2 lesions in articular cartilage. Serial sagittal T2-weighted fast spin-echo images show a focal T2 lesion (arrow) in the femoral cartilage adjacent to the posterior horn of the lateral meniscus at baseline (A). Follow-up imaging 9 months later (B) shows a partial-thickness (Grade 2.0) defect at that exact location. (From Peterfy, CG. Scratching the surface: articular cartilage disorders in the knee. MRI Clin N Am 8(2): 409–430.)

over time with multi-echo imaging (Fig. 13). Limitations of this approach include technical tradeoffs between imaging time and the number of echoes, spatial resolution and signal-to-noise ratio attainable. Further development is needed here.

Significantly less work has been done with magnetization transfer as a marker of collagen integrity in articular cartilage. Theoretically, this marker could be used almost exactly the same way that cartilage T2 is used. However, even less is known about its diagnostic accuracy, responsiveness to disease and therapy, dynamic range, and measurement precision. Accordingly, further characterization is needed.

As mentioned above, methods for evaluating the integrity of the proteoglycan matrix by probing regional variations in fixed negative charged density in articular cartilage have recently been developed (Fig. 14). The histological and biochemical validity of this approach has been well demonstrated by a number of groups [22–24]. Using cartilage-nulling inversion recovery sequences at high spatial resolutions and high field strength, Bashir et al. [22] demonstrated strong histological correlation between the distribution of anionic Gd-DTPA^{2-} and perichondrocytic glycosaminoglycan (GAG) depletion following incubation of cartilage explains with IL-1 (interleukin-1). Subsequent studies have demonstrated a linear correlation

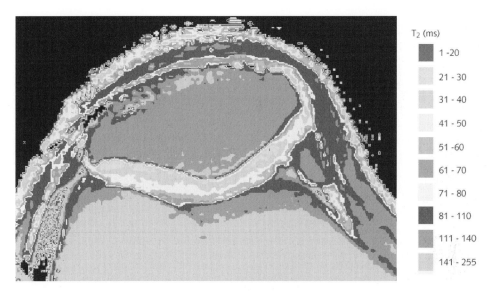

Figure 13
T2 relaxation of normal adult articular cartilage. T2 map generated from multi-slice, multi-echo (11 echoes: TE = 9,18,... 99 ms) spin-echo images acquired at 3 T shows increasing T2 towards the articular surface. (Courtesy of B.J. Dardzinski, Ph.D. University of Cincinnati College of Medicine).

between T1 changes associated with Gd-DTPA^{2-} and cartilage GAG concentration ranging from 10 mg/ml to 70 mg/ml as measured biochemically [25]. In a study by Trattnig et al. [24] areas of abnormal Gd-DTPA^{2-} uptake in cartilage specimens harvested at total knee replacement surgery all corresponded to sites of collagen loss based on azan-stained histology. Unfortunately, this study did not report the correlation with areas of abnormal T2, if any were present. The study also reported marked inter-individual variation in the pattern of Gd-DTPA^{2-} uptake in eight normal volunteers who were examined, as well as marked differences in the diffusion times observed for cartilages of different thickness. Accordingly, while Gd-DTPA^{2-} uptake appears to be a valid method for quantifying GAG concentration and distribution in articular cartilage, with good dynamic range properties relative to GAG concentration, the relationship of this marker to cartilage T2 has yet to be reported. Does abnormal Gd-DTPA^{2-} uptake precede abnormal T2 temporally? What is the relative performance of these two markers in terms of sensitivity, specificity, responsiveness to disease and therapy, dynamic range, predictive power for subsequent cartilage loss, other structural changes associated with osteoarthritis, and clinical outcomes of osteoarthritis? Finally, what is the optimal *in vivo* acquisition

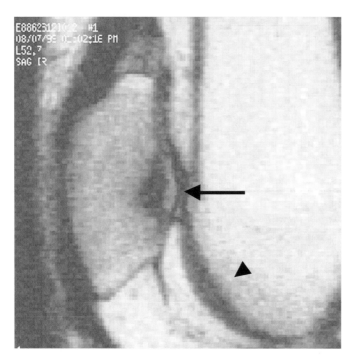

Figure 14
Imaging cartilage matrix damage. Sagittal inversion-recovery image of a knee following i.v. administration of Gd-DTPA shows a region of high signal intensity (arrow) in the patellar cartilage indicative of abnormal uptake of anionic Gd-DTPA, and therefore, local proteoglycan depletion. Cartilage in the trochlear groove (arrowhead) shows low signal intensity indicative of repulsion of Gd-DTPA by negatively charged proteoglycans. (Courtesy of Synarc, Inc., with permission.)

technique for cartilage Gd-DTPA^{2-} uptake as a marker? In this regard, cartilage T2 may be a simpler marker to use. Of course, advantages in convenience and cost typically rank lower than performance and reliability in most clinical trials.

Conclusion

Powerful techniques are being developed for evaluating articular cartilage with MRI. Much of this development is being driven by the pharmaceutical and biotechnology industries searching for novel therapeutic solutions to the growing problem of arthritis in our aging society. Accordingly, the imaging tools that ultimately will be used to direct patients to specific therapies and then to monitor treatment effectiveness and safety are currently being refined and validated in rigorous multi-center and multinational clinical trials aimed at gaining regulatory approval of these new therapies. This process represents not only a new pathway for innovation in medical imaging, but one that allows radiology to advance in sync with the rest of medicine.

References

1 Vignon E, Conrozier T, Piperno M, Richard S, Carrillon Y, Fantino O (1999) Radiographic assessment of hip and knee osteoarthritis. Recommendations: recommended guidelines. *Osteoarthritis Cartilage* 7: 434–436
2 Peterfy C (1997) Imaging techniques. In: J Klippel, P Dieppe (eds): *Rheumatology*, 2nd ed. 1 vol. Mosby, Philadelphia, 14.1–14.18
3 Buckland-Wright JC, Clarke GS, Chikanza IC, Grahame R (1993) Quantitative microfocal radiography detects changes in erosion area in patients with early rheumatoid arthritis treated with myocrisine. *J Rheumatol* 20: 243–247
4 Buckland-Wright JC, Macfarlane DG, Jasani MK, Lynch JA (1994) Quantitative microfocal radiographic assessment of osteoarthritis of the knee from weight bearing tunnel and semiflexed standing views. *J Rheum* 21: 1734–1741
5 Mazzuca SA, Brandt KD, Buckland-Wright J, Buckwalter KA, Katz BP et al (1999) Field test of the reproducibility of automated measurements of medial tibiofemoral joint space width derived from standardized knee radiographs. *J Rheumatol* 26: 1359–1365
6 Peterfy C, Li J, Duryea J, Lynch J, Miaux Y, Genant H (1998) Nonfluoroscopic method for flexed radiography of the knee that allows reproducible joint-space width measurement. *Arthritis Rheum* 41: S361
7 Buckland-Wright J, Wolfe F, Ward R, Flowers N, Hayne C (1999) Substantial superiority of semiflexed (MTP) views in knee osteoarthritis: a comparative radiographic study, without fluoroscopy, of standing extended, semiflexed (MTP), and Schuss views. *J Rheumatol* 26: 2664–2674
8 Bredella MA, Tirman PF, Peterfy CG, Zarlingo M et al (1999) Accuracy of T2-weighted fast spin-echo MR imaging with fat saturation in detecting cartilage defects in the knee: comparison with arthroscopy in 130 patients. *Am J Roentgenol* 172: 1073–1080
9 Recht MP, Kramer J, Marcelis S, Patkria MN et al (1993) Abnormalities of articular cartilage in the knee: analysis of available MR techniques. *Radiology* 187: 473–478
10 Disler DG, McCauley TR, Wirth CR, Fuchs MC (1995) Detection of knee hyaline articular cartilage defects using fat-suppressed three-dimensional spoiled gradient-echo MR imaging: comparison with standard MR imaging and correlation with arthroscopy. *AJR* 165: 377–382
11 Peterfy CG, White D, Tirman P, Lu Y, Miaux Y, Leff R, Genant HK (1999) Whole-organ evaluation of the knee in osteoarthritis using MRI. European League Against Rheumatism. Glasgow, Scotland

12 Wildy K, Zaim S, Peterfy C, Newman B, Kritchevsky S, Nevitt M (2001) Reliability of the Whole-Organ review MRI scorning (WORMS) method for knee osteoarthritis (OA) in a multicenter study. 65th Annual Scientific Meeting of the American College of Rheumatology. San Francisco, CA, Nov. 11–15, 2001
13 Peterfy CG, van Dijke CF, Janzen DL, Gluer CC, Namba R, Majumdar S, Lang P, Genant HK (1994) Quantification of articular cartilage in the knee by pulsed saturation transfer and fat-suppressed MRI: optimization and validation. *Radiology* 192: 485–491
14 Eckstein F, Sitteck H, Gavazzenia A, Milz S, Putz R, Reiser M (1995) Assessment of articular cartilage volume and thickness with magnetic resonance imaging (MRI). *Trans Orthop Res Soc* 20: 194
15 Pilch L, Stewart C, Gordon D, Inman R, Parsons K, Pataki I, Stevens J (1994) Assessment of cartilage volume in the femorotibial joint with magnetic resonance imaging and 3D computer reconstruction. *J Rheum* 21: 2307–2321
16 Peterfy C, White D, Zhao J, Van Dijke C, Genant H (1998) Longitudinal measurement of knee articular cartilage volume in osteoarthritis. American College of Rheumatology. San Diego
17 Cohen ZA, McCarthy DM, Ateshian GA, Kwak SD, Peterfy CG, Alderson P, Grelsamer RP, Henry JH, Mow VC (1997) *In vivo* and *in vitro* knee joint cartilage topography, thickness, and contact areas from MRI. Orthopaedic Research Society. San Francisco, February 1997
18 Eckstein F, Gavazzeni A, Sittek H, Haubner M, Losch A, Milz S, Englmeier KH, Schulte E, Putz R, Reiber M (1996) Determination of knee joint cartilage thickness using three-dimensional magnetic resonance chondro-Crassometry (3D MR-CCM). *Magn Reson Med* 36: 256–265
19 Xia Y, Farquhar T, Burton-Wurster N, Lust G (1997) Origin of cartilage laminae in MRI. *JMRI* 7: 887–894
20 Xia Y, Moody JB, Burton-Wurster N, Lust G (2001) Quantitative *in situ* correlation between microscopic MRI and polarized light microscopy studies of artcular cartilage. *Osteoarthritis Cartilage* 9: 393–406
21 Zaim S, Lynch JA, Li J, Genant HK, Peterfy CG (2001) MRI of early cartilage degeneration following meniscal surgery: a three-year longitudinal study. International Society for Magnetic Resonance in Medicine. Glasgow, Scotland, April 2001
22 Bashir A, Gray ML, Burstein D (1996) Gd-DTPA as a measure of cartilage degradation. *Magn Reson Med* 36: 665–673
23 Bashir A, Gray ML, Hartke J, Burstein D (1999) Nondistructive imaging of human cartilage glycosaminoglycan concentration by MRI. *Magn Reson Med* 41: 857–865
24 Trattnig S, Mlynarkck V, Breilenseher M, Huber M, Zembsch A, Rand T, Imhof H (1999) MR visualization of proteoglycan depletion in articular cartilage *via* intravenous injection of Gd-DTPA. *Magn Reson Imaging* 17: 577–583
25 Bashir A, Gray ML, Hartke J, Burstein D (1998) Validation of gadolinium-enhanced MRI for GAG measurement in human cartilage. International Society of Magnetic Resonance in Medicine. Philadelphia, PA

Diffraction enhanced x-ray imaging of articular cartilage

Carol Muehleman[1], Jürgen A. Mollenhauer[2], Matthias F. Aurich[2], Klaus E. Kuettner[3], Zhong Zhong[4], Ada Cole[1] and Dean Chapman[5]

[1]Departments of Anatomy and Biochemistry, Rush Medical College, 600 South Paulina, Chicago, IL 60612, USA; [2]Department of Orthopedics, University of Jena, Waldkrankenhaus "Rudolf-Elle", Klosterlausnitzer Strasse 81, 07607 Eisenberg, Germany; and Department of Biochemistry, Rush Medical College, 1653 W. Congress Parkway, Chicago, IL 60612, USA; [3]Department of Biochemistry and Department of Orthopedics, Rush University, Rush-Presbyterian-St. Luke's Medical Center, 1653 W. Congress Parkway, Chicago, IL 60612, USA; [4]NSLS Brookhaven National Laboratory, Upton, NY 11973, USA; [5]Biological, Chemical and Physical Sciences Department, Illinois Institute of Technology, 3101 South Dearborn, Chicago, IL 60616, USA

Introduction

The detection of articular cartilage and its abnormalities through non-invasive means is of importance for the early diagnosis of the degenerative joint disease, osteoarthritis (OA). The development of treatment strategies depends largely on the ability to identify early pathological changes in the articular cartilage and monitor the progression of its degeneration. Because this tissue has little or no x-ray absorptive contrast, it cannot be visualized with conventional radiography, thus joint abnormalities can be assessed only through bony changes and the joint space narrowing that occurs at late stage disease. However, because of the short wavelength nature of x-rays they should be capable of detecting a higher level of tissue information than other modes of imaging. Here, we describe a novel x-ray technology in its early stages of development, Diffraction Enhanced Imaging (DEI), for the high contrast imaging of articular cartilage of synovial joints [1].

Diffraction enhanced imaging

DEI is a computed imaging process that allows the acquisition of numerous images from a single exposure without increasing the total radiation exposure over conventional radiography. Although the x-ray source is currently a synchrotron, the technique is not inherently dependent upon synchrotron x-rays and efforts are underway by one of the co-authors (Chapman) toward the development of an x-ray tube based DEI system.

The details of the DEI methodology have been previously published [2]. Briefly, DEI utilizes the x-ray beam at the National Synchrotron Light Source at Brookhaven National Laboratory, where a polychromatic beam is diffracted into a nearly monochromatic, highly collimated, single energy beam by two matching crystals. Once this beam passes through the subject, a third crystal (analyzer crystal) of the same orientation, and using the same reflection, is introduced.

If the analyzer crystal is rotated about an axis, the crystal will rotate through the Bragg condition for diffraction, and the diffracted intensity will trace out a profile or "rocking curve" [3]. X-ray diffraction from a perfect crystal, with its narrow reflection angular width provides the tools necessary to prepare and analyze x-ray beams traversing a specimen on the microradian scale [3]. The rocking curve is observed as a variation in intensity as the analyzer crystal is rotated in the horizontal direction. The character of the subject image, therefore, changes depending on the setting of the analyzer crystal.

To extract refraction information, the analyzer is typically set to the half intensity points on the low and high angle sides of the rocking curve. For optimal extinction sensitivity, the analyzer is typically set to the peak of the rocking curve. To image the x-rays scattered by the sample, the analyzer is set in the wings of the rocking curve.

The x-ray beam transmitted through the subject can be either imaged directly as in conventional radiography or following diffraction in the vertical plane by the analyzer crystal. The detector can be radiographic film or an image plate.

Materials and methods

The specimens that were DE imaged included intact and disarticulated human knee and ankle joints (obtained within 24 h of death of the donor through the Regional Organ Bank of Illinois) and osteoarthritic knee cartilage and cartilage/subchondral bone pieces from knee replacement patients from the Department of Orthopedic Surgery at Rush-Presbyterian-St. Luke's Medical Center. All specimens were categorized in terms of level of cartilage degeneration using the criteria of our previously published scale [4] and then x-rayed and DE imaged in a posterior to anterior direction. Each specimen was subsequently prepared for histological examination.

Results

For each sample, either no cartilage, or just a faint image of cartilage, could be seen with ordinary radiography. However, with DEI, the cartilage was clearly visible. Furthermore, the DE image of each cartilage specimen appeared to be reflective of its gross visual appearance. For samples displaying no gross visual cartilage degen-

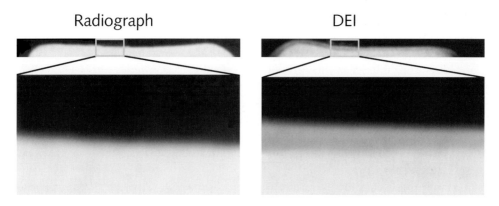

Figure 1
Images of a human talar dome showing only the bone in the radiograph on the left and cartilage and bone in the DE image on the right.

eration, the DE image of the cartilage appeared homogeneous with an intact, uninterrupted surface. The cartilage thickness could be determined and the surface and bone/cartilage interface were readily visible. A representative sample of an ordinary x-ray (left) and a DE image (right) of the talus of an ankle joint displaying no degenerative signs are shown in Figure 1. The DE image was taken at 18 keV and at +1.8 mrad on the rocking curve.

Specimens displaying signs of gross visual and histological degeneration such as fibrillation, fissuring and erosion down to subchondral bone had DE images that showed contrast heterogeneities in comparable locations. The severity of the gross visual and histological cartilage degeneration corresponded to the extent of contrast heterogeneities within the DE images. The DE images of the specimens varied in contrast characteristics depending upon the location on the rocking curve. At the present time, the optimal location on the rocking curve for DE imaging has not been determined. It is possible that a composite of several images from different locations on the rocking curve may best depict the morphology of the articular cartilage. For intact knee and ankle joints, the articular cartilage and surrounding soft tissues were also visible. In the knee joint, the boundaries between articular cartilage, menisci and peripheral soft tissues were clearly demarcated and the integrity of the articular surface was apparent.

In summary, we have shown that the articular cartilage of disarticulated and intact human knee and ankle joints can be radiographically imaged through DEI. Furthermore, cartilage pathologies such as are typical of OA can be observed in DE images of such specimens. We believe that with further optimization and development into a clinical device, DEI can be utilized for the radiographic imaging of joint cartilages for the early diagnosis of degenerative joint disease.

References

1 Mollenhauer J, Aurich M, Muehleman C, Zhong, Kuettner K, Cole A, Chapman D (2001) Using diffraction enhanced x-ray imaging to detect articular cartilage. *Proceedings of the annual meeting of the Orthopedic Research Society*
2 Chapman D, Thomlinson W, Johnston RE, Washburn D, Pisano E, Gmur N, Zhong Z, Menk R, Arfelli F, Sayers D (1997) Diffraction enhanced x-ray imaging. *Phys Med Biol* 42: 2015–2025
3 Zachariasen WH (1945) *Theory of x-ray diffraction in crystals*. John Wiley & Sons, New York, chapter 4
4 Muehleman C, Bareither D, Huch K, Cole A, Kuettner KE (1997) Prevalence of degenerative morphological changes in the joints of the lower extremity. *J Osteoarthritis Cart* 5: 1–15

Bone changes in osteoarthritis (OA)

Iain Watt

Department of Clinical Radiology, Bristol Royal Infirmary, Bristol BS2 8HW, UK

Introduction

The role and relevance of bone in joint deformity in patients with osteoarthritis (OA) was emphasised nearly 200 years ago [1]. One hundred years ago it was conventional to categorise arthritis as "inflammatory" or "degenerative". In a seminal publication two conclusions were drawn. Firstly, each was a reaction to a different form of insult, and secondly different types of "degenerative" disease were described for the first time [2]. Later, hypertrophic arthritis of the finger joints was shown to have a strong hereditary tendency [3], and OA was described as a generalised condition, associated with Heberden's nodes [4]. Further, attention was drawn to the concept that such joints went through phases of disease evolution. Lawrence and others [5] emphasised the fact that x-ray changes, whilst predisposing to symptoms, did not equate to them – a still extremely relevant axiom.

Thus to summarise, it has been long recognised that the subchondral bone response to OA and hyaline cartilage degradation is not uniform. It may be hypertrophic or atrophic; has a hereditary and familial tendency, and is part of a generalised condition that is not necessarily symptomatic.

Osteoarthritis is not a single disease

Osteoarthritis is age-related with obvious separate clinical and radiological subsets. These include posterior hip OA, primary elbow joint OA [6], erosive OA [7, 8] and disorders that are "secondary" to a spectrum of pre-existing conditions which have in some way disrupted normal hyaline cartilage. These conditions include juvenile arthritis, Perthes disease and rheumatoid arthritis. Further, epidemiological studies suggest different incidences and associations between individual joints afflicted by OA. For example, knee OA occurs at an earlier age and in more female patients than OA of the hip, which is an older and more male disease.

Whilst OA is clearly age related, it must be distinguished from normal age-related changes in synovial joints. For example, an age-related correlation exists in finger joints concerning apparent synovial swelling, joint margin spurs and intra-articular loose bodies [9]. Increasing joint congruity in older persons may contribute to hyaline cartilage thinning, perhaps due to reduced cartilage nutrition [10]. An ageing synovial joint may go through a series of phases starting with increasing congruity, reduced synovial fluid circulation, deficient hyaline cartilage nutrition, subsequent hyaline cartilage thinning and decreased joint stability associated with reduced muscle function, and reducing trabecular bone mass. Under these circumstances, minor marginal osteophytes or spurs develop that should not be confused with OA.

Are crystals the controllers of subchondral bone response in OA?

Deposition of calcium pyrophosphate dihydrate (CPPD) is an age-related finding in hyaline cartilage. Episodes of crystal shedding are associated with the clinical manifestation of pseudogout. In 1970, Martel [11] drew attention to a roentgenologically distinctive arthropathy in some patients with pseudogout syndrome. He noted that the arthropathy was severe and associated with marked new bone formation in a somewhat unusual distribution, with an abundance of subchondral cysts. Later Resnick and others [12] coined the term "pyrophosphate arthropathy" describing a markedly hypertrophic bone response to joint insult. The concept that the hypertrophic bone response was due to the presence of CPPD has not stood the test of time. Normal human femoral head hyaline cartilage contains crystals in the 50 μ range, in all specimens and at all ages. They are located superiorly in the femoral head in the predominantly weight-bearing area [13]. Larger aggregates may even be beneficial to cartilage resistance to sheer stress loading [13, 14]. Detailed work by Dieppe and co-workers in Bristol has suggested strongly that no clear causal relationship exists between the presence of CPPD crystals in a joint and the development of a hypertrophic bone response. From this and other work, the conclusion has been drawn that the delineation between "crystal and non-crystal" OA is artificial [15, 16]. Differences in crystal morphology do exist between patients suffering from pseudogout and those suffering from hypertrophic OA, perhaps accounting for the more inflammatory nature of pseudogout as compared to "pyrophosphate arthropathy" [16]. Thus it may be considered, perhaps simplistically, that the hypertrophic bone response to joint failure is due to overactive chondrocytes producing an abundance of CPPD. Other work has demonstrated that crystal deposition may be enhanced by joint damage, for example, previous surgical meniscectomy [17]. However, the extent of knee OA in such patients is strongly correlated with the presence of OA elsewhere in the skeleton – for example in the hands [18]. In other words, this implies a systemic predisposition to OA enhancing the effect of local injury. Finally, long-term follow up of patients with hypertrophic OA associated with CPPD demonstrated that whilst chondrocalcinosis may increase with time, and episodes of acute pseudogout occur, the hypertrophic OA appears benign in terms of the need for surgical intervention [19].

At the other end of the OA bone spectrum lies atrophic destructive OA. These elderly, usually female, patients have large joint effusions, with considerable joint instability associated with severe bone attrition and virtually no secondary bone response. Joint fluid in such patients is rich in basic calcium phosphates, which predominately comprise "bone dust" in association with joint destruction. Basic calcium phosphate (BCP) received attention when it was described as the cause of rapidly destructive shoulder joint OA, the so-called Milwaukee shoulder [20]. Subsequent analysis of joint fluid in such patients demonstrated that abundant BCP crystals were not associated with secondary high lytic enzyme concentrations alleged to cause the bone destruction in the initial paper [21]. Indeed, it is important to emphasise that some BCP is a normal component of joint fluid in most people.

Thus, again it may be concluded that at least two more processes are operating in so-called primary OA. These are a hypertrophic response to joint injury, a marker of which is chondrocalcinosis due to CPPD, and an atrophic joint failure with increased BCP in joint fluid. Further, high BMI patients tend to fall into the former category and elderly osteoporotic patients into the latter. Thus the response of a joint to insult may reflect not only local factors, but also generalised skeletal mineralisation. Hence, the bone response in OA is multifaceted. Crystals are likely to be markers of OA subsets rather than a cause.

How else may this spectrum of bone responses be analysed radiologically?

Magnetic resonance imaging (MRI) research has focused, almost exclusively, on hyaline cartilage estimation and measurement and also on synovitis and synovial diseases. Yet in a routine spectrum of patients with OA, considerable differences in subchondral bone response may be shown. This varies from osteophytosis, either marginal or articular surface ("stud" osteophytes), to complete effacement of hyaline cartilage with no secondary bone response. Little or no work has been done on marrow and subchondral bone changes in experimental OA save for that of Nolte-Ernsting [22]. This publication evaluated developing marrow inflammatory changes secondary to acute joint instability, with increasing fibrosis and sclerosis as the lesions matured. Increased signal in subchondral marrow, particularly on STIR or T2W sequences, has been dismissed often as "marrow oedema". This view has been challenged recently showing that it corresponds to inflammatory tissue and subsequent fibrosis [23]. Recent correlation between plain knee radiographs and MRI [24] has shown that the presence of an osteophyte on a plain film is always associated with a hyaline cartilage defect on MRI.

Skeletal scintigraphy, using routine bone-imaging radio-pharmaceuticals, has proved extremely rewarding in the understanding of OA. In late phase bone scintigraphy (approximately 3 h from injection) activity occurs at sites of both bone formation and resorption [25]. In experimental OA the earliest feature of joint disease was the development of scintigraphic activity on the margins of joints, where a chondrophyte was developing. These changes antedated hyaline cartilage degradation. Hutton and others, studying OA of the hands, demonstrated that bone scintigraphy had strong positive and negative predictive values for future change on hand x-rays [26]. Dieppe and others demonstrated similar predictive properties in a 5-year follow up of scintigraphy in knee OA [27]. Petersson and others [28] showed a strong correlation between serum COMP and BSP in patients who had scintigraphically positive knee OA, as opposed to those without. Thus, scintigraphy indicates patients who have raised metabolic function in subchondral bone and predicts those who will and will not progress on x-ray over time. Clearly it is uncertain as to whether the abnormal subchondral bone activity antedates, post-dates or coexists with hyaline cartilage change.

Scintigraphic subsets. Work by McCrae and others [29] demonstrated that a number of scintigraphic patterns emerged in knee OA. The two most important they termed "tram-line", where activity was focused along the joint line, and the "extended pattern", where scintigraphic abnormality extended well below the immediate subchondral zone. Radiographically, the tramline pattern corresponds closely to marginal osteophytosis but with relatively preserved hyaline cartilage thickness, whereas the extended pattern correlated with diffuse subchondral sclerosis and a greater degree of hyaline cartilage loss. McAlindon and others [30] correlated these scintigraphic changes with MRI findings. They demonstrated an association between the tramline pattern and abundant osteophyte formation on the margins of the joint and between those patients exhibiting the extended scintigraphic pattern and diffuse subchondral signal change on MRI. Thus, akin to the spectrum of hypertrophic *versus* atrophic OA on plain film, MRI and scintigraphy also reflect polar changes.

Fractal signature analysis in isolated femoral heads compared with DEXA changes

Femoral heads which exhibit eburnation and sclerosis appear to have a normal fractal signature, but increased subchondral bone density, whereas femoral heads which are predominately osteophytic have an altered compressive (load bearing) fractal signature, but normal subchondral bone density [31]. Three important conclusions were drawn from this work. Firstly, an inverse relationship existed between these two features, that is subchondral sclerosis and osteophytosis are not parallel bone responses. Secondly, hypertrophic osteophytosis has a reorganised subchondral trabecular anatomy and normal bone density, perhaps suggesting a healing response to joint instability. Thirdly, the maintenance of normal trabecular orientation but increasing density may represent joint failure in the form of sclerosis and attrition.

Why are some joints at risk of OA, whereas others are not?

It has been suggested that our recent emergence as a species in evolutionary terms may dictate the joints that are at risk of OA. The argument goes that they are the latest to evolve as we have adopted a purely upright posture [32]. These joints include the hips (from the sailor's gate in the chimpanzee to a true hominid bipedal stance); the knee (from a roller ball joint to a flattened articular surface to enable a locked standing position); the cervical and lumbar facet joints (the spine now being a loaded vertical structure rather than a bridge-like span), and the hand, particularly the thumb carpometacarpal joints and all the terminal interphalangeal joints (due to the development of a pinch grip). Thus, the concept of a dysplastic joint, such as the hip, has been taken a step further, implying that it represents underdevelopment in the upright ape.

Conclusion

In the evaluation of OA, considerable attention has been focused on hyaline cartilage, its physiology, biochemistry, visualisation and quantification. Yet clearly subchondral bone is a major player in the evolution of OA and has been relatively neglected by laboratory science. The systemic predisposition to the OA disorders has been noted for nearly 200 years, the generalised nature of OA being emphasised 50 years ago. OA is clearly not a single disease entity either clinically or radiologically. Too many obvious subsets exist where categorisation may be based purely on the distribution of joint involvement. Further, rates of progression or degrees of inflammatory change must be taken into account as in rapidly progressive OA or erosive OA. It is obvious that bone response varies dramatically from the markedly hypertrophic to the essentially atrophic. This spectrum is associated with crystals in joint fluid and in hyaline cartilage, the hypertrophic being associated with CPPD on the one hand, and atrophic with BCP on the other. Despite orthodox teaching to the contrary, it is highly likely that these crystals are markers, or co-expressions of disease, not the prime pathogenesis, as suggested by Martel in his original paper more than 30 years ago [11]. Further, two other major differences have emerged with regard to the response to joint failure in human, or perhaps all mammalian joints. These comprise the reorganisation of trabecular orientation and alterations in bone mineral density changes in subchondral bone. These are associated scintigraphically with changes in subchondral bone blood flow. Thus, one must conclude that the aetiopathology of OA is truly multifactorial, including genetic, developmental and other risk factors known and unknown. Perhaps OA does start with hyaline cartilage disease, but if so, it may be the secondary bone response which dictates how a joint will respond to this failure. Perhaps marginal osteophytosis on a radiograph, the tramline scintigraphic pattern, and "bright" osteophytes on a MRI scan reflect a benign healing response to joint instability. Conversely, plain x-ray bone attrition and subchondral sclerosis, the extended scintigraphic pattern, and increased marrow signal on a STIR sequence, may indicate a failure of trabecular reorganisation and mechanical effectiveness. Lastly, it has been speculated that the way that a joint responds in OA reflects overall skeletal well being. Broadly, the postulate exists that OA of a hypertrophic nature is associated with normal or perhaps raised mineral density in the spine, whereas the converse applies to atrophic destructive OA. Some data support this suggestion [33]. If so, the way a joint responds to trauma or other insults may simply be a reflection of skeletal mineralisation and the way that the rest of skeleton responds to mechanical loading and/or instability.

References

1   Heberden W (1803) Commentaries on the history and cure of diseases. Printed for T Payne, Mewsgate, London, 2nd ed.
2   Nichols EH, Richardson FL (1909) Arthritis deformans. *J Med Research* 21: 149–221
3   Stecher RM (1941) Heberden's nodes: heredity in hypertrophic arthritis of the finger joints. *Am J Med Sci* 201: 801–809
4   Kellgren JH, Moore R (1952) Generalised osteoarthritis and Heberden's nodes. *Br Med J* 1: 181–187
5   Lawrence JS, Bremner JM, Bier F (1966) Osteo-arthrosis. Prevalence in the population and relationship between symptoms and x-ray changes. *Ann Rheum Dis* 25: 1–24
6   Doherty M, Preston B (1989) Primary osteoarthritis of the elbow. *Ann Rheum Dis* 48: 743–747
7   Utsinger PD, Resnick D, Shapiro RF, Wiesner KB (1978) Roentgenologic, immunologic and therapeutic study of erosive (inflammatory) osteoarthritis. *Arch Intern Med* 138: 693–697
8   Cobby M, Cushnaghan J, Creamer P, Dieppe P, Watt I (1990) Erosive OA: Is it a separate disease entity? *Clin Rad* 42: 258–263
9   Makela P, Virtama P, Dean PB (1979) Finger joint swelling: correlation with age, gender and manual labour. *Am J Roentgenology* 132: 939–943
10 Bullough PG (1981) The geometry of diarthrodial joints, its physiologic maintenance, and the possible significance of age-related changes in geometry-to-load distribution and the development of osteoarthritis. *Clin Orthop* 156: 61–66
11 Martel W, Champion CK, Thompson GR, Carter TL (1970) A roentgenologically dis-

tinctive arthropathy in some patients with the pseudogout syndrome. *Am J Roentgenology* 109: 587–607
12 Resnick D (1997) *Diagnosis of bone and joint disorders*, 4th ed. WB Saunders, Philadelphia
13 Scotchford CA, Greenwald S, Ali SY (1992) Calcium phosphate crystal distribution in the superficial zone of human femoral head articular cartilage. *J Anat* 181: 293–300
14 Hayes A, Clift SE, Miles AW (1997) An investigation of the stress distribution generated in articular cartilage by crystal aggregates of varying material properties. *Med Eng Phys* 19: 242–252
15 Swan A, Chapman B, Heap P, Seward H, Dieppe P (1994) Submicroscopic crystals in osteoarthritic synovial fluid. *Ann Rheum Dis* 53: 467–470
16 Swan A, Heywood B, Chapman B, Seward H, Dieppe P (1995) Evidence for a causal relationship between the structure, size and load of calcium pyrophosphate dihydrate crystals, and attacks of pseudogout. *Ann Rheum Dis* 54: 825–830
17 Doherty M, Watt I, Dieppe P (1982) Localised chondrocalcinosis in post-meniscectomy knees. *Lancet* I: 1207–1210
18 Doherty M, Watt I, Dieppe P (1983) Influence of primary generalised osteoarthritis on development of secondary osteoarthritis. *Lancet* 2 (8340): 8–11
19 Doherty M, Dieppe P, Watt I (1993) Pyrophosphate arthropathy – a prospective study. *Br J Rheumatol* 32: 189–196
20 Halverson PB, McCarty DJ, Cheung HS (1984) Milwaukee shoulder syndrome. *Semin Arthritis Rheum* 14: 36–44
21 Dieppe P, Cawson T, Mercer E, Campion G, Hornby J, Hutton C, Doherty M, Watt I, Wolfe A, Hazleman B (1988) Synovial fluid collagenase in patients with destructive arthritis of the shoulder joint. *Arthritis Rheum* 31: 882–890
22 Nolte-Ernsting C, Adam G, Bühne M, Prescher A, Gunther RW (1996) MRI of degenerative bone marrow lesions in experimental osteoarthritis of canine knee joints. *Skeletal Radiol* 25: 413–420
23 Zanetti M, Bruder E, Romero J, Hodler J (2000) Bone marrow edema pattern in osteoarthritic knees: correlation between MR imaging and histologic findings. *Radiology* 215: 835–840
24 Boegard T, Rudling O, Petersson IF, Jonsson K (1998) Correlation between radiographically diagnosed osteophytes and magnetic resonance detected cartilage defects in the patellofemoral joint. *Ann Rheum Dis* 57: 395–400
25 Christensen SB (1985) Osteoarthrosis. Changes of bone, cartilage and synovial membrane in relation to bone scintigraphy. *Acta Orthop Scand* 214: 1–43
26 Hutton C, Higgs E, Jackson P, Watt I, Dieppe P (1986) ^{99m}technetium HMDP bone scanning in generalised nodal osteoarthritis-1. Comparison of the standard radiograph and 4-hour bone scan image. *Ann Rheum Dis* 45: 617–621. Also 2. The 4-hour bone scan image predicts radiographic change. *Ann Rheum Dis* 45: 622–626
27 Dieppe PA, Cushnaghan J, Young P, Kirwan J (1993) Prediction of the progression of

joint space narrowing in osteoarthritis of the knee by bone scintigraphy. *Ann Rheum Dis* 52: 557–563
28 Petersson IF, Boegard T, Dahlstrom J, Svensson B, Heinegard D, Saxne T (1998) Bone scan and serum markers of bone and cartilage in patients with knee pain and osteoarthritis. *Osteoarthritis Cartilage* 6: 33–39
29 McCrae F, Shouls J, Dieppe P, Watt I (1992) The scintigraphic assessment of osteoarthritis of the knee joint. *Ann Rheum Dis* 51: 938–942
30 McAlindon T, Watt I, McCrae F, Goddard P, Dieppe PA (1991) Magnetic resonance imaging of OA of the knee: correlation with radiographic and scintigraphic findings. *Ann Rheum Dis* 50: 14–19
31 Sharma S, Rogers J, Watt I, Buckland-Wright C (1997) Bone mineral density and fractal signature analysis in hip osteoarthritis – a study of a post-mortem and postoperative population. *Clin Radiol* 52: 872 (quoted in abstract)
32 Hutton CW (1987) Generalised osteoarthritis: an evolutionary problem? *Lancet* 27: 1463–1465
33 Dequeker J (1999) The inverse relationship between osteoporosis and osteoarthritis. *Adv Exp Med Biol* 455: 419–422

Radiographic joint space width (JSW): A marker of disease progression in OA of the hip

Joel A. Block

Section of Rheumatology, Rush Medical College, Chicago, IL 60612, USA

Introduction

Osteoarthritis (OA) of the hip is a significant source of pain and disability. As with OA involving other joints, there are few reliable surrogate markers which permit prognostication regarding disease progression. Moreover, in contradistinction to OA of the knee, there are few longitudinal analyses of progression of hip OA. The qualitative assessment of narrowing of the radiographic joint space width (JSW) has long been recognized in OA as a significant marker of disease severity, although clinical correlations with radiographic severity have never been highly specific or sensitive. Moreover, as these schemes rely on subjective assessments rather than on quantitative measurements, it is often difficult to discern subtle signs of progression; this difficulty may be magnified in hip OA because, whereas radiographic classification schemes such as that described by Kellgren and Lawrence [1] emphasize the presence of osteophytes, narrowing of the JSW may in fact be more relevant to clinically symptomatic disease [2, 3]. We have previously described a quantitative method to evaluate radiographic joint space width of the hip utilizing standard anteroposterior (AP) radiographs and hand-held calipers to assess the visible JSW of the hips [4]. This technique does not require advanced imaging technology and has the advantage of being available in the routine clinical setting. By applying this technique to radiographs obtained longitudinally in several distinct patient cohorts, we have evaluated the utility of quantitative radiographic JSW measurements as a marker of disease progression in OA of the hip.

Radiographic JSW of the hip

For quantification, the radiographic JSW is defined as the narrowest point between the cortical surface of the acetabulum and the bone contour of the femoral head and is measured using precision dial calipers. The method has been validated for interobserver and intra-observer variability by evaluating a series of radiographs in triplicate in blinded order by separate investigators in sessions separated by several weeks [4], and has been demonstrated to be highly reproducible among trained individuals who are geographically widely scattered; the overall inter- and intra-observer components of variance (0.003531 and 0.004035, respectively) accounted for < 1% of the total variation in JSW measurements; the rater-to-rater variation was 0.0619 mm and replicate-to-replicate variation was 0.0635 mm. After technical validation of the methodology, quantification of JSW was applied to assess the long-term progression of hip OA in several well-characterized populations.

The rate of narrowing of JSW is linear in hip OA

Evaluation of a cohort of 99 consecutive patients who underwent unilateral total hip arthroplasty (THA) for advanced OA of the hip was performed after long-term follow-up. The primary parameter assessed was the change in JSW of the contralateral, non-operated hip. Subjects were restricted to those without systemic inflammatory arthritides and those who did not undergo contralateral THA within the first 12 months after the index surgery. The group consisted of 49 females and 50 males of mean age 55 years. Annual clinical and radiographic follow-up was performed for a mean duration of 104 months (range 12–149 months). The overall median initial JSW of the contralateral hip was 3.48 mm (inter-quartile range 1.55). A total of 619 AP radiographs of the hips, consisting of the perioperative and annual follow-up examinations of the subject population, were available for analysis. Evaluation of the longitudinal narrowing of the JSW in the contralateral hips demonstrated that the rate of narrowing in each case was linear with a population median decline of 0.1 mm/year. Analyses of the aggregate disparity between the predicted y-intercept and the observed initial JSW for each individual confirmed that the rate of decline in JSW remained constant and linear throughout the period of study (Spearman $r_s = -0.69$, $p < 0.0001$). Moreover, in light of the concordance between the predicted and the measured intercept, it was clear that there was no threshold width beyond which the rate of JSW decline increased; thus, if true disease modifying therapies were to be identified for OA, it may be possible to arrest progression even relatively late in the disease course.

Confirmation that the rate of decline in JSW is linear throughout the course of OA suggests that evaluation of JSW early in the disease course may be predictive of the rapidity of disease progression. A test of this hypothesis in this cohort demonstrated that the observed rate of decline in the first 20 months of observation remained constant throughout the followup period ($p < 0.001$). Also, during follow-up, the cohort segregated into two distinct subpopulations of JSW narrowing in the contralateral hip: those with slow decline in JSW, ≤2 mm/year, accounted for approximately 85% of the population, and those with rapid decline, > 2 mm/year, comprised approximately 15%. These subpopulations did not differ with regard to initial JSW ($p = 0.10$), Kellgren-Lawrence radiographic grade ($p = 0.14$), or body mass index (BMI) ($p = 0.72$); however, by 20 months of follow-up, the pattern of JSW narrowing was clearly distinguishable among individuals.

To demonstrate that these findings were representative, an independent cohort was identified [5]. The second group consisted of 28 patients who underwent THA for OA and who had asymptomatic contralateral hips with mild or absent radiographic evidence of contralateral OA (Kellgren-Lawrence grade 2 or less). Eight of the 28 subjects had narrowing of the contralateral JSW at rates > 2 mm/year in the first 2 years, comparable to the earlier cohort. Moreover, analyses of bone mineral density (BMD) by dual-photon X-ray absorptiometry demonstrated that although those subjects destined to undergo rapid JSW narrowing had indistinguishable JSWs at baseline from the slow-narrowing group, their BMD at the hip, femoral neck, and lumbar spine were all significantly higher than the group of slow narrowers ($p < 0.05$), suggesting that elevated BMD may either be a marker of, or predispose to, progression of hip OA.

JSW is the best predictor of future contralateral THA

If the rate of narrowing of JSW is indeed constant in patients who have OA of the hip, then regardless of whether the patient is destined to be in the "rapid narrowing" or the "slow narrowing" group, it could be hypothesized that the magnitude of the JSW should have prognostic significance. To test this, a consecutive group of 170 patients undergoing unilateral THA for OA of the hip was followed longitudinally. The initial JSW of the contralateral hip was determined and radiographic grades, clinical parameters, and questionnaires were assessed in annual follow-up. This cohort consisted of 90 males and 80 females with an average age of 57.7 ± 11.7 years (mean ± SD) at the time of THA. With a median follow-up of 96.4 months, there were 34 contralateral THAs performed, for an overall risk of 31% (95% CI 20–40%). However, in a search for significant risk factors predictive of the need for contralateral THA, log rank statistics suggested that age ($p = 0.9$), gender ($p = 0.6$), and original side of THA ($p = 0.5$) were not related to future contralateral THA [6]. Similarly, the Cox proportional hazards model suggested that BMI was not a significant risk factor ($p = 0.42$). The most significant predictive value of the need for future contralateral THA was initial JSW. The overall contralateral initial JSW of the cohort was normal, with a value of 3.53 ± 1.09 mm (mean ± SD). For those subjects with initial contralateral JSW < 2 mm, the median duration to THA was 71 months; if the initial JSW was 2–4 mm, only 29% had undergone contralateral THA by 120 months; and for those with initial JSW wider than 4 mm, there were no contralateral THAs performed (Fig. 1). Furthermore, the Cox proportional haz-

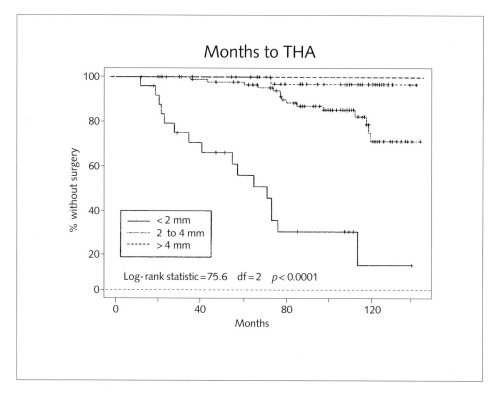

Figure 1
Kaplan-Meier plot illustrating risk of contralateral THA after the initial THA for OA of the hip. No subjects with an initial contralateral JSW of >4 mm underwent contralateral THA. In contrast, if the initial JSW was <2 mm, there was a high risk of contralateral THA, with a median duration to surgery of 71 months.

ards model predicted that the risk was continuous and progressive; thus for each additional 1 mm JSW at baseline, the model predicted a 74% reduction in relative risk of THA ($p = 0.001$).

Conclusions

The radiographic JSW of the hip is easily, precisely, and reproducibly quantifiable using standard clinical AP radiographs; this is in contradistinction to JSW measurements of the knees which require special radiographic techniques. The ability to apply this methodology without special technology may permit its use in the routine clinical setting. Moreover, it is clear that quantification of the radiographic JSW of the hip provides important prognostic information both longitudinally and cross-sectionally. Longitudinal evaluations of JSW provide an estimate of the rate of decline in JSW, and this rate has been demonstrated to be linear and constant over several years during the progression of hip OA; this may provide the basis for assessing putative disease-modifying therapies as well as permit planning for long-term disease management. Similarly, cross-sectionally, a single assessment of JSW may provide information regarding the risks of progression to advanced OA over the succeeding several years. Hence, the quantification of radiographic JSW of the hip may be considered as a surrogate marker of OA of the hip. Its relevance to subjects without clinical OA remains to be determined.

References

1   Kellgren JH, Lawrence JS (1957) Radiographical assessment of osteoarthrosis. *Ann Rheum Dis* 16: 494–502
2   Altman RD, Fries JF, Bloch DA, Carstens J, Cooke TD, Genant H, Gofton P, Groth H, McShane DJ, Murphy WA (1987) Radiographic assessment of progression in osteoarthritis. *Arthritis Rheum* 30: 1214–1225
3   Croft P, Cooper C, Wickham C, Coggon D (1990) Defining osteoarthritis of the hip for epidemiologic studies. *Am J Epid* 132: 514–522
4   Goker B, Doughan AM, Schnitzer TJ, Block JA (2000) Quantification of progressive joint space narrowing in osteoarthritis of the hip: longitudinal analysis of the contralateral hip after total hip arthroplasty. *Arthritis Rheum* 43: 988–994
5   Goker B, Sumner DR, Hurwitz D, Block JA (2000) Bone mineral density varies as a function of the rate of joint space narrowing in the hip. *J Rheumatol* 27: 735–738
6   Doughan AM, Goker B, Schnitzer TJ, Kull L, and Block JA (1997) Time course of contralateral total hip arthroplasty (THA) after hip replacement for osteoarthritis (OA). *Orthopaedic Transactions* 21: 826

Small proteoglycans in knee and ankle cartilage

Gabriella Cs-Szabo[1,2], Deborah Ragasa[1], Richard A. Berger[2] and Klaus E. Kuettner[1,2]

Departments of [1]Biochemistry and [2]Orthopedic Surgery, Rush Medical College at Rush-Presbyterian-St. Luke's Medical Center, 1653 W. Congress Parkway, Chicago, IL 60612, USA

Introduction

Clinical and epidemiological data provide evidence that the prevalence of symptomatic osteoarthritis (OA) is higher in knee than in ankle joints. Several factors can contribute to this phenomenon, including a difference in biochemical composition and gene expression levels. This difference in composition may provide the ankle tissue with more resistance to progressive tissue degeneration or with an increased potential for repair following damage. In addition to the major structural components of the cartilage matrix (aggrecan and collagens), several other molecules can contribute to tissue development and repair. One of these players is the group of the small proteoglycans (PGs), members of the family of leucin-rich repeat proteins. These molecules, which are ideal for protein-protein interactions, can indeed influence matrix assembly. Several members of the small PG family, biglycan, decorin, fibromodulin and lumican, are present in articular cartilage. One or more of them were shown to be involved in collagen fibrillogenesis and interactions with other extracellular matrix components, e.g. growth factors (TGFβ, BMP-7 (unpublished observation)), fibronectin and thrombospondin (for summary see [1]). Thus, their presence in appropriate concentrations may be crucial in developing and/or regenerating cartilage.

We have demonstrated elevated levels of the small PGs, biglycan, decorin, fibromodulin as well as lumican, in OA knee cartilage samples when compared to normal controls [2, 3]. Now we follow changes affecting two of the small PGs, biglycan and decorin, at different stages of cartilage degeneration in knee and ankle cartilage. This information may help to understand why knee cartilage is more prone to develop OA than ankle cartilage, as well as to reach our ultimate goal of defining the role small PGs play in the development of OA and/or the repair process following cartilage damage.

Samples

As both biglycan and decorin undergo age-related changes [4], cartilages from donors between 50 and 80 years old were used. Patients donating OA cartilage after total knee replacement surgery come from the same age group. Additionally, we have shown that the levels of small PGs change as a function of tissue load [5], thus cartilage was harvested only from the load bearing surfaces of the knee and the ankle joints. These cartilage surfaces were visually graded for tissue degeneration using a scale developed by Collins which was adapted to the knee and the ankle cartilage by our research group [6]. According to this scale, grade 0 cartilage is normal (without degenerative changes) and grades 1–4 indicate increasing severity in cartilage tissue degeneration ranging from surface fibrillation (grade 1) to loss of more than 30% of the articular cartilage from the subchondral bone along with the development of osteophytes (grade 4). Since osteophytic cartilage represents a less mature phenotype, this cartilage was not harvested. Finally, an appropriate number of tissue samples was collected to generate statistically significant data. This was not an easy task, since in this age group, knee samples show more pronounced tissue damage (more grade 3 and 4 than grade 0 or 1) than ankle cartilage (more grade 0 and 1 than grade 3 and 4).

Results

While the macroscopic appearances of knee and ankle cartilage of the same Collins grade are similar to each other, there are differences in tissue composition regarding the small PGs.

Biglycan core protein undergoes degradation and fragments are generated in both knee and ankle cartilage tissues. These fragments are processed mainly from the N-terminal of the core protein. Interestingly, these fragments do stay in cartilage even in heavily degraded tissue, suggesting that the major binding site on the core protein is farther towards the C-terminal. There is no detectable difference in the fragment pattern between knee and ankle cartilage; however, more fragments are detectable in grade 3 and 4 cartilages than in grade 0–2 samples. The number of fragments further increase in OA knee cartilage, thus in this respect, a grade 4 cartilage, even if visually similar looking, cannot be considered undiagnosed OA. The combined (intact and fragmented) core protein level of biglycan is slightly decreased in grade 1 and steadily increases in more severely degenerated knee cartilage (Tab. 1). On the contrary, in ankle cartilage, biglycan content (both intact and degraded) declines with increasing damage severity (Tab. 1). Biglycan message increases in parallel with the severity of cartilage degeneration in knees, but declines in ankle cartilage (Tab. 2) following the pattern of change on the protein level. Thus,

Table 1 - Change of small PG core protein levels in degenerating cartilage (normalized to cartilage wet weight and grade 0)*

Grade			0	1	2	3	4
Biglycan	Knee	↑	1	0.8	1.3	1.5	1.8
	Ankle	↓	1	0.8	0.6	0.7	0.5
Decorin	Knee	↑	1	1.2	1.3	1.5	1.8
	Ankle	↓	1	0.9	0.6	0.8	0.7

*measured as relative densities on Western blots

Table 2 - Change of small PG message levels in degenerating cartilage (normalized to GAPDH)*

Grade			0	1	2	3	4
Biglycan	Knee	↑	0.4	0.5	0.6	0.65	0.75
	Ankle	↓	0.7	0.65	0.6	0.5	0.4
Decorin	Knee	↑	1.0	5.0	3.0	2.5	2.5
	Ankle	↓	1.5	1.2	1.0	0.8	0.6

*measured as relative densities in gels

it is most likely that the expression of biglycan is regulated at the transcriptional level.

Decorin does not go through intensive degradation during tissue degeneration. Only one slightly shorter core protein fragment was detected in all of our samples, which is processed from the N-terminal. This fragment content slightly increases in more heavily degenerated samples but still represents only a minor fraction of the total. There are few additional shorter core protein species present in OA knee cartilage, however, these still add up to a small fraction of the intact core protein. Decorin core protein content increases proportionally to increasing grade of cartilage degeneration in the knee (Tab. 1). On the contrary, the protein content decreases in ankle cartilage as a function of tissue damage (Tab. 1). Decorin mRNA in knee cartilage increases significantly in grade 1 tissue and then declines in heavily degenerated samples, while in ankle cartilage, a continuous decline in the message for decorin is detected (Tab. 2).

Conclusion

We demonstrate here the change of expression of small PGs in relation to cartilage degeneration. Interestingly, this change shows opposite directions in a joint that frequently develops into OA (knee) and in another joint in which the occurrence of OA is rare (ankle). These results may imply that abnormal accumulation of small PGs might be responsible for the failure of proper repair processes in knee cartilage. Moreover, the difference in the response of small PGs to damage in knee and ankle cartilages may be an indication of the difference in their functions in cartilage homeostasis and in the process of repair.

Acknowledgements
This work was supported by the NIH (AG-16024 and 2P50 AR-39239) and through a research agreement by the GlaxoSmithKline Pharmaceuticals. Collaboration with the Regional Organ Bank of Illinois and A. Margulis, M.D. is greatly appreciated.

References

1   Iozzo R (1999) The biology of the small leucine-rich proteoglycans. *J Biol Chem* 274: 18843–18846
2   Cs-Szabo G, Roughley PJ, Plaas AH, Glant TT (1995) Large and small proteoglycans of osteoarthritic and rheumatoid articular cartilage. *Arthritis Rheum* 38: 660–668
3   Cs-Szabo G, Melching LI, Roughley PJ, Glant TT (1997) Changes in messenger RNA and protein levels of proteoglycans and link protein in human osteoarthritic cartilage samples. *Arthritis Rheum* 40: 1037–1045
4   Roughley PJ, Melching LI, Reckliens AD (1994) Changes in the expression of decorin and biglycan in human articular cartilage with age and regulation by TGF-beta. *Matrix Biol* 14: 51–59
5   Cs-Szabo G, Liu W, Chubinskaya S, Roughley PJ, Buzás E, Jacobs JJ, Glant, TT (1999) Loading and osteoarthritic changes influence the levels and distribution of small proteoglycans in human knee joints. *Trans ORS* 24: 400
6   Muehleman C, Bareither D, Huch K, Cole AA, Kuettner KE (1997) Prevalence of degenerative morphological changes in the joints of the lower extremity. *Osteoarthritis Cartilage* 5: 1–15

Discussion

Introduction

Q: *Stefan Lohmander*: If you regard the OA process as a continuum, then the distinction is artificial and perhaps dependent on the technology available to us for diagnosing and assessing joint damage. Is that not right?

A: *Paul Dieppe*: I agree that it is artificial to dichotomize the continuum of early change in joint structure through to severe joint destruction. However, several different studies have come to similar conclusions – i.e. that there are different risk factors for initiation and progression of OA. What this suggests to me is that there are several different aspects to the OA disease process, and that as joint damage progresses, we move from dominance of some of these mechanisms to dominance of others, and that there are different risk factors for these different types of process. Put more simply, it may be that, for example, synovitis and inflammation are unimportant in early joint damage, but important drivers of progression.

Q: *Stefan Lohmander*: What was the cut-off used to differentiate OA from non-OA in the different studies that you mention?

A: *Paul Dieppe*: The cut-off we used for initiation of OA was going from a completely normal radiograph to any grade of change on the Kellgren and Lawrence scale, including grade 1. Others have not regarded grade 1 as indicative of OA. Similarly, there have been slightly different definitions used for progression in different studies. Your point is a perfectly valid one; we are currently dependent on the very blunt instrument of changes in Kellgren and Lawrence grades on an x-ray to make these differentiations.

Q: *Stefan Lohmander*: In my view we should think about where we are in the continuum of the OA process, and where and when different risk factors kick in, as this may give us new insights into what is going on.

A: *Paul Dieppe*: Yes, I agree.

Q: *Klaus Kuettner*: Paul, in your lecture you didn't use the concept of osteophytosis. Was that on purpose or was it an oversight?

A: *Paul Dieppe*: That was an oversight, really, Klaus. It is an important point, because the grade 1 on the Kellgren and Lawrence scale, which I just referred to, means osteophytes without any evidence of cartilage loss. I think that lone osteophyte formation, without cartilage damage, may be a separate clinical and pathological condition, distinct from OA. In my clinical practice I use the term "painful osteophytosis", because as osteophytes grow, they can cause pain by periosteal elevation, even if the rest of the joint is entirely normal.

Q: *Richard Loeser*: In your presentation you avoided discussing the role of inflammation (of either synovium or cartilage) in either the initiation or the progression of OA, would you be willing to make a few comments on that?

A: *Paul Dieppe*: There is certainly evidence of conventional inflammation in the synovium, cartilage and subchondral bone in advanced OA, so it is clearly part of the OA process. In general, the more advanced the OA the more inflammation you are likely to see. That is not surprising, as inflammation can be defined as the response of living tissues to injury, so if you injure the joint it will become inflamed. However, it is difficult to know what part this sort of inflammation may be playing in driving further damage. It seems likely that it makes some contribution, both to symptoms such as pain and stiffness, and to joint damage. I do not know how much, but I suspect relatively little.

Q: *Frank Luyten*: I hope we agree that we are talking about local inflammation here. In OA you do not have any increase in ESR or CRP. If you see that then it is not OA anymore.

A: *Paul Dieppe*: Absolutely, the inflammation in OA is limited to the joint and is not systemic. We do see increased CRP levels in OA, Frank, but it's a very minor increase, and like you, I have never seen a raised ESR as a result of OA alone.

Q: *Eugene Thonar*: I have found that very low levels of cytokines can have a profound influence on chondrocytes and bone, so even the low level of CRP might be important. What we need to do is to redefine what we consider inflammation in OA. If we cannot define that we cannot determine how important it is.

A: *Paul Dieppe*: Points well taken Eugene. I think Frank's point is that in the clinical arena we have to distinguish between disorders like rheumatoid arthritis and infection, where there are major systemic manifestations of inflammation, and disorders like OA, where the inflammatory process is largely opaque at the systemic level, and confined to the joint.

Q: *Robin Poole*: It is striking that in Mary Fran Sower's study on the initiation of OA in a female population, she found up-regulation of CRP in the very early stages. We have to consider what this reflects – is it inflammation or is it changes in cytokine regulation accompanying changes in connective tissue turnover and chondrocyte activity? It could result from stimulation by interleukin-6 and other cytokines within the musculoskeletal tissues. In our studies of ageing, that we are doing with Ada Cole, Matthias Aurich and Klaus Kuettner, we find striking differences in collagen synthesis in people with OA, which is not reflected by any joint pathology, suggesting that there may be fundamental differences in skeletal tissue turnover in OA populations. I would urge people to look differently and critically at inflammation in the light of what Paul is talking about as turnover of tissues in the OA process, rather than the sort of inflammation that we see in rheumatoid synovitis.

A: *Paul Dieppe*: I agree with that. I guess the finding of susceptibility to OA in people with increased bone density would also speak to the sort of concept that you are putting forward.

Q: *Thomas Andriacchi*: In your clinical observation have you noticed a difference between weight bearing and non-weight bearing joints in the rate of stabilization or progression? What motivates that question is the potential to modify loading on the lower extremities during locomotion, whereas in upper extremities, there may be other important factors.

A: *Paul Dieppe*: That's a question I find difficult to answer, partly because I feel that we don't really understand what aspects of loading of the different joints matters. Many joints go through a cycle of change followed by stabilization, and in my experience, the interphalangeal joints of the hand do this the most rapidly. In just 2 or 3 years they may change from normal to severe OA, and then stop changing their anatomy for ever. In general the knee and hip joints seem to change more slowly, although the hip is more likely to undergo both phases of rapid destructive change and spontaneous healing than the knee. I think most of these changes must be mechanically driven, but I don't know how.

Imaging cartilage changes in osteoarthritis

Q: *Frank Luyten*: We are getting very nice pictures from contrast CT. In our hands it gives better information on cartilage than the MRI data.

A: *Charles Peterfy*: You can delineate the articular cartilage if you inject some contrast into the joint, then you can see the surface of the cartilage. That is a potential approach. One could compare that technique with others for specific applications. If you have CT, but not MRI, you could certainly get a lot out of it.

Q: *Frank Luyten*: Even when we compare CT with MRI, the contrast CT provides a little sharper image than the MRI. Has that also been your experience?

A: *Charles Peterfy*: If you make a straightforward measurement like contrast, you are right, CT gives a good result. But it's really about what endpoint you want to measure; do you want cartilage thickness or volume for example?

Q: *Matthias Aurich*: I was very impressed with the high-resolution MRI pictures you showed. They seemed to show black stripes on a white background in the cartilage. Do these stripes correlate with any structure inside the cartilage?

A: *Charles Peterfy*: One study from Iain Watt and Paul Dieppe did show histological correlates to this banding pattern. What you are seeing is expression of T2 relaxation, which correlates with collagen content. But there is also a "magic angle phenomenon" having to do with anisotropy of T2 relaxation in cartilage, which can produce artifactual banding if the orientation is inappropriate. One has to take this into account, particularly if you are trying to obtain quantitative data from T2 relaxation times.

Q: *Matthias Aurich*: So, you think it's related to the collagen structure?

A: *Charles Peterfy*: Yes, I do.

Robin Poole: We have done quite large studies looking at the relationships between MRI of articular cartilage and collagen, and could never see any correlations. It may be a much more complicated picture, dependent on molecular interactions and water content, and involving proteoglycans as well as collagen.

Q: *Jenny Tyler*: The interesting idea of acquiring quantitative T2 relaxation rates from cartilage in the knee depends on a two-dimensional slice. What is involved in repositioning that slice for a longitudinal study?

A: *Charles Peterfy*: You have to contend with the usual errors of two-dimensional slicing and partial volume averaging on either side. The problem with the conventional T2 sequence that one would basically have to use for this is that it has a poor signal to noise ratio, so you often have to take slightly thicker slices. The one good thing is that these T2 lesions tend to be relatively large and cover quite a bit of territory, so the challenges to spatial resolution are less than they perhaps would be for tiny focal defects.

Q: *Jenny Tyler*: What are the prospects of getting T2 information in a three-dimensional data set?

A: *Charles Peterfy*: There is DEFT which is a T2-weighted 3D sequence that is being developed by Garry Gold in Stanford. It will be some time before that is available in conventional scanners, and it is not clear how soon it might be available for clinical trials, but there is the capacity to do it. You could do three-dimensional fast spin echoes as an alternative.

Bone change in osteoarthritis

Q: *Johannes Flechtenmacher*: How do early bone scan abnormalities compare to STIR sequence MR images?

A: *Iain Watt*: There is an extremely good correlation between perfusion scan abnormalities and fat-suppressed or T2 weighted MRI sequences showing marrow edema. It is so good that I no longer do very many bone scans.

Q: *Paul Dieppe*: Do you think that the bone change comes before the cartilage change?

A: *Iain Watt*: Yes. For example, I have pictures of Cathy Carlson's work on early OA in monkeys, which shows evidence of major changes occurring in the subchondral bone and well-preserved articular cartilage. I believe that the bone is crucial to the development of OA.

Q: *Frank Luyten*: You mentioned that in a number of cases, the subchondral bone is responsible for the pain – if so, how should we treat these patients – with pure analgesics rather than anti-inflammatory agents?

A: *Iain Watt*: I'm a radiologist, not a clinician, but I would suggest to you that the vast majority of pain that we ascribe to radiographic change is actually coming from structures that we cannot see on the x-rays – the capsule, ligaments, tendons, muscle, or whatever.

Paul Dieppe: Frank, I would just add that we've got to go beyond describing both pain and radiographic changes as present or absent. We need to understand the different features, patterns, severity and associations of each phenomenon. That work is just starting but really needs to go much further.

Q: *Rick Sumner*: Given all these bone changes, is it a good strategy or a bad strategy to turn off bone remodeling as a cartilage protective agent? Or does it depend on the type of OA?

A: *Iain Watt*: Instinctively, I think it has to be bad to turn off a bone response. I would not want to inhibit callus formation or fracture repair.

Rick Sumner: However, there is a poster at this meeting showing that in experimental models agents that inhibit bone remodeling can protect cartilage.

Q: *Ernst Hunziker*: When you see thickening of the joint space, it may not be due to the presence of hyaline cartilage, it may be fibrous cartilage or some other tissue. When we look at late OA specimens or those with a so-called repair response, we find that they are pathologically very active, that the repair tissue comes from the bone, and that it is not hyaline cartilage that is forming on the surface of the joint. Similarly if you stimulate repair with a microfracture technique, it is fibro-cartilage that forms.

A: *Iain Watt*: I want to make it absolutely clear that when I said repair, I did not mean the restitution of normal anatomy. What I meant was a containment exercise – a fallback position for the joint. Radiographically and scintigraphically we see this. The problem is that I need to know where you get your specimens from because if these are operative, then by definition the disease is not quiescent. If you think about the analogy with a fracture repair, we know that this process can continue to change for at least 2 years. So I am not at all surprised to hear that you see continuing pathological changes in a joint that is considered to be quiescent, or in *post mortem* specimens.

Radiographic joint space width (JSW): A marker of disease progression in osteoarthritis of the hip

Paul Dieppe: I should have mentioned when I was talking about bone density and OA progression, that I was only referring to the knee joint. Joel's data from the hip, showing a different type of relationship, highlights the general point that different joints may be responding in unique ways to insults, and that the OA process may be under different influences at the varying joint sites affected.

Q: *Johannes Flechtenmacher*: Do you have any idea about what diagnosis leads to the first implant? Is there a difference between those with hip dysplasia and idiopathic OA?

A: *Joel Block*: We stratified patients diagnostically as having primary or secondary OA, and by the cause if secondary. When we do this we find that those with secondary OA are at less risk of having a rapid decline in the contralateral hip. This makes sense, as many of those in the secondary group had unilateral trauma as the cause of their hip OA.

Q: *Stefan Lohmander*: I noticed in your first cohort that the average age was something like 55, and I thought that was a fairly young age for having hip replacement.

A: *Joel Block*: The average age of the overall group is not relevant because a huge number of the younger people have inflammatory arthritis. However, we were also struck by the apparent young age of the OA cohort, whose age ranged from about 35 to 85. A lot of the younger ones had OA secondary to disorders such as a dysplasia. We did the analyses both including and excluding these groups, and got similar results.

Small proteoglycans in knee and ankle cartilage

Q: *Robin Poole*: As you point out, these molecules are directly or indirectly associated with collagen fibrils, so it is very important to relate the data to the content of the different types of collagen.

A: *Gabriella Cs-Szabo*: I know that there is an increase in Type I collagen level and that biglycan, decorin and fibromodulin can bind to it. Also, we can measure an increase in Type VI collagen, which is the major binding partner for biglycan. I have not studied type II collagen yet, but I intend to do some more work in this area.

Q: *Chris Handley*: Can you tell us a little bit about the degradation of decorin and biglycan and where the cleavage sites are? Also, have you got any information about the glycosylation of these molecules? In some work we've been doing with ligament, there is quite a large population of non-glycosylated core proteins.

A: *Gabriella Cs-Szabo*: We don't know the actual cleavage sites, but the degradation is coming from the N terminal, and the fragments are short. There is more cleavage of biglycan than decorin, and fibromodulin is intermediate. I did not look for glycosylation in this project, but when I worked with normal vs. OA samples, I could see a lot of unglycosylated biglycan and decorin in OA tissue.

Biomechanics and cartilage metabolism

Introduction

Alan Grodzinsky

Continuum Electromechanics Group, Center for Biomedical Engineering, Massachusetts Institute of Technology, M.I.T. Room 38-377, Cambridge, MA 02139, USA

During the past decade, increasing attention has focused on the ability of cells and tissues to respond to mechanical forces and other physical stimuli in their environment. Investigators have studied cellular mechanotransduction in a broad range of soft and hard connective tissues, epithelial and endothelial tissues, and muscle. This Section focuses on the effects of mechanical loading on cartilage metabolism and the resulting chondrocyte-mediated synthesis, assembly and degradation of this tissue. Articular cartilage is subjected to a wide range of static and dynamic mechanical loads in human synovial joints. The ability of cartilage to withstand these compressive, tensile, and shear loads depends on the composition and structural integrity of its extracellular matrix (ECM). In turn, the maintenance of a functionally intact ECM requires that chondrocytes respond to mechanical forces as well as biological signals as regulators of matrix synthesis, assembly, and turnover.

It is now well accepted that mechanical stimuli in the microenvironment of the chondrocytes can significantly affect cartilage metabolism. However, the cellular transduction mechanisms that regulate chondrocyte response to mechanical stimuli are not well understood. Recent data suggest that there are multiple regulatory pathways by which chondrocytes sense and respond to mechanical stimuli, including upstream signaling pathways and mechanisms that may lead to direct changes at the levels of transcription, translation, and post-translational modifications. Correspondingly, there may be multiple pathways by which physical stimuli can alter not only the rate of matrix production, but the quality and molecular structure of newly synthesized proteoglycans, collagens, and other ECM molecules. In this manner, specific mechanical loading regimens may either enhance or compromise the ultimate biomechanical function of cartilage.

Much progress has been made in understanding the magnitude and distribution of physical forces and flows in the neighborhood of the chondrocyte within loaded cartilage. Experiments have shown that compression of cartilage causes deformation of cells and matrix, hydrostatic pressure gradients, fluid flow, streaming potentials and currents, and physicochemical changes including altered matrix water content, fixed charge density, mobile ion concentrations, and osmotic pressure. Any of these mechanical, chemical, or electrical signals may modulate cellular metabolism. As a result, our interpretation of the observed effects of dynamic compression on chondrocyte behavior is complicated by the fact that many of these physical signals will occur simultaneously within the tissue during compression.

The first study reported in this Section was motivated in part by the complexity of interpreting the response of chondrocytes to mechanical compression. Jin et al. focused on the effects of dynamic tissue shear deformation on chondrocyte biosynthesis. The conditions of their experiments highlight the role of cell and matrix deformation produced by tissue shear, while minimizing the presence of intratissue fluid flow, hydrostatic pressure gradients, and streaming potentials. (These experiments are in marked contrast to other complementary studies in the literature in which isolated cells plated onto substrates are subjected to applied fluid shear). Jin et al. observed that tissue shear in the absence of serum could stimulate protein and proteoglycan synthesis in a dose-dependent manner with increasing tissue shear strain. In the presence of serum, tissue shear preferentially stimulated protein synthesis (predominantly collagen synthesis in this immature bovine cartilage system) over proteoglycan synthesis. Taken together, the data suggest that tissue shear deformation can stimulate chondrocyte biosynthesis even in the absence of significant fluid flow and the associated fluid transport of soluble factors that is typically induced by intratissue fluid flow. In a related set of experiments, Jin et al. showed that nanoscale electrostatic repulsive interactions between charged GAG chains within the ECM play an important role in providing shear stiffness of cartilage during shear deformation of the tissue.

Urban and coworkers then emphasized that GAG chains are also responsible for the high osmotic swelling pressure of cartilage that gives the tissue increased stiffness in compression. At the same time, extracellular osmolarity and pH, which are regulated by the concentration of GAGs within the tissue, constitute important physicochemical signals that regulate chondrocyte synthesis of GAG-rich aggrecan molecules. Thus, there is an important feedback loop between tissue GAG content and GAG synthesis by chondrocytes. This loop appears linked to the relation between GAG fixed charge density and the resulting osmolarity of the ECM. Urban et al. showed that changes in osmolarity produced *in vitro* by alterations in bath ionic strength or by applied mechanical compression can play a key role in regulating transport of ions and water across the chondrocyte cell membrane which, in turn, can activate MAP kinase pathways and membrane transporters. This concept provides an important link between the extracellular environment and the intracellular milieu of the chondrocyte within intact tissue. Changes in intracellular ion concentrations can thereby have a direct and rapid influence on GAG synthesis.

The study by Sah and coworkers focuses on changes in the biomechanical properties of cartilage that occur with growth, age and degeneration. Based on measurements in tension and compression, significant changes in tissue properties are identified that are related to the underlying composition and structure of the ECM.

These changes are related, in turn, to the changing mechanical environment seen by the chondrocyte during tissue growth and remodeling from fetal to post-natal, childhood, and adult stages of life. The compressive modulus of cartilage increases from fetus to adult, with an accompanying decrease in ECM hydraulic permeability. Together, these changes result in an improvement in load-carrying characteristics of cartilage. In contrast, tissue sites with a high predilection for osteoarthritis exhibit low tensile strength in young adults or a decrease in tensile strength with age. Both normal aging and osteoarthritic degeneration contribute to the biomechanical dysfunction of cartilage. Sah relates these observed changes in tissue properties to underlying changes in molecular composition and the integrity of the collagen network and the collagen-glycosaminoglycan composition within the different layers of the tissue.

The potentially important role of mechanical signaling in cartilage growth and degeneration are contrasted in the studies by Aurich et al. and Kisiday et al. Investigators have shown previously that the prevalence of OA degeneration is lower in the ankle than the knee joint. Aurich et al. emphasize that there are biochemical as well as biomechanical differences in the cartilages from these joints, and that chondrocytes from these tissues may exhibit an intrinsically different metabolic response to inflammatory insults. As an example, they compared the response of knee and ankle chondrocytes cultured in alginate gel beads to treatment with IL-1β, which inhibits synthesis of sulfated glycosaminoglycans. Interestingly, they found that the 50% inhibitory concentration of IL-1β was five times lower in knee compared to ankle chondrocytes; that is, the ankle chondrocytes are more resistant to this catabolic stimulus. The possibility that such a response is linked to the different microbiomechanical environment of the two cell populations provides an important theme for the integration of biological and biomechanical hypotheses relating to cartilage metabolism. Kisiday et al. applied dynamic mechanical compression directly to bovine chondrocytes seeded into a self-assembling peptide gel scaffold as a model for mechanical conditioning of cartilage tissue engineering constructs. They found that unloaded chondrocytes synthesize and assemble a cartilage-like ECM within this peptide gel material. Interestingly, dynamic compression induced a 20–30% increase in the accumulated GAG content during the ~2 week loading period, and thereby increased the mechanical functionality of the tissue. Taken together, these studies demonstrate that the regulation of chondrocyte anabolic and catabolic pathways appears to be sensitively tuned to the cell's biomechanical environment.

Fixed charge density and cartilage biomechanics

Robert J. Wilkins, Bethan Hopewell and Jill P.G. Urban

Physiology Laboratory, Oxford University, Parks Rd, Oxford OX1 3PT, UK

Introduction

The biomechanical properties of cartilage depend to a large extent on its ability to maintain hydration and tissue thickness under the mechanical stresses encountered during normal activities. The resistance to fluid expression under pressures which can rise to 20 MPa (200 atmospheres) [1] has long been known to depend on the swelling pressure imparted by the osmotic properties of aggrecan. Aggrecan osmotic pressure is largely ionic in origin [2] and arises from the difference between the ionic composition of cartilage and that of bathing medium or synovial fluid. Swelling pressure thus increases as the mobile-ion concentrations in cartilage rise or as medium ionic strength falls. The importance of the ionic contribution to cartilage stiffness can be shown by compressing cartilage in media of different ionic strengths; to compress a cartilage plug to 600 microns required 200 kPa at low ionic strength (0.01 M saline) but only 50 kPa at high ionic strengths (1 M saline) [3].

The concentration of mobile ions in cartilage is dictated by the concentration of negative fixed charges on the glycosaminoglycans (GAGs) [4]; there is thus a direct relationship between cartilage stiffness and GAG content [5]. For cartilage to fulfil its biomechanical role it is therefore vital that GAG concentrations are kept at appropriate levels, i.e., that chondrocyte activity maintains rates of GAG production and degradation in balance. Many factors can affect these rates; one not often considered is the extracellular ionic environment itself. Feedback from local ion concentrations or osmolality could thus provide a rapid and sensitive mechanism for chondrocytes to sense and react to local GAG concentrations. In this chapter we will discuss the effects of extracellular osmolality and pH, both of which are determined to a large extent by the fixed charge density of the extracellular matrix, on GAG synthesis and on other aspects of chondrocyte behaviour.

Extracellular osmolality

Fixed charge density and cartilage osmolality

The extracellular osmolality of cartilage principally arises from mobile ion concentrations [2], which depend directly upon the fixed charge density of GAGs. In general cartilage osmolality lies in the range 350–450 mOsm but varies from joint to joint and throughout the depth of cartilage as it follows the gradients in fixed charge density [2]. Osmolality is lowest in the surface zone, which has the lowest GAG content. It is also lower in osteoarthritic than normal cartilage because of the fall in GAG concentration through dilution and loss [7].

Extracellular osmolality and chondrocyte metabolism

When cartilage is compressed under static loads, GAG synthesis falls in a load-dependent manner and Jones et al. [8] first suggested that this fall in synthesis was mediated by the consequences of fluid expression. Several studies have since found that the fall in GAG synthesis seen when extracellular osmolality is increased at constant hydration, parallels that seen after increase in extracellular osmolality consequent on fluid expression [9], indicating that chondrocytes appear to respond to the increase in ionic strength and osmolality rather than loss of fluid *per se*. While the effects of fluid expression have been well documented, the effects of swelling and GAG dilution are not easily seen in normal cartilage which swells only slightly. However responses to hypo-osmolarity have been studied in articular chondrocytes isolated from the matrix (and intervertebral disc, which swells dramatically) and show that synthesis also falls after exposure to hypo-osmotic conditions [10]. Thus synthesis rates fall in a dose-dependent manner after both hyper and hypo-osmotic shock [9], and maximal synthesis rates under experimental conditions are thus at osmolalities close to the average experienced *in situ* (~380 mOsm) (Fig. 1).

While short-term studies with intact cartilage or isolated chondrocytes show dramatic falls in synthesis if osmolality is changed from 380 to 450 mOsm (Fig. 1), chondrocytes *in situ* are exposed to a wide range of extracellular osmolalities, from c.350 mOsm at the surface to c.500 mOsm in the deep zone. Thus *in vivo* at least, the chondrocyte population must adapt to a wide range of extracellular osmolalities. Evidence from long-term culture studies indicates that chondrocytes are indeed able to adapt to hyperosmolality. Following an initial suppression of GAG synthesis in response to the onset of hyperosmotic conditions, chondrocyte synthesis rates then appear to recover and even surpass original levels after 12–24 h culture [9, 11] (Fig. 1). Adaptation to hyperosmolality has also been studied in renal cells where activation of members of the mitogen activated protein (MAP) kinase family appear responsible for the adaptive changes to osmotic stress [12], which involve upregu-

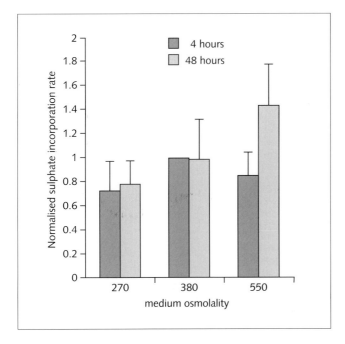

Figure 1
The effect of extracellular osmolality on sulphate incorporation by isolated bovine articular chondrocytes cultured in alginate beads. Chondrocytes were cultured in medium whose osmolality was altered by sodium addition and rates of synthesis measured using ^{35}S-sulphate at 4 and 48 h after change in osmolality (adapted from [13]).

lation of transporters for neutral solutes; similar pathways appear activated in chondrocytes [13, 14] (Fig. 2). In contrast, in response to hypo-osmotic conditions (280 mOsm; the standard osmolality of tissue culture medium), GAG synthesis remained suppressed throughout 24 h of culture [13]. Chondrocytes thus appear unable to adapt hypo-osmotic conditions, which could have consequences in early stages of osteoarthritis when tissue hydration increases and interstitial osmolality falls.

Extracellular pH

Extracellular pH in cartilage

Another ion which has a large effect on cellular behaviour is H^+. GAG concentrations regulate the distribution of H^+, as other cations, according to Gibbs-Donnan

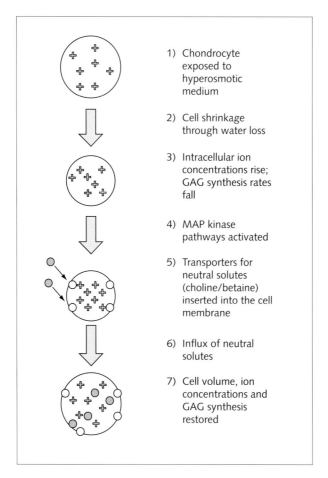

Figure 2
Schematic diagram showing changes in intracellular composition after hyperosmotic shock. (1–3) Immediately (seconds) after hyperosmotic shock, chondrocyte lose water and volume, intracellular solutes concentrate to maintain iso-osmotic conditions across the cell membrane, and GAG synthesis falls [9]. (4–6) The rise in intracellular ions initiates MAP kinase pathways (minutes) which result in the insertion of neutral solute membrane transporters into the cell membrane (hours) [11–14]. (7) Neutral intracellular solute composition rises, cell volume increases and intracellular ion concentrations and GAG synthesis are restored (> 12 h) [11:13].

equilibria; titration studies indicate that pH in cartilage is around 0.5 pH units lower than in the synovial fluid on account of the fixed negative charge density of the tissue [15]. However H^+ concentrations are also affected by another mechanism,

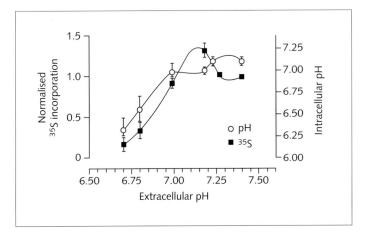

Figure 3
Relationship between intra and extracellular pH in isolated chondrocytes. Isolated chondrocytes were incubated in media of controlled extracellular pH. Intracellular pH was measured using the fluorescent dye BCECF [22], and GAG synthesis was determined from rates of ^{35}S-sulphate incorporation. (Adapted from [21]).

viz. cell energy metabolism. Chondrocytes obtain their energy predominantly through breakdown of glucose to produce lactic acid [16]. The resulting hydrogen ions leave the chondrocyte through membrane-bound H$^+$-lactate- cotransport carrier proteins [17] and diffuse through the matrix to the synovial fluid or subchondral blood supply where they are removed into the systemic circulation. Gradients of pH through cartilage would thus depend on GAG concentration gradients and on factors regulating energy metabolism, and thus the rate of lactic acid production. Concentration gradients of hydrogen ions produced by metabolism also depend critically on pathways for removal; the pH in the deep zone will be considerably less acidic if the subchondral plate is permeable than if the only route for diffusion is through the cartilage surface into the synovial fluid (Selard, unpublished observations). In all cases however, pH would be less acidic at the surface than in the deep or middle layers of cartilage and also more acidic around the cell than in the inter-territorial matrix.

Effects of extracellular pH on chondrocyte metabolism

Extracellular pH, like osmolality, has a powerful influence on the rate of proteoglycan synthesis [18] with rates 30–40% greater at pH 7.1 than at pH 7.4, and then falling steeply as extracellular falls below pH 6.8 (Fig. 3). Extracellular pH has other effects on cell metabolism. Rates of glycolysis fall as pH becomes acidic leading to a decrease in intracellular ATP levels [17]. Acidic pH levels affect actions of agents such as growth factors; the upregulation of proteoglycan synthesis induced by TGFβ in media at pH 7.4, disappears as pH becomes acidic (Razaq, unpublished observations).

Influence of extracellular ions on the intracellular milieu

The sensitivity of cartilage matrix turnover to changes in its extracellular ionic or osmotic environment undoubtedly reflects the pronounced effects which the various components of load have on chondrocyte "housekeeping". This housekeeping involves the provision of nutrients, removal of wastes, and general maintenance of a constant intracellular milieu in order to sustain viable cellular function. Regulation of intracellular composition is effected by an array of proteins in the cell membrane, which operate as transport pathways [19]. They range from simple channels, through which solutes diffuse down electrochemical gradients, to complex ATP-hydrolysing carriers, which bind solutes and perform concentrative transport energised by the ATP.

Regulation of intracellular pH

As for all cells, intracellular pH is determined by the balance between inward leak of H^+ ions moving down their electrochemical gradient and acid extrusion from the cell [20]. This acid extrusion is essential to maintain near-neutral pH, and can be mediated by a variety of transporters which either eject H^+ ions or import HCO_3^- ions to buffer protons. The importance of maintenance of intracellular pH is illustrated by experiments in which for cells suspended in acidic solutions, H^+ ion extrusion becomes overrun by the inward leak. The intracellular acidification which results closely correlates with the inhibition of matrix synthesis recorded in acidic solutions (Fig. 3) [21]. Bovine and human chondrocytes rely heavily on H^+ ion carriers, in particular $Na^+ \times H^+$ exchange (NHE), which uses influx of Na^+ ions down their electrochemical gradient to energise the extrusion of H^+ ions [22]. While an H^+-ATPase also extrudes H^+ ions, and a monocarboxylate carrier (MCT) transports H^+ and lactate-ions out of the cell, it is the $Na^+ \times H^+$ exchanger which is the keystone transporter in chondrocytes.

Responses to osmotic perturbations

Osmotic water movements across the cell membrane affect intracellular ion concentrations directly in relation to fluid shifts. As these in general disturb cellular homeostasis, membrane transporters may be activated in an attempt to minimise or even reverse the alterations to cell volume which ensue [23]. Studies in bovine and human chondrocytes have shown that a Na$^+$ driven uptake of K$^+$ and Cl$^-$ ions is activated to increase intracellular osmolyte content and offset cell shrinkage. Likewise, following cell swelling, K$^+$ and Cl$^-$ ions, along with taurine, are discarded as they diffuse down electrochemical gradients through volume-activated channels [24].

Effect of changes in extracellular composition or mechanical stress on the intracellular milieu

Intracellular composition depends both on the activity of the transporters and also on transmembrane gradients. Changes in extracellular ionic composition can affect both, as can mechanical load. Dynamic pressurisation and hyperosmolarity associated with static fluid expression have both been shown to augment Na$^+\times$H$^+$ exchange activity, with resultant modifications to intracellular pH [25, 26]. The transporters which regulate recovery from osmotic perturbations may also be affected by hydrostatic pressure [27]. It is apparent that the responses are not simple, e.g., in chondrocytes and other cells it has been observed that a decrease in extracellular pH will increase intracellular H$^+$ concentrations, activate Na$^+$/H$^+$ exchange and hence increase intracellular Na$^+$ concentration (Browning and Wilkins, unpublished observations). This however has subsequent knock-on influences such as increasing Na$^+$/K$^+$-ATPase activity thus returning intracellular Na$^+$ towards normal values but increasing intracellular K$^+$ [28]. In chondrocytes it has been shown that changes in intracellular pH also influence levels of intracellular Ca^{2+} [29]. Thus changes in one transporter or ion gradient can influence intracellular concentrations of others and the integrated activity of individual transport processes will determine chondrocyte intracellular composition.

Conclusions

Changes in the extracellular ionic environment arising from changes in GAG concentration *in vivo*, or from chondrocyte isolation and culture under non-physiological conditions and ionic concentrations, or from application of loads which alter activity of ion transporters, all may affect the intracellular milieu. The consequent changes in intracellular ion concentrations in particular, as well as activating second-messenger cascades and gene expression, appear to have a direct and rapid influence on GAG synthesis through mechanisms at present unknown. Thus changes in the fixed charge density-regulated ionic composition provide a feedback mechanism for regulating GAG concentrations in the matrix.

Acknowledgements

The work was supported by the Arthritis Research Campaign (U0511, W064) BH is supported by an MRC-CASE studentship with AstraZeneca.

References

1 Hodge WA, Fuan RS, Carlson KL, Burgess RG, Harris WH, Mann RW (1986) Contact pressures in the human hip joint measured *in vivo*. *Proc Natl Acad Sci USA* 2879–2883
2 Urban JPG, Maroudas A, Bayliss MT, Dillon J (1979) Swelling pressure of proteoglycans at the concentrations found in cartilaginous tissues. *Biorheol* 16: 447–464
3 Eisenberg SR, Grodzinsky AJ (1985) Swelling of articular cartilage and other connective tissues: electromechanochemical forces. *J Orthop Res* 3: 148–159
4 Maroudas A, Muir H, Wingham J (1969) The correlation of fixed negative charge with glycosaminoglycan content of human articular cartilage. *Biochim Biophys Acta* 177: 492–500
5 Kempson GE, Muir H, Swanson SA, Freeman MA (1970) Correlations between stiffness and the chemical constituents of cartilage on the human femoral head. *Biochim Biophys Acta* 215: 70–77
6 Maroudas A, Evans H (1974) A study of ionic equilibria in cartilage. *Conn Tiss Res* 1: 69–79
7 Kempson GE, Spivey CJ, Swanson SA, Freeman MA (1971) Patterns of cartilage stiffness on normal and degenerate human femoral heads. *J Biomech* 4: 597–609
8 Jones IL, Klamfeldt A, Sandstrom T (1982) The effect of continuous mechanical pressure upon the turnover of articular cartilage proteoglycans *in vitro*. *Clin Orthop* 165: 283–289
9 Urban JPG, Hall AC, Gehl KA (1993) Regulation of matrix synthesis rates by the ionic and osmotic environment of articular chondrocytes. *J Cell Physiol* 154: 262–270
10 Ishihara H, Warensjo K, Roberts S, Urban J-PF (1997) Proteoglycan synthesis in the intervertebral disk nucleus: The role of extracellular osmolality. *Am J Physiol* 272: C1499–C1506
11 Borghetti P, Dellsalda L, De Angelis E, Maltarello MC, Petronini PG, Cabassi E, Marcato PS, Maraldi NM, Borghetti AF (1995) Adaptive cellular response to osmotic-stress in pig articular chondrocytes. *Tissue & Cell* 27: 173–183
12 Berl T, Siriwardana G, Ao L, Butterfield LM, Heasley LE (1997) Multiple mitogen-activated protein kinases are regulated by hyperosmolality in mouse IMCD cells. *Am J Physiol* 272: F305–F311
13 Hopewell B, Urban JPG (2002) Adaptation of chondrocytes to long-term changes in extracellular osmolality. *Biorheology; in press*
14 De Angelis E, Petronini PG, Borghetti P, Borghetti AF, Wheeler KP (1999) Induction of betaine-gamma-aminobutyric acid transport activity in porcine chondrocytes exposed to hypertonicity. *J Physiol* 518 (Pt 1): 187–194

15 Grodzinsky AJ (1983) Electromechanical and physicochemical regulation of cartilage strength and metabolism. *CRC Crit Rev Bioeng* 9: 133–199
16 Ysart GE, Mason RM (1994) Responses of articular cartilage explant cultures to different oxygen tensions. *Biochim Biophys Acta* 1221: 15–20
17 Lee RB, Wilkins RJ, Razaq S, Urban JPG (2002) The effect of mechanical stress on cartilage energy metabolism. *Biorheology* 39: 133–143
18 Gray M, Pizzanelli A, Grodzinsky A, Lee R (1988) Mechanical and Physiochemical Determinants of the Chondrocyte Biosynthetic Response. *J Orthop Res* 6: 777–792
19 Wilkins RJ, Browning JA, Ellory JC (2000) Surviving in a matrix: membrane transport in articular chondrocytes. *J Membr Biol* 177: 95–108
20 Thomas RC (1986) Intracellular pH. In: R Hainsworth (ed): *Acid-base balance*. Manchester University Press, Manchester, 50–74
21 Wilkins RJ, Hall AC (1995) Control of matrix synthesis in isolated bovine chondrocytes by extracellular and intracellular pH. *J Cell Physiol* 164: 474–481
22 Wilkins RJ, Hall AC (1992) Measurement of intracellular pH in isolated bovine articular chondrocytes. *Exp Physiol* 77: 521–524
23 Lang F, Busch GL, Ritter M, Volki H, Waldegger S, Gulbins E, Haussinger D (1998) Functional significance of cell volume regulatory mechanisms. *Physiol Rev* 78: 247–306
24 Bush PG, Hall AC (2001) Regulatory volume decrease (RVD) by isolated and *in situ* bovine articular chondrocytes. *J Cell Physiol* 187: 304–314
25 Browning JA, Walker RE, Hall AC, Wilkins RJ (1999) Modulation of $Na^+ \times H^+$ exchange by hydrostatic pressure in isolated bovine articular chondrocytes. *Acta Physiol Scand* 166: 39–45
26 Yamazaki N, Browning JA, Wilkins RJ (2000) Modulation of $Na(+) \times H(+)$ exchange by osmotic shock in isolated bovinearticular chondrocytes. *Acta Physiol Scand.* 169: 221–228
27 Hall AC (1999) Differential effects of hydrostatic pressure on cation transport pathways of isolated articular chondrocytes. *J Cell Physiol* 178: 197–204
28 Mendoza SA (1988) The role of ion transport in the regulation of cell proliferation. *Pediatr Nephrol* 2: 118–123
29 Browning JA, Wilkins RJ (2000) Effects of extracellular alkalinisation on Ca^{2+} homeostasis in articular chondrocytes. *J Physiol* 526: 105P

Influence of tissue shear deformation on chondrocyte biosynthesis and matrix nano-electromechanics

Moonsoo Jin[1], Alan J. Grodzinsky[1], Thomas H. Wuerz[1], Gregory R. Emkey[1], Marcy Wong[2], and Ernst B. Hunziker[2]

[1]Continuum Electromechanics Group, Center for Biomedical Engineering, Massachusetts Institute of Technology, M.I.T. Room 38-377, Cambridge, MA 02139, USA; [2]M.E. Mueller Institute for Biomechanics, University of Bern, Bern, Switzerland

Introduction

Articular cartilage provides lubrication and load bearing functions during the motion of synovial joints. Such a specialized biomechanical function is enabled by the mechanical and electromechanical properties of cartilage extracellular matrix (ECM) and the interaction between cartilage and synovial fluid. Within cartilage matrix, highly charged aggrecan molecules are embedded within a dense collagen fibrillar network. The proteoglycan-associated repulsive forces are restrained by tensile forces within the collagen network. At the molecular level, these repulsive or swelling stresses are mostly due to electrical double layer repulsion associated with the negative fixed charges on glycosaminoglycan (GAG) chains, in addition to the elastic and entropic interactions between GAG macromolecules.

In vivo loading of cartilage causes coupled electromechanical and physicochemical changes that are known to modulate chondrocyte metabolism *via* multiple pathways including upstream signaling, transcriptional and translational regulation, post-translational modification, intracellular vesicular transport, and extracellular processing [1, 2]. Low levels of oscillatory strain (1–5% dynamic strain amplitude at frequencies ranging from 0.01–1.0 Hz) can stimulate chondrocyte biosynthesis of proteoglycans, collagens, and other important matrix proteins [1, 3–6]. Higher levels of strain and strain rate can cause cell death and matrix damage [7–9]. As the magnitude of strain [10] and strain rate in response to physiological loading depend in part on the mechanical properties of cartilage tissue, it is important to understand both the biophysical and biological stimuli that regulate the coordinated synthesis and degradation of ECM by the chondrocytes.

In this chapter we first review the regulation of cartilage metabolism by mechanical loading, especially the effect of tissue shear deformation on mRNA regulation and matrix synthesis. We then describe experimental and theoretical studies of certain nano-electromechanical interactions within cartilage ECM that govern the macroscopic mechanical behavior of cartilage.

Physical regulation of cartilage metabolism

The development and maintenance of cartilage *in vivo* is regulated in part by mechanical loading, including compressive and shear deformations, and concomitant mechanical and physicochemical forces and flows [1]. Soluble factors including vitamins, hormones, growth factors, and cytokines are known to be important regulators of chondrocyte biosynthesis and differentiation [2, 11]. Therefore, investigators have studied the regulation of cartilage metabolism *in vitro*, by utilizing soluble factors combined with mechanical or physicochemical stimuli [11–14]: static and cyclic hydrostatic pressure, fluid-induced shear, dynamic tissue deformation, changes in osmolarity and pH. The cellular mechanisms that underly mechanotransduction responses are not well understood. Signaling mechanisms may include stretch-activated ion channels [15], ligand-cell surface receptor interactions, and integrin-cytoskeleton machinery [16] that can trigger kinase cascades leading to changes in transcriptional activity. Mechanical stresses may also affect chondrocyte biosynthesis at the level of translation and post-translational modification by changing the structure of organelles such as endoplasmic reticulum and Golgi apparatus [1].

The influence of tissue shear loading on chondrocyte biosynthesis

Dynamic compression of cartilage explants causes cell and matrix deformation as well as fluid flow within the extracellular matrix (ECM) in the environment of the cells, mimicking these aspects of loading *in vivo*. Previous studies suggested that the increase in proteoglycan and protein synthesis caused by dynamic compression *in vitro* was associated with intratissue fluid flow, streaming potential, and cell deformation [17, 18]. These studies modeled the spatial profiles of biophysical phenomena within cylindrical explants, including fluid flow and hydrostatic pressure, and compared them to the measured spatial profiles of newly synthesized proteoglycans using quantitative autoradiography. Possible effects of fluid flow on cartilage metabolism include (1) increased availability of nutrients and growth factors due to the convective transport, (2) streaming potentials, and (3) flow-induced shear stress. It has been speculated that each of these possible mechanisms can regulate chondrocyte biosynthesis.

In a recent study [19], we examined the effect of cell and matrix deformation (in the absence of significant fluid flow) on mRNA regulation and matrix biosynthesis using applied tissue shear deformation. Since tissue shear causes little or no volumetric deformation or intratissue pressure gradients, there is minimal intratissue fluid flow. Cartilage disks (3 mm diameter by 1 mm thick) were obtained from the femoropatellar groove of 1–2-week-old calves. Sinusoidal shear deformation of 0.5–6% dynamic strain amplitude at frequencies between 0.01–1.0 Hz was applied

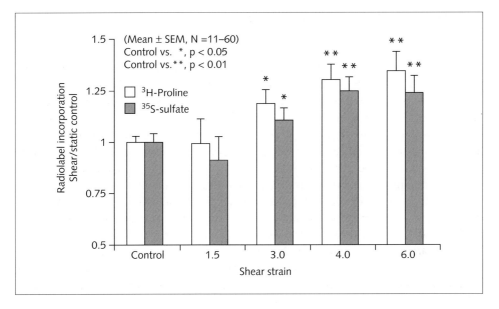

Figure 1
Radiolabel incorporation (pmol/µg DNA/h) in dynamically sheared disks was normalized to that in control disks where disks were maintained within tissue culture media without FBS. Synthesis of total protein and proteoglycans was significantly increased by shear deformation above 1.5% dynamic strain amplitude at a frequency of 0.1 Hz. The increase at 6% shear was similar to that observed previously in disks incubated in the presence of 10% FBS and subjected to 3% shear strain [19].

to groups of disks using an incubator-housed biaxial apparatus [39]; anatomically matched control disks were maintained at the same static offset compression (i.e., the cut thickness of the disks) but not subjected to shear. During the entire loading period, disks were incubated in DMEM ± 10% FBS, with ^{35}S-sulfate and ^{3}H-proline incorporation used as measures of proteoglycan and total protein synthesis, respectively [19].

Shear loading above 1.5% strain amplitude applied for 24 h at a frequency of 0.1 Hz with no FBS caused significant stimulation of total protein and proteoglycan synthesis over static controls (Fig. 1). By 6% strain amplitude, proline and sulfate incorporation had increased by ~40% and ~25%, respectively (Fig. 1), which was close to the stimulatory level caused by 3% shear strain in the presence of 10% FBS in our previous study [19]. In the presence of 10% FBS [19], tissue shear deformation stimulated biosynthesis at strain amplitudes as low as 1%.

In a separate series of experiments (using 10% FBS), the spatial profiles of radiolabel incorporation into newly synthesized matrix molecules were further analyzed by quantitative autoradiography. Cartilage disks were distributed into four groups ($n = 6$ each): a dynamic shear and control group maintained in medium containing 10 µCi/ml ^{35}S-sulfate, and another dynamic shear and control group with 20 µCi/ml ^{3}H-proline. Both shear groups were subjected to 3% dynamic shear strain amplitude at 0.1 Hz for 24 h. After loading in the presence of label, the disks were prepared for the quantitative autoradiography analysis [19]. Sub-sectional images (100 µm × 80 µm) selected from center ($r = 0$ to 0.3 mm), middle ($r = 0.6$ to 0.9 mm), and edge regions ($r = 1.2$ to 1.5 mm) of the cross-section (3 mm wide by 1 mm high by 1 µm thick) of cylindrical cartilage explants (Fig. 2A) were used to analyze radial variations in proline and sulfate grain densities.

Grain densities of dynamically sheared disks were normalized to the static control disks (Fig. 2B). Overall, the proline grain density increased significantly by 50% due to shear loading ($p < 0.001$) and was relatively uniform with increasing radius from the center of the disk ($p > 0.6$, by two-way ANOVA). In contrast, the sulfate grain densities increased with a trend towards significance ($p = 0.076$), and there was a significant trend in the radial variation of sulfate grains ($p < 0.05$). In previous studies of the effects of dynamic compression on cartilage explants [17, 18], the relative increase in sulfate grains was higher near the peripheral edges of the disk where chondrocytes were subjected to a higher level of intratissue fluid flow. Therefore it is interesting to note that the increase in sulfate incorporation in response to tissue shear deformation was found to be greater in the center region. Together, these results may be associated with the influence of tissue shear deformation on local (cell-level) facilitated transport or diffusion of growth factors and other macromolecules throughout the entire region of cylindrical disks, in the absence of significant fluid flow.

The effect of tissue shear loading on changes in mRNA expression of type II collagen and aggrecan core protein were further investigated using RT-PCR (data not shown). For these experiments, cartilage disks were subjected to 3% shear strain amplitude at 0.1 Hz for 0.5, 2, 6, and 24 h. Compared to the level of matched static controls, the expression of type II collagen mRNA in response to tissue shear increased ~50% by 0.5 h. This level of stimulation was maintained over 24 h. However, there was no significant increase in the expression of aggrecan core protein mRNA by shear deformation.

Matrix nano-electromechanics

Experimental and theoretical studies of the mechanical and electromechanical properties of cartilage can be classified into macroscopic-tissue-scale *versus* nano-molecular-scale approaches. Macroscopic approaches focus on the material properties of cartilage tissue and the ability of cartilage to respond to complex joint loading *in vivo*. Measurements of tissue properties have utilized testing configurations to quan-

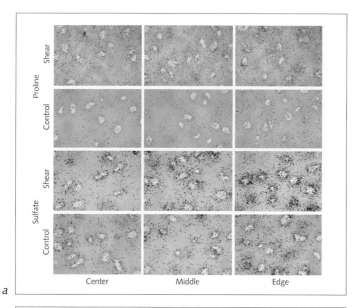
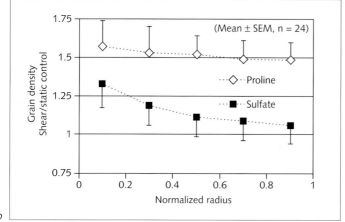

Figure 2
(A) Autoradiographic appearance of chondrocytes with proline and sulfate grains from center, middle, and edge regions of cartilage explants. Proline and sulfate grain density appeared higher in dynamically sheared disks than in static controls (white bar at the bottom right = 10 μm). (B) The radial variation in proline and sulfate grain densities of the shear group was normalized to that of the control group. Overall proline grain density increased by 50% due to shear loading ($p < 0.001$) and was relatively uniform with disk radius ($p < 0.6$, two-way ANOVA). In contrast, the relative increase in the sulfate grain densities due to shear was marginally significant ($p < 0.076$) and there was a significant trend in the radial variation of sulfate grains ($p < 0.05$) with higher stimulation near center region of cylindrical disk.

tify the compressive, shear, and tensile behavior of the ECM [20–27]. The tensile strength of the collagen network within native tissue has been determined using the force balance between an applied stress, the proteoglycan (PG) swelling pressure, and the collagen tensile stiffness [23]. This study showed that the ability of collagen fibrils to limit the hydration of tissue and, thus, maintain a high PG concentration in normal cartilage was significantly compromised in osteoarthritic tissue. Measurements of tissue-level mechanical properties have also been performed before and after enzymatic degradation of proteoglycan consituents [21], and after changes in bath ionic concentration and pH [21, 22, 24]. These studies have revealed the critically important contribution of GAG-associated electrostatic interactions to the compressive and shear stiffness of cartilage under equilibrium as well as time varying conditions.

Theoretical models have been developed to describe the macroscopic behavior of cartilage tissue, addressing the interaction between fluid and solid matrix, and the intrinsic properties of the constituent macromolecules [28]. Poroelasticity and mixture theories have predicted the observed stress relaxation, creep, and electrokinetic behavior of cartilage under volumetric deformation [25, 26]. The intrinsic viscoelastic properties of the solid matrix have been incorporated into a poroviscoelasticity theory, thus addressing combined relaxation from the intrinsic properties of macromolecules and the interaction between solid and fluid phases [27].

The investigation of nano-molecular interactions and biophysical properties of constituents of the ECM has enhanced our understanding of the origin of the macroscopic properties of cartilage. For example, aggrecan is non-covalently bound to hyaluronan, stabilized by link protein. The viscosity and shear modulus of solutions of aggrecan with link protein were found to be higher than those of link-free aggrecan solutions, highlighting the stabilizing effect of link protein in PG aggregates [29]. Nano-scale structural visualization of aggrecan and collagen fibrils has been performed by electron microscopy [30, 31] and, recently, by atomic force microscopy (AFM), which enables the imaging of macromolecules in ambient air or in near physiological fluids [32, 33]. In addition, intermolecular and intramolecular electrostatic repulsion forces between GAG chains end grafted onto a gold-coated silicon wafer were quantified using high resolution force spectroscopy [34]. The resulting force per area occupied by a single GAG-chain was found to be on the order of the known macrocopic Donnan swelling stress of cartilage tissue [20, 23], giving further support to the molecular level origin of electrostatic swelling forces within the tissue. The Poisson-Boltzmann (PB) mean field theory has been used to model such electrostatic repulsion interactions between charged polyelectrolyte molecules in colloidal systems [35]. Of relevance to our study, the contribution of GAG electrostatic interactions to the osmotic swelling pressure of proteoglycan solutions and the equilibrium compressive modulus of cartilage was predicted quantitatively using the PB theory applied to the unit cell model of GAG [36].

Effect of electrostatic interactions between GAGs on the shear modulus of cartilage

The mechanical properties of cartilage derive from distinct electrical (charge-dependent) and non-electrical (charge-independent) contributions. The electrical contributions are mainly associated GAG electrostatic interactions, while the non-electrical contributions are associated with the electrically neutral collagen fibrils as well as with the elastic forces due to the steric and entropic effects within the ECM. As previously described, GAG molecules play an important role in the shear behavior of cartilage by inflating the collagen network, causing a tensile prestress that enables the collagen-aggrecan matrix to resist shear deformation [21, 23]. The shear modulus of cartilage changed significantly after extraction of aggrecan [21]. In addition, it has been inferred from measurement of the confined compression modulus of cartilage tissue [37] that increased ionic strength could decrease the shear modulus of cartilage. However, we know of no previous studies of the possible mechanisms at the nano-molecular scale by which GAG electrostatic interactions may contribute directly to the tissue shear stiffness.

To explore this, we measured the dynamic and equilibrium shear modulus of cartilage disks (9.65 mm diameter by 1 mm thick) in a torsional configuration (at 10% compressive offset) and varied the NaCl concentration at neutral pH to modulate the electrostatic interactions between GAGs [22]. At physiological pH, the properties of the collagen fibrils do not change significantly with ionic strength in the range of 0.01–1.0 M [38]. For measurement of the equilibrium shear modulus, a ramp-and-hold shear strain of 1.5% was applied, resulting in an initial increase and subsequent relaxation of the shear stress. This sequence was repeated four times, and the slope of the relaxed equilibrium stress and strain was used to compute the equilibrium modulus. After returning the specimen to 0% shear strain, a 0.8% amplitude sinusoidal shear strain was applied at 0.5 Hz. This sequence of equilibrium and dynamic shear tests at 0.15 M NaCl was then repeated sequentially after re-equilibration in 0.05 M, 0.01 M, 0.5 M, and 1.0 M NaCl, and corresponding moduli were calculated at each concentration.

The equilibrium shear stress recorded after 10–30 min of relaxation increased linearly with applied shear strain at all ionic concentrations. At each shear strain, the equilibrium shear stress decreased monotonically with increasing NaCl due to the shielding of electrostatic repulsive forces between GAGs [19]. The equilibrium shear modulus, computed by the linear regression from the stress-strain curves, decreased monotonically with increasing NaCl concentration (Fig. 3). The dynamic shear modulus also decreased with increasing NaCl concentration in a similar manner (Fig. 3). These data suggest that electrostatic interactions between GAG chains contribute significantly to the shear properties of cartilage.

We then hypothesized that cartilage's resistance to shear deformation was provided, in part, by changes in the electrostatic forces between neighboring GAG

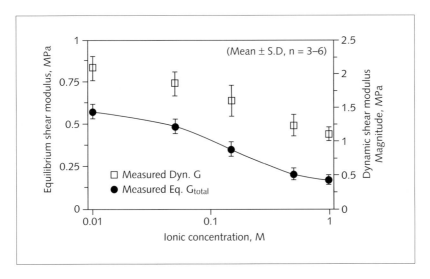

Figure 3
The equilibrium and dynamic shear modulus increased monotonically as the ionic concentration decreased, suggesting the important role of GAG electrostatic interactions in the shear properties of cartilage. The solid line is the predicted electrical component of equilibrium shear modulus (G_e) using two fitting parameters: the GAG concentration (C_{GAG}) and ionic strength-independent shear modulus (G_{ne}). The fitting parameters were determined as C_{GAG} = 5.7% by wet weight and G_{ne} = 173 kPa which are similar to values from independent measurements.

chains caused by the nano-scale rearrangement of GAGs that occurs during macroscopic tissue shear deformation. To test this hypothesis, we modeled GAG segments as charged rods with the outer boundary representing the average distance between neighboring GAGs (unit cell model). This unit cell approach has been used previously to model streaming potentials [40], compression-induced changes in hydraulic permeability [41], and GAG electrostatic interactions in compression [36]. The orientation of the unit cell axis and the vectors describing the unit cell boundary, motivated by the model of Quinn et al. [41], were varied in a manner consistent with the applied macroscopic shear deformation. Theoretical predictions of the electrical contribution to the equilibrium shear modulus (G_e) were obtained using an energy method by differentiating the changes in the electrostatic free energy with respect to the applied macroscopic deformation [22]. The Poisson-Boltzmann equation was incorporated into the unit cell model (PB unit cell) to predict the nano-scale electric potential and mobile ion distribution, which enabled the calculation of the free energy. The overall change in the free energy in response to macroscopic shear deformation was calculated by probabilistically averaging over all GAG orientations, using the approach of Quinn et al. [41]. The shear modulus (G_{total}) was modeled as the sum of an electrical contribution, G_e, and an ionic strength-independent (non-electrical) contribution, G_{ne}, e.g., associated with the collagen network ($G_{total} = G_e + G_{ne}$). The solid line of Figure 3 is the predicted electrical component G_e based on the best fit of two adjustable parameters: the GAG concentration (C_{GAG} = 5.7% by wet weight at 0% compressive offset) and G_{ne} = 173 kPa, where the chi-square function was minimized. For comparison, the measured GAG concentration in newborn calf femoropatellar groove cartilage has been reported previously to be 5.6 ± 0.6% by wet weight [3] which is very close to the best fit value obtained from the present model. The value of G_{ne} is close to the value obtained by extrapolating the measured equilibrium shear modulus to that obtained at 1.0 M NaCl concentration, by which electrostatic interactions are essentially screened (Fig. 3). The good comparison between experimental measurements and theoretical predictions (Fig. 3) strongly suggests that the nano-structural rearrangement of GAG molecules during shear deformation is an important determinant in the shear properties of cartilage.

Conclusions

Articular cartilage is a unique material that can support high loads and deformations during joint loading and simultaneously exhibit extremely low levels of friction. Biomechanical studies of cartilage have focused on native tissue as well as individual molecular components of the ECM. Various macroscopic continuum theories have been developed to describe the material behavior of cartilage during compression, shear, and tensile deformation. At the molecular scale, the properties of aggrecan and collagen fibrils have been studied with attention to electromechanical, physicochemical, and rheological characteristics of these components. Here we described the contribution of nano-scale GAG electrostatic interactions to the shear properties of cartilage, and we suggested that nano-scale rearrangement of GAG molecules is an important mechanism underlying the shear stiffness of tissue during *in vivo* shear loading.

At the same time, mechanical loading forces concomitant with biophysical changes appear to regulate matrix biosynthesis and turnover across multiple pathways including upstream signaling, transcription, translation, post-translational modification, and intracellular vesicular transport. Tissue shear deformation, which induces little or no fluid flow, was found to stimulate protein and proteoglycan synthesis, and the increase in protein synthesis was accompanied by an increase in expression of type II collagen mRNA. The specific mechanisms by which the multiple regulatory pathways interact and lead to the changes in chondrocyte metabolism are under continued study.

Acknowledgments

This research was supported by NIH Grant AR 33236. The authors are indebted to the contributions of Drs. Eliot Frank and Thomas Quinn, and thank Elke Berger, Veronique Gaschen, and Prasanna Perumbuli for technical contributions.

References

1. Grodzinsky AJ, Levenston ME, Jin M, Frank EH (2000) Cartilage tissue remodeling in response to mechanical forces. *Annu Rev Biomed Eng* 2: 691–713
2. Hering TM (1999) Regulation of chondrocyte gene expression. *Front Biosci* 4: D743–761
3. Sah RL, Kim YJ, Doong JY, Grodzinsky AJ, Plaas AH, Sandy JD (1989) Biosynthetic response of cartilage explants to dynamic compression. *J Orthop Res* 7(5): 619–636
4. Kim YJ, Sah RL, Grodzinsky AJ, Plaas AH, Sandy JD (1994) Mechanical regulation of cartilage biosynthetic behavior: physical stimuli. *Arch Biochem Biophys* 311(1): 1–12
5. Parkkinen JJ, Lammi MJ, Helminen HJ, Tammi M (1992) Local stimulation of proteoglycan synthesis in articular cartilage explants by dynamic compression *in vitro*. *J Orthop Res* 10(5): 610–620
6. Wong M, Siegrist M, Cao X (1999) Cyclic compression of articular cartilage explants is associated with progressive consolidation and altered expression pattern of extracellular matrix proteins. *Matrix Biol* 18(4): 391–399
7. Newberry WN, Garcia JJ, Mackenzie CD, Decamp CE, Haut RC (1998) Analysis of acute mechanical insult in an animal model of post-traumatic osteoartrosis. *J Biomech Eng* 120: 704–709
8. Loening AM, James IE, Levenston ME, Badger AM, Frank EH, Kurz B, Nuttall ME, Hung HK, Blake SM, Grodzinsky AJ, Lark MW (2000) Injurious mechanical compression of bovine articular cartilage induces chondrocyte apoptosis. *Arch Biochem Biophys* 381: 205–212
9. Kurz B, Jin M, Patwari P, Cheng DM, Lark MW, Grodzinsky AJ (2001) Biosynthetic response and mechanical properties of articular cartilage after injurious compression. *J Orthop Res* 19(6): 1140–1146
10. Wong M, Wuethrich P, Buschmann MD, Eggli P, Hunziker E (1997) Chondrocyte biosynthesis correlates with local tissue strain in staticallycompressed adult articular cartilage. *J Orthop Res* 15(2): 189–196
11. Bonassar LJ, Grodzinsky AJ, Srinivasan A, Davila SG, Trippel SB (2000) Mechanical and physicochemical regulation of the action of insulin-like growth factor-I on articular cartilage. *Arch Biochem Biophys* 1379(1): 57–63
12. Urban JP, Hall AC, Gehl KA (1993) Regulation of matrix synthesis rates by the ionic and osmotic environment of articular chondrocytes. *J Cell Physiol* 154(2): 262–270
13. Parkkinen JJ, Ikonen J, Lammi MJ, Laakkonen J, Tammi M, Helminen HJ (1993)

Effects of cyclic hydrostatic pressure on proteoglycan synthesis in cultured chondrocytes and articular cartilage explants. *Arch Biochem Biophys* 300(1): 458–465

14   Sah RL, Trippel SB, Grodzinsky AJ (1996) Differential effects of serum, insulin-like growth factor-I, and fibroblast growth factor-2 on the maintenance of cartilage physical properties during long-term culture. *J Orthop Res* 14(1): 44–52

15   Wright M, Jobanputra P, Bavington C, Salter DM, Nuki G (1996) Effects of intermittent pressure-induced strain on the electrophysiology of cultured human chondrocytes: evidence for the presence of stretch-activated membrane ion channels. *Clin Sci (Lond)* 90(1): 61–71

16   Millward-Sadler SJ, Wright MO, Davies LW, Nuki G, Salter DM (2000) Mechanotransduction *via* integrins and interleukin-4 results in altered aggrecan and matrix metalloproteinase 3 gene expression in normal, but not osteoarthritic, human articular chondrocytes. *Arthritis Rheum* 43(9): 2091–2099

17   Kim YJ, Bonassar LJ, Grodzinsky AJ (1995) The role of cartilage streaming potential, fluid flow and pressure in the stimulation of chondrocyte biosynthesis during dynamic compression. *J Biomech* 28(9): 1055–1066

18   Buschmann MD, Kim YJ, Wong M, Frank E, Hunziker EB, Grodzinsky AJ (1999) Stimulation of aggrecan synthesis in cartilage explants by cyclic loading is localized to regions of high interstitial fluid flow. *Arch Biochem Biophys* 366(1): 1–7

19   Jin M, Frank EH, Quinn TM, Hunziker EB, Grodzinsky AJ (2001) Tissue shear deformation stimulates proteoglycan and protein biosynthesis inbovine cartilage explants. *Arch Biochem Biophys.* 395(1): 41–48

20   Eisenberg SR, Grodzinsky AJ (1987) The kinetics of chemically induced nonequilibrium swelling of articular cartilage and corneal stroma. *J Biomech Eng* 109(1): 79–89

21   Zhu W, Mow VC, Koob TJ, Eyre DR (1993) Viscoelastic shear properties of articular cartilage and the effects of glycosidase treatments. *J Orthop Res* 11(6): 771–781

22   Jin M, Grodzinsky AJ (2001) Effect of electrostatic interactions between glycosaminoglycans on the shear stiffness of cartilage: A molecular model and experiments. *Macromolecules* 34(23): 8330–8339

23   Basser PJ, Schneiderman R, Bank RA, Wachtel E, Maroudas A (1998) Mechanical properties of the collagen network in human articular cartilage as measured by osmotic stress technique. *Arch Biochem Biophys* 351(2): 207–219

24   Frank EH, Grodzinsky AJ (1987) Cartilage electromechanics – II. A continuum model of cartilage electrokinetics and correlation with experiments. *J Biomech* 20(6): 629–639

25   Frank EH, Grodzinsky AJ (1987) Cartilage electromechanics – I. Electrokinetic transduction and the effects of electrolyte pH and ionic strength. *J Biomech* 20(6): 615–627

26   Mow VC, Kuei SC, Lai WM, Armstrong CG (1980) Biphasic creep and stress relaxation of articular cartilage in compression: Theory and experiments. *J Biomech Eng* 102(1): 73–84

27   Setton LA, Zhu W, Mow VC (1993) The biphasic poroviscoelastic behavior of articular cartilage: role of the surface zone in governing the compressive behavior. *J Biomech* 26(4–5): 581–592

28. Li LP, Buschmann MD, Shirazi-Adl (2000) A fibril reinforced nonhomogeneous poroelastic model for articular cartilage: inhomogeneous response in unconfined compression. *J Biomech* 33(12): 1533–1541
29. Zhu W, Lai WM, Mow VC (1991) The density and strength of proteoglycan-proteoglycan interaction sites in concentrated solutions. *J Biomech* 24(11): 1007–1018
30. Buckwalter JA, Roughley PJ, Rosenberg LC (1994) Age-related changes in cartilage proteoglycans: quantitative electron microscopic studies. *Microsc Res Tech* 28(5): 398–408
31. Holmes DF, Gilpin CJ, Baldock C, Ziese U, Koster AJ, Kadler KE (2001) Corneal collagen fibril structure in three dimensions: Structural insights into fibril assembly, mechanical properties, and tissue organization. *Proc Natl Acad Sci USA* 98(13): 7307–7312
32. Chen CH, Hansma HG (2001) Basement membrane macromolecules: insights from atomic force microscopy. *J Struct Biol* 131(1): 44–55
33. Sun HB, Smith GN, Hasty KA, Yokota H (2000) Atomic force microscopy-based detection of binding and cleavage site of matrix metalloproteinase on individual type II collagen helices. *Anal Biochem* 283(2): 153–158
34. Seog J, Dean D, Plaas A, Wong-Palms S, Grodzinsky AJ, Ortiz C (2002) Direct measurement of glycosaminoglycan intermolecular interactions via high-resolution force spectroscopy. *Macromolecules* 35(14): 5601–5615
35. Katchalsky A (1971) *Pure Applied Chemistry* 26: 327–373
36. Buschmann MD, Grodzinsky AJ (1995) A molecular model of proteoglycan-associated electrostatic forces in cartilage mechanics. *J Biomech Eng* 117(2): 179–192
37. Bursac P, McGrath CV, Eisenberg SR, Stamenovic D (2000) A microstructural model of elastostatic properties of articular cartilage in confined compression. *J Biomech Eng* 122(4): 347–353
38. Ripamonti A, Roveri N, Braga D, Hulmes DJ, Miller A, Timmins PA (1980) Effects of pH and ionic strength on the structure of collagen fibrils. *Biopolymers* 19(5): 965–975
39. Frank EH, Jin M, Loening AM, Levenston ME, Grodzinsky AJ (2000) A versatile shear and compression apparatus for mechanical stimulation of tissue explants. *J Biomech* 33: 1523–1527
40. Eisenberg S, Grodzinsky AJ (1988) Electrokinetic micromodel of extracellular matrix and other polyelectrolyte networks. *Physicochemical Hydrodynamics* 10(4): 517–539
41. Quinn TM, Dierickx P, Grodzinsky AJ (2001) Glycosaminoglycan network geometry may contribute to anisotropic hydraulic permeability in cartilage under compression. *J Biomech* 34: 1483–1490

The biomechanical faces of articular cartilage in growth, aging, and osteoarthritis

Robert L. Sah

Professor Robert L. Sah, Department of Bioengineering, Mail Code 0412, University of California-San Diego, 9500 Gilman Drive, La Jolla, CA 92093-0412, USA

Introduction

Articular cartilage ideally functions over a lifetime as a low-friction wear-resistant load-bearing material that covers the ends of long bones. During the fetal, postnatal, and childhood stages of life and into adulthood, articular cartilage grows and adapts to varying mechanical demands. However, with advancing age or traumatic injury, cartilage degeneration and mechanical dysfunction occur frequently, especially at particular sites and joints. During the stages of growth up to early adulthood, articular cartilage normally exhibits a variety of biomechanical properties. With aging in the adult as well as the osteoarthritic degeneration that occurs commonly in advanced age, articular cartilage shows a different set of biomechanical properties. Thus, articular cartilage has many biomechanical "faces" during the life stages of growth, aging, and osteoarthritis.

This chapter describes ways of viewing the biomechanical faces of cartilage, some biomechanical faces that cartilage exhibits, the compositional and structural bases for these faces, and ways of analyzing changes in the biomechanical faces of cartilage.

Ways to look at the biomechanical faces of articular cartilage

There are a variety of ways to view the biomechanical faces of cartilage as a load-bearing tissue, even for a particular cartilage specimen at any particular point in time. These faces depend primarily on the viewpoint and objectives of the observer. One such objective is to gain a better understanding of tissue biomechanical function *in vivo*, and parameters such as mechanical stress and strain [20]. In this case, determination of multiple biomechanical properties that fully describe cartilage material in a geometry-independent fashion, as well as the complex loading conditions of the tissue, are desirable [4]. Another objective is to obtain a non-destructive biomechanical index of tissue function, such as for diagnosing disease or determin-

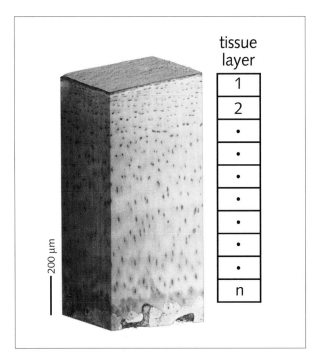

Figure 1
Inhomogeneous and anisotropic structure of adult human articular cartilage, and representation of depth-varying biomechanical properties. Digital volumetric image of a human articular cartilage sample (left), and schematic of cartilage layer properties as a function of depth from the articular surface (right).

ing the efficacy of a putative therapy. In this case, one or more physical properties of cartilage that can be determined non-destructively, such as indentation stiffness, may be desirable [22]. Still another objective is to use a biomechanical measure as a sensitive marker of cartilage dysfunction. For example, the tensile properties of cartilage measured in isolated samples, are sensitive to age-associated changes [26, 27].

The biomechanical faces that are discerned are also dependent on the length scale at which cartilage is examined (Fig. 1). Cartilage can be analyzed at length scales ranging from intact joints to full-thickness regions of cartilage over the joint surface, layers of cartilage tissue from different depths, cellular and extracellular regions of cartilage tissue, molecular constituents within these regions, and beyond. One set of biomechanical faces can be observed if articular cartilage is assumed to be homogeneous [36]. Another set of very different faces can be observed if cartilage is analyzed at the finer length scale of tissue layers at varying depths from the articular surface [45]. The length scale can be even finer, for example, to focus on the indwelling chondrocytes, which are relatively soft compared to the matrix [24], or the glycosaminoglycan component of the matrix [14]. This chapter focuses on tissue-scale biomechanical properties of cartilage, as well as relationships between biomechanical properties at different length scales.

Biomechanical faces of articular cartilage during growth, aging, and osteoarthritis

The way in which the biomechanical faces of articular cartilage differ normally, at different stages of maturation from fetus to adulthood as well as in various joint locations, may have significant implications for the development of osteoarthritis as well as the repair or replacement of degenerate cartilage. Certain growth-related diseases, such as congenital hip dysplasia [59], pose an increased risk for development of osteoarthritis, and it has been suggested that the development of osteoarthritis in a high percentage of individuals is due to subtle predisposing developmental abnormalities. With advancing age, there is an increasing prevalence of osteoarthritis in adults at particular joint locations. Thus, the cumulative change in cartilage properties, from the initial stages of development through stages of growth and then adulthood, may pose an increased site-specific risk for the development of osteoarthritis. In addition, an understanding of the basis for the biomechanical faces in the immature state may be relevant to osteoarthritis, since it has been postulated that certain aspects of cartilage development and growth are trigerred aberrantly in osteoarthritis, leading to tissue breakdown [57]. Finally, delineation of the normal evolution of biomechanical properties serves as a basis for evaluating the degree of success of a biomimetic regeneration strategy since efforts to repair cartilage by tissue engineering techniques often attempt to regenerate articular cartilage by recapitulating normal development and growth processes.

During fetal and postnatal growth, there is a marked evolution of the mechanical properties of articular cartilage. The compressive modulus of articular cartilage increases from the fetus to the adult, in a manner that varies between joints [60, 62]. The increase in compressive modulus (a measure of the stiffness of the solid matrix) is accompanied by a decrease in hydraulic permeability (a measure of the ease with which pressure gradients can drive fluid to flow through the tissue). The maturation-associated changes result in an improvement of the load-carrying characteristics of cartilage, in that the same compressive stress induces a lower compressive strain in adult cartilage than in fetal cartilage. In addition, the tensile modulus and strength of knee cartilage increases to an even greater degree from the fetus to adult [61]. Thus, maturation-associated changes in cartilage serve to enhance the overall integrity of the tissue. The biomechanical properties of fetal cartilage may also be associated with a cartilage structure that facilitates matrix accretion and tissue expansion.

Shortly after birth and into adulthood, articular cartilage has biomechanical properties that vary markedly with depth from the articular surface (Fig. 1). Compared to deeper zones, the superficial zone of normal articular cartilage is relatively soft in compression [16, 33, 45, 49, 50, 58, 60] and strong in tension [25]. These mechanical properties of superficial cartilage may serve a number of critical functions. The compressive characteristics of superficial cartilage promote efficient load distribution over a broad area of the articular surface. The tensile and cohesive characteristics of superficial cartilage provide a high degree of tissue integrity. In addition, the chondrocytes in the superficial zone of cartilage and specialized molecules of the articular surface also contribute to the lubricating function of cartilage, as described elsewhere in this book [47].

In the young adult and with normal aging, articular cartilage also exhibits mechanical properties that vary between joints and joint locations. At sites that have a relatively high predilection for osteoarthritis, the tensile strength of cartilage appears to attain relatively low values in young adults or perhaps diminishes markedly with age. Age-related weakening of cartilage in tension is evident not only in the distal femur, but also in the femoral head [26, 42], but weakening is not evident in the talus [27]. In the young adult, bovine cartilage of the femoral condyle has a lower tensile strength than that of patellofemoral groove [61], and human cartilage has tensile strength that is increasingly higher in the patella, medial femoral condyle, and lateral femoral condyle [53]. The site-associated weakening or weakness in adult cartilage has a pattern similar to the incidence of cartilage degeneration and osteoarthritis, which is increasingly less frequent at the patella, medial femoral condyle, lateral femoral condyle, femoral head, and talus [12, 39]. These maturation- and aging-associated variations in cartilage biomechanical properties may be particularly important if, in osteoarthritis, areas of cartilage damage expand from weak areas and coalesce.

Normal aging and osteoarthritic degeneration appear to each contribute to the biomechanical dysfunction of cartilage. Most of the biomechanical studies of cartilage in human osteoarthritis have compared tissue samples that are osteoarthritic and aged with others that are normal and young, and thus have age as a confounding variable. The compressive modulus of full thickness cartilage of osteochondral cores from the lateral facet of the patella decreases with aging-associated degeneration [6]. The cartilage of osteoarthritic and aged humans is weaker in tension than that of normal and young adults, for tissue from various regions of the femoral condyle and patella [3, 5]. While part of these differences are likely due to normal aging as described above, other differences appear to be associated with osteoarthritic degeneration processes. The tensile strength and stiffness of the superficial cartilage layer, even with normal gross appearance, decreases with increasing proximity to degenerate areas [28, 42]. Since deep layer cartilage from the same samples is not markedly different, the superficial layer of cartilage appears to be affected in early stages of degeneration. However, in more advanced aging [26, 27] as well as aging and osteoarthritis [3], all layers of cartilage appear to be affected. Thus, the superficial, middle, and deep zones of cartilage may be affected characteristically and distinctly at different times during aging- and osteoarthritis-associated degeneration.

The compositional and structural bases for a particular biomechanical face of cartilage

The biomechanical faces of articular cartilage are related to the biological components of the tissue as well as the specialized way in which such components are organized. More than two decades ago, Maroudas proposed that the swelling and load-bearing behavior of cartilage [32] was determined primarily by the balance between the electrostatic swelling tendency of fixed charge components of the tissue, and the restraining properties of the collagen network component of the tissue. Basser and Maroudas formalized this description with a theoretical model for the case of osmotically-applied compression, and used it with experimental data to determine properties of the collagen network [10], and recent work is described elsewhere in this book [11]. A similar approach can be used to assess the contribution of the collagen network in mechanically-applied compression tests [17].

The organization of the molecules of cartilage also contributes to the biomechanical properties of the tissue. The tensile properties of cartilage have traditionally been attributed to its collagen network [25, 37]. In tensile tests of tissues cut parallel to the articular surface, cartilage is much stiffer and stronger in the superficial region than the deeper regions. This appears to be due to variations in tissue structure rather than composition, since the collagen content of cartilage varies little with depth from the articular surface, and exhaustive extraction of non-collagen components does not alter tensile strength [48]. Rather, the organization of the collagen network appears primarily responsible, as it varies markedly from the superficial zone where relative thin fibrils are oriented parallel to the surface and carry tensile load very effectively, to the deep zone where relatively thick fibrils are oriented perpendicular to the surface and are much less efficient at bearing tensile load.

The variation in compressive properties of cartilage during growth appear to be due, in large part, to variations in tissue composition. A number of studies have analyzed the growth-associated change in cartilage composition [9, 40, 43, 52, 54]. The solid content of the cartilaginous epiphysis of the bovine distal femur and distal humerus increases progressively, between the fetal, newborn, young adult, and old adult stages. Most of this increase is due to an increase in collagen, which accounts for an increasing portion of the dry mass of cartilage and which is accompanied by an increase in pyridinoline crosslink content. By comparison, the glycosaminoglycan content of epiphyseal or full-thickness adult articular cartilage changes little, especially in the articular surface layer. The variability in compressive modulus and hydraulic permeability in cartilage across all stages of development is associated with variations in tissue matrix components; however, the development-associated increase in modulus and decrease in permeability are associated primarily with an increase in collagen content [60, 61], as are changes in other compressive properties [62]. These results implicate changes in the collagen component of articular cartilage as having important functional consequences during normal development and growth.

The variation in cartilage compressive properties with depth from the articular surface and between various locations also appears to governed primarily by the glycosaminoglycan and collagen components. While the compressive modulus [5] and hydraulic permeability [33, 37] of cartilage have long been attributed primarily to the glycsoaminoglycan or fixed charge constituent, it also appears that during growth as well as in the adult, collagen also contributes to compressive properties [44, 55]. On the other hand, other components or organizational factors, such as the orientation of collagen fibrils, also contribute to the compressive properties since the contents of glycosaminoglycan and collagen are relatively constant in the middle and deep layers of cartilage [7], but compressive modulus continues to increase with depth in these regions [16, 45, 46].

The biomechanical weakening of articular cartilage in aging and osteoarthritis also appears due, at least in part, to structural alterations. Adult aging leads to accumulation of non-enzymatic glycation endproducts in cartilage, and such products are associated with an increase in tissue stiffness and brittleness (cartilage failure at a lesser strain) [7, 15, 56]. Also, with aging, disruption of the articular surface is evident from application of India ink [35]. The surface disruption is characterized at early stages by wear in the form of linear striations that run generally in the direction of motion of one joint surface over another [34] and that can be viewed in structural detail (Fig. 2). Such patterns are related to histopathological indices of osteoarthritis, as sections oriented perpendicular to the striations revealed multiple vertical fissures, whereas sections parallel to the striations revealed relatively few fissures. The macroscopic grade of cartilage degeneration, determined after staining with India Ink, has been correlated with biomechanical dysfunction in old/osteoarthritic cartilage [6]. In osteoarthritis, the degradation of the collagen network of cartilage [19] has been associated with a functional softening of the collagen network [8].

Analysis of the changing faces of articular cartilage

The biomechanical faces of articular cartilage also change dynamically during the stages of skeletal growth, aging in the adult, and osteoarthritis. While the compositional and structural properties of cartilage are the basis for its various biomechanical faces at any instant of time, cartilage metabolism and remodeling are major

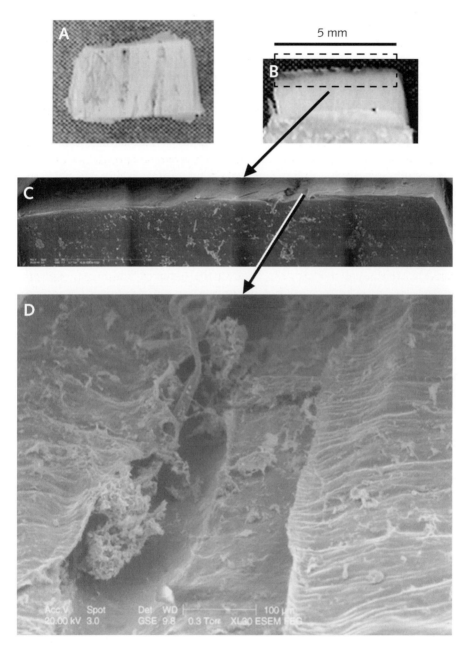

Figure 2
Surface structure of adult human articular cartilage. (A) Top- and (B) side-view photographs of ink-stained osteochondral block. (C) Low- and (D) high-magnification environmental scanning electron micrographs of cartilage surface.

determinants of the time-dependent evolution of cartilage properties over a lifetime. When a physical or chemical regulator acts upon cartilage, the physical properties of the tissue determine how this stimulus translates into a mechanical or chemical change in the cellular and extracellular microenvironments (Fig. 3A). This microenvironment, along with biological factors such as age and genetic make-up, govern both cell metabolism and cell fate and, consequently, can lead to changes in the composition, structure, and hence the mechanical properties of the tissue. Models that describe the relationships between mechanical and biochemical stimuli, cell metabolism, and tissue adaptation may lead to a better understanding of the processes of cartilage growth, degeneration, and repair, and may ultimately become predictive and useful for designing treatment strategies.

Such models can be quantitative or qualitative in nature. Quantitatively, the time-dependent adaptation of cartilage has been modeled using two main approaches. The first approach describes the metabolism of cartilage components. These models include single- and multi-compartmental models [1, 21, 31] as well as spatially-varying continuum models [18, 30, 38], the latter of which can explicitly describe processes such as biosynthesis, diffusion, and reaction. The second approach addresses cartilage adaptation as a mechanical process [51], incorporating growth laws that describe the amount and orientation of deposition of various molecules in the tissue [29, 30]. These models are not necessarily exclusive, and are likely to converge in the future.

Qualitatively, an aging-associated model leading to the development of osteoarthritis can be hypothesized (Fig. 3B). The initial stage (I → II) of wear and fatigue leads to weak or weakening of cartilage, especially in the superficial layer [26, 41, 53]. This is followed by a loss (II → III) and then activation (III → IV) of cells, the latter being associated with more severe matrix fragmentation, severe loss of mechanical integrity, and tissue swelling. The basis for age-associated tensile weakening of cartilage in aging and osteoarthritis remains to be established, but may involve mechanical wear [34] or enzymatic degradation of the collagen network [8, 23]. Whether cells die before the aging-associated decrease in cellularity [53] or in osteoarthritis [13] is controversial [2], although theoretically, a slow change in cartilage cellularity due to cell death could certainly occur with only very few dead cells being apparent in the tissue at any instant of time. When cartilage age reaches the state characterized by weakened tensile properties and diminished cellularity, either by normal aging, following injury, or in early osteoarthritis, it may be highly susceptible to progress to focal erosion and osteoarthritic progression.

Acknowledgments

I thank Won Bae, Kyle Jadin, Kelvin Li, Kori Rivard, Barbara Schumacher, Michele Temple, Ben Wong, Resolution Sciences Corporation, and FEI Company for processing samples and providing the figures.

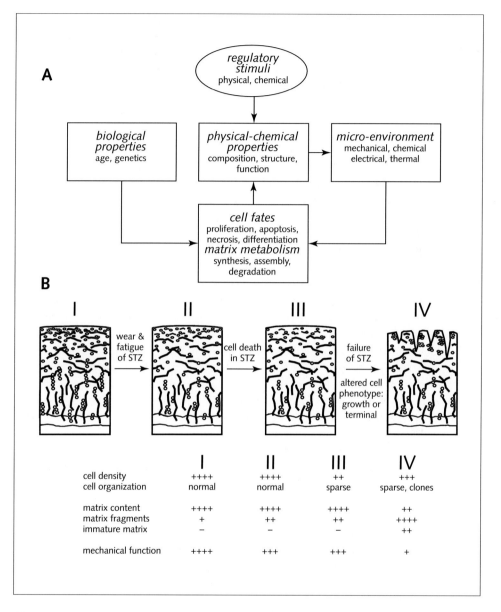

Figure 3
Models of cartilage dynamics. (A) Block diagram of influence of environmental regulatory stimuli and biological properties on cartilage remodeling, which may lead to cartilage growth, degeneration, or repair. (B) Hypothetical sequence of changes of human articular cartilage, focusing on changes in the superficial tangential zone (STZ), during aging (I → II → III) and osteoarthritis (III → IV), leading to mechanical failure.

References

1 Ahsan T, Harwood FH, Amiel D, Sah RL (2000) Kinetics of collagen crosslinking in adult bovine articular cartilage. *Trans Orthop Res Soc* 25: 111
2 Aigner T, Hemmel M, Neureiter D, Gebhard PM, Zeiler G, Kirchner T, McKenna L (2001) Apoptotic cell death is not a widespread phenomenon in normal aging and osteoarthritic human articular knee cartilage. *Arthritis Rheum* 44: 1304–1312
3 Akizuki S, Mow VC, Muller F, Pita JC, Howell DS, Manicourt DH (1986) Tensile properties of human knee joint cartilage: I. influence of ionic conditions, weight bearing, and fibrillation on the tensile modulus. *J Orthop Res* 4: 379–92
4 Almeida ES, Spilker RL (1997) Mixed and penalty finite element models for the non-linear behavior of biphasic soft tissues in finite deformation. Part I-alternative formulations. *Comp Meth Biomechanics Biomed Engng* 1: 25–46
5 Armstrong CG, Mow VC (1982) Biomechanics of normal and osteoarthrotic articular cartilage. In: PD Wilson, LR Straub (eds): *Clinical trends in orthopaedics*. Thieme-Stratton, New York, 189–97
6 Armstrong CG, Mow VC (1982) Variations in the intrinsic mechanical properties of human articular cartilage with age, degeneration, and water content. *J Bone Joint Surg* 64–A: 88–94
7 Bank RA, Bayliss MT, Lafeber FPJG, Maroudas A, Tekoppele JM (1998) Ageing and zonal variation in post-translational modification of collagen in normal human articular cartilage. The age-related increase in non-enzymatic glycation affects biomechanical properties of cartilage. *Biochem J* 330: 345–51
8 Bank RA, Soudry M, Maroudas A, Mizrahi J, TeKoppele JM (2000) The increased swelling and instantaneous deformation of osteoarthritic cartilage is highly correlated with collagen degradation. *Arthritis Rheum* 43: 2202–2210
9 Barone-Varelas J, Schnitzer TJ, Meng Q, Otten L, Thonar E (1991) Age-related differences in the metabolism of proteoglycans in bovine articular cartilage explants maintained in the presence of insulin-like growth factor-1. *Connect Tissue Res* 26: 101–120
10 Basser PJ, Schneiderman R, Bank R, Wachtel E, Maroudas A (1998) Mechanical properties of the collagen network in human articular cartilage as measured by osmotic stress technique. *Arch Biochem Biophys* 351: 207–219
11 Ben Zaken C, Schneiderman R, Kaufmann H, Maroudas A (2002) Age-dependent changes in some physico-chemical properties of articular cartilage. In: VC Hascall, KE Kuettner (eds): *The many faces of osteoarthritis*. Birkhäuser, Basel, 249–254
12 Bennett GA, Waine H, Bauer W (1942) Changes in the knee joint at various ages with particular reference to the nature and development of degenerative joint disease. The Commonwealth Fund, New York, 97
13 Blanco FJ, Guitian R, Vazquez-Martul E, De Toro FJ, Galdo F (1998) Osteoarthritis chondrocytes die by apoptosis: a possible pathway for osteoarthritis pathology. *Arthritis Rheum* 41: 284–849

14 Buschmann MD, Grodzinsky AJ (1995) A molecular model of proteoglycan-associated electrostatic forces in cartilage mechanics. *J Biomech* Eng 117: 179–192
15 Chen AC, Temple MM, Ng DM, Richardson CD, DeGroot J, Verzijl N, TeKoppele JM, Sah RL (2001) Age-related crosslinking alters tensile properties of articular cartilage. *Trans Orthop Res Soc* 26: 128
16 Chen SS, Falcovitz YH, Schneiderman R, Maroudas A, Sah RL (2001) Depth-dependent compressive properties of normal aged human femoral head articular cartilage. *Osteoarthritis Cartilage* 9: 561–569
17 Chen SS, Sah RL (2001) Contribution of collagen network and fixed charge to the confined compression modulus of articular cartilage. *Trans Orthop Res Soc* 26: 426
18 DiMicco MA, Sah RL (2002) Dependence of cartilage matrix composition on biosynthesis, diffusion, and reaction. *Transport in Porous Media; in press*
19 Dodge GR, Poole AR (1989) Immunohistochemical detection and immunochemical analysis of type II collagen degradation in human normal rheumatoid, and osteoarthritis articular cartilage and in explants of bovine articular cartilage cultured with interleukin 1. *J Clin Invest* 83: 647–661
20 Fung YC (1993) *Biomechanics: Mechanical properties of living tissues*, 2nd ed. Springer-Verlag, New York
21 Hascall VC, Luyten FP, Plaas AHK, Sandy JD (1990) Steady-state metabolism of proteoglycans in bovine articular cartilage. In: K Kuettner, A Maroudas (eds): *Methods in cartilage research*. Academic Press, San Diego, 108–112
22 Hayes WC, Keer LM, Herrmann KG, Mockros LF (1972) A mathematical analysis for indentation tests of articular cartilage. *J Biomech*anics 5: 541–551
23 Hollander AP, Heathfield TF, Webber C, Iwata Y, Bourne R, Rorabeck C, Poole AR (1994) Increased damage to type II collagen in osteoarthritic articular cartilage detected by a new immunoassay. *J Clin Invest* 93: 1722–1732
24 Jones WR, Ting-Beall HP, Lee GM, Kelley SS, Hochmuth RM, Guilak F (1999) Alterations in the Young's modulus and volumetric properties of chondrocytes isolated from normal and osteoarthritic human cartilage. *J Biomech* 32: 119–127
25 Kempson GE (1979) Mechanical properties of articular cartilage. In: MAR Freeman (ed): *Adult articular cartilage*. Pitman Medical, Tunbridge Wells, 333–414
26 Kempson GE (1982) Relationship between the tensile properties of articular cartilage from the human knee and age. *Ann Rheum Dis* 41: 508–511
27 Kempson GE (1991) Age-related changes in the tensile properties of human articular cartilage: a comparative study between the femoral head of the hip joint and the talus of the ankle joint. *Biochim Biophys Acta* 1075: 223–230
28 Kempson GE, Muir H, Pollard C, Tuke M (1973) The tensile properties of the cartilage of human femoral condyles related to the content of collagen and glycosaminoglycans. *Biochim Biophys Acta* 297: 456–472
29 Klisch SM, Chen SS, Masuda K, Thonar EJ-MA, Hoger A, Sah RL (2001) Application of a growth and remodeling mixture theory to developing articular cartilage. *Trans Orthop Res Soc* 26: 316

30. Klisch SM, DiMicco MA, Hoger A, Sah RL (2002) Bioengineering the growth of articular cartilage. In: F Guilak, D Butler, D Mooney, S Goldstein (eds): *Functional tissue engineering*. Springer-Verlag, New York; in press
31. Lohmander LS, Kimura J (1986) Biosynthesis of cartilage proteoglycan. In: K Kuettner, R Schleyerbach, VC Hascall (eds): *Articular cartilage biochemistry*. Raven Press, New York
32. Maroudas A (1976) Balance between swelling pressure and collagen tension in normal and degenerate cartilage. *Nature* 260: 808–809
33. Maroudas A (1979) Physico-chemical properties of articular cartilage. In: MAR Freeman (ed): *Adult articular cartilage*. Pitman Medical, Tunbridge Wells, England, 215–290
34. Meachim G (1972) Light microscopy of Indian ink preparations of fibrillated cartilage. *Ann Rheum Dis* 31: 457–464
35. Meachim G, Emery IH (1974) Quantitative aspects of patello-femoral cartilage fibrillation in Liverpool necropsies. *Ann Rheum Dis* 33: 39–47
36. Mow VC, Kuei SC, Lai WM, Armstrong CG (1980) Biphasic creep and stress relaxation of articular cartilage in compression: theory and experiment. *J Biomech Eng* 102: 73–84
37. Mow VC, Ratcliffe A (1997) Structure and function of articular cartilage and meniscus. In: VC Mow, WC Hayes (eds): *Basic orthopaedic biomechanics*. Raven Press, New York, 113–718
38. Obradovic B, Meldon JH, Freed LE, Vunjak-Novakovic G (2000) Glycosaminoglycan deposition in engineered cartilage: Experiments and mathematical model. *AICHE J* 46: 1860–7181
39. Oliveria Sa, Felson DT, Reed JL, Cirillo PA, Walker AM (1995) Incidence of symptomatic hand, hip, and knee osteoarthritis among patients in a health maintenance organization. *Arthritis Rheum* 38: 1134–1141
40. Pal S, Tang L-H, Choi H, Habermann E, Rosenberg L, Roughley P, Poole AR (1981) Structural changes during development in bovine fetal epiphyseal cartilage. *Collagen Rel Res* 1: 151–176
41. Rivard KL, Temple MM, Bae WC, Sah RL (2002) Aging-associated biomechanical weakening of human patellar cartilage: relationship to tissue structure and composition. *Trans Orthop Res Soc* 27: 400
42. Roberts S, Weightman B, Urban J, Chappell D (1986) Mechanical and biochemical properties of human articular cartilage in osteoarthritic femoral heads and in autopsy specimens. *J Bone Jt Surg* 68-B: 278–288
43. Sah RL, Chen AC, Grodzinsky AJ, Trippel SB (1994) Differential effects of IGF-I and bFGF on matrix metabolism in calf and adult bovine cartilage explants. *Arch Biochem Biophys* 308: 137–147
44. Sah RL, Trippel SB, Grodzinsky AJ (1996) Differential effects of serum, IGF-I, and FGF-2 on the maintenance of cartilage physical properties during long-term culture. *J Orthop Res* 14: 44–52

45 Schinagl RM, Gurskis D, Chen AC, Sah RL (1997) Depth-dependent confined compression modulus of full-thickness bovine articular cartilage. *J Orthop Res* 15: 499–506

46 Schinagl RM, Ting MK, Price JH, Sah RL (1996) Video microscopy to quantitate the inhomogeneous equilibrium strain within articular cartilage during confined compression. *Ann Biomed Eng* 24: 500–512

47 Schmid T, Su JL, Lindley KM, Soloveychik V, Madsen L, Block JA, Kuettner KE, Schumacher BL (2002) Superficial zone protein is an abundant glycoprotein in human synovial fluid with lubricating properties. In: VC Hascall, KE Kuettner (eds): *The many faces of osteoarthritis*. Birkhäuser, Basel, 161–164

48 Schmidt MB, Mow VC, Chun LE, Eyre DR (1990) Effects of proteoglycan extraction on the tensile behavior of articular cartilage. *J Orthop Res* 8: 353–363

49 Schneiderman R, Keret D, Maroudas A (1986) Effects of mechanical and osmotic pressure on the rate of glycosaminoglycan synthesis in the human adult femoral head cartilage: an *in vitro* study. *J Orthop Res* 4: 393–408

50 Setton LA, Zhu W, Mow VC (1993) The biphasic poroviscoelastic behavior of articular cartilage: role of the surface zone in governing the compressive behavior. *J Biomech* 26: 581–592

51 Skalak R, Gasgupta G, Moss M, Otten E, Dullemeijer P, Vilmann H (1982) Analytical description of growth. *J Theor Biol* 94: 555–577

52 Strider W, Pal S, Rosenberg L (1975) Comparison of proteoglycans from bovine articular cartilage. *Biochim Biophys Acta* 379: 271–281

53 Temple MM, Bae WC, Rivard KL, Sah RL (2002) Age- and site-associated biomechanical weakening of human articular cartilage of the femoral condyle: relationship to cellularity and wear. *Trans Orthop Res Soc* 27: 84

54 Thonar EJ-M, Sweet MBE (1981) Maturation-related changes in proteoglycans of fetal articular cartilage. *Arch Biochem Biophys* 208: 535–547

55 Treppo S, Koepp H, Quan EC, Cole AA, Kuettner KE, Grodzinsky AJ (2000) Comparison of biomechanical and biochemical properties of cartilage from human knee and ankle pairs. *J Orthop Res* 18: 739–748

56 Verzijl N, DeGroot J, Zaken CB, Braun-Benjamin O, Maroudas A, Bank RA, Mizrahi J, Schalkwijk CG, Thorpe SR, Baynes JW et al (2002) Crosslinking by advanced glycation end products increases the stiffness of the collagen network in human articular cartilage. *Arthritis Rheum* 46: 114–123

57 von der Mark K, Kirsch T, Aigner T, Reichenberger E, Nerlich A, Weseloh G, Stob H (1992) The fate of chondrocytes in osteoarthritic cartilage: regeneration, dedifferentiation, or hypertrophy? In: KE Kuettner, R Schleyerbach, JG Peyron, VC Hascall (eds): *Articular cartilage and osteoarthritis*. Raven Press, New York, 221–234

58 Vunjak-Novakovic G, Martin I, Obradovic B, Treppo S, Grodzinsky AJ, Langer R, Freed LE (1999) Bioreactor cultivation conditions modulate the composition and mechanical properties of tissue-engineered cartilage. *J Orthop Res* 17: 130–139

59 Weinstein SL (1992) Congenital hip dislocation. *Clin Orthop* 281: 69–74

60   Williamson AK, Chen AC, Sah RL (2001) Compressive properties and function-composition relationships of developing bovine articular cartilage. *J Orthop Res* 19: 1113–1121
61   Williamson AK, Drake D, Sah RL (2002) Topographical variation in tensile mechanical properties and collagen network of articular cartilage during growth and development. *Trans Orthop Res Soc* 27: 81
62   Wong M, Ponticiello M, Kovanen V, Jurvelin JS (2000) Volumetric changes of articular cartilage during stress relaxation in unconfined compression. *J Biomech* 33: 1049–1054

Cartilage tissue engineering using a new self-assembling peptide gel scaffold

John D. Kisiday[1], Moonsoo Jin[2], Bodo Kurz[4], Han-Hwa Hung[4], Carlos Semino[4], Shuguang Zhang[4] and Alan J. Grodzinsky[1,2,3,4]

[1]Division of Biological Engineering, Departments of Mechanical[2] and Electrical[3] Engineering, [4]Center for Biomedical Engineering, Massachusetts Institute of Technology, Cambridge, MA 02139-4307, USA

Introduction

Emerging therapies for repair of articular cartilage include delivery of cells or cell-seeded scaffolds to a defect site to initiate de novo tissue regeneration. Biocompatible scaffolds assist in providing a template for cell distribution and extracellular matrix accumulation in a three-dimensional defect geometry. A variety of scaffolds have been investigated for cartilage repair in tissue culture and/or in animals. However, no scaffold-based cartilage construct is yet available for clinical application. In this study, we have explored the use of a novel self-assembling peptide hydrogel as a three-dimensional tissue engineering scaffold for cartilage repair [1].

Self-assembling peptides are characterized by amino acid sequences of alternating hydrophobic and hydrophilic side groups, with sequences of charged amino acid residues including alternating positive and negative charges [2–4]. The proposed model for self-assembly is outlined in Figure 1 [5]. When dissolved in water, the peptides assemble into stable beta sheets one peptide molecule thick containing two distinct surfaces of either hydrophobic or hydrophilic side chains. Hydrophobic bonding orients the beta sheet into a twisted tape configuration, creating nanofibers 10–20 nm thick. Exposure to electrolyte solution then initiates beta sheet assembly into an interwoven nanofiber network, creating the hydrogel structure (Fig. 1).

Previous cell culture studies have been conducted using self-assembling peptide material by seeding cells onto the surface of pre-assembled peptide hydrogel. Diverse mammalian cell types were found to attach and proliferate on the peptide membrane-like surface [2–4]. In particular, neuronal cells were found to attach, differentiate, and undergo extensive neurite outgrowth and synapse formation on the surface of the 16 amino acid peptide (RADA)$_4$ gel [4]. Additionally, two peptides were tested for immunogenicity in rats. Injection of (EAKA)$_4$ and (RADA)$_4$ into leg muscle of Fisher 344 rats resulted in no detectable toxic reaction after 9 days and 5 weeks, respectively [4]. These studies suggested that self-assembling peptide hydrogels are suitable for cell and animal compatibility studies relevant to cartilage tissue repair.

Practical applications to cartilage repair require the distribution of chondrocytes in a three-dimensional geometry for implantation into chondral defects. We hypothesized that a self-assembling peptide hydrogel would provide an appropriate environment for retention of chondrocyte phenotype and synthesis of cartilage extracellular matrix (ECM). In this study a method was developed to encapsulate chondrocytes within the peptide hydrogel during self-assembly. The time-dependent evolution of ECM biosynthesis, accumulation (including identification of collagens as an indicator of phenotypic expression) and mechanical functionality was quantified during subsequent *in vitro* culture. Chondrocyte proliferation was evaluated during early timepoints in culture. We also compared chondrocyte behavior in self-assembling peptide scaffolds to that in agarose hydrogel culture as a reference for chondrogenic potential in a well-established gel culture system [6, 7].

Methods

The amino acid sequence AcN-KLDLKLDLKLDL-CNH$_2$ (KLD12) was synthesized and lyophilized to a powder. A 0.5% peptide casting solution was obtained by dissolving KLD12 in an aqueous solution of 295 mM sucrose. Primary chondrocytes from 1–2 week-old bovine femoropatellar groove cartilage were re-suspended in the casting solution at a concentration of 15×10^6 cells/ml. The suspension was injected into a casting frame and placed into a 1X PBS bath for 20 min to initiate self-assembly into a 1.6-mm-thick slab [1]. A control agarose hydrogel was seeded in a similar manner. Peptide and agarose gels were cultured in DMEM supplemented

---

Figure 1 (top of next page)
Model of peptide self-assembly (adapted from [5]).

Figure 2 (bottom of next page)
Matrix accumulation in chondrocyte-seeded peptide hydrogel. (A) Total glycosaminoglycan (GAG) accumulation in cell-seeded peptide hydrogel cultured in FBS and ITS/FBS medium. Chondrocyte-seeded agarose hydrogel analyzed in parallel as a well-defined reference chondrocyte culture system. (B) Toluidine blue staining of chondrocyte-seeded peptide hydrogel, day 15. (C) Immunohistochemical staining for type II collagen in cell-seeded peptide hydrogel, day 15. D) SDS-PAGE of collagens extracted from chondrocyte-seeded peptide hydrogel, day 35. Standards (kindly donated by Prof. P. Bruckner, Münster): Chick cartilage for collagen II and XI banding pattern; mouse skin for collagen I alpha helix 2, indicative of collagen expression of a de-differentiated, fibroblastic phenotype.

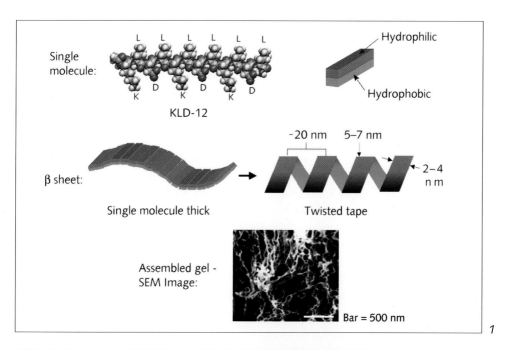

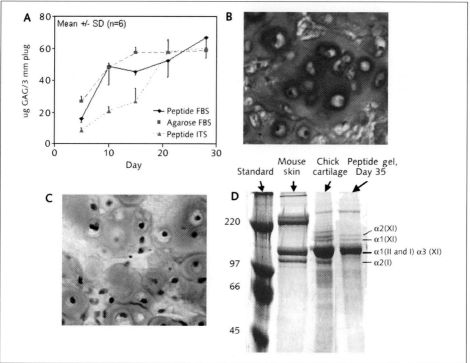

with 10% FBS (FBS). A second peptide gel preparation was seeded similarly and maintained in DMEM plus 1% ITS (insulin, transferrin and selenium) and 0.2% FBS (ITS). The following measurements were performed on both chondrocyte seeded peptide and agarose hydrogels: (1) quantification of the rates of biosynthesis of proteoglycans and proteins *via* ^{35}S-sulfate and ^{3}H-proline radiolabel incorporation, respectively, (2) GAG accumulation (DMMB dye binding), and (3) cell proliferation as determined by viable cell content (MTS viable cell assay, Promega). In order to interpret MTS values, a calibration curve for 3-D cultures was first established using agarose cultures. Agarose gels were seeded at various cell concentrations and analyzed during the first 5 days of culture. Groups of plugs were punched and analyzed for MTS output, or digested and evaluated for DNA content *via* Hoechst dye analysis [8]. Mean MTS output was plotted against mean DNA content to establish the calibration curve. MTS data were then obtained for seeded peptide plugs, and viable cell counts were determined using the calibration curve. BrdU incorporation in chondrocyte seeded peptide hydrogels was used to further explore cell proliferation. Chondrocyte seeded peptide hydrogels were also characterized for GAG accumulation (histological sections stained with toluidine blue), collagen type (immunohistochemical techniques and SDS-PAGE of extracted collagen), and mechanical stiffness (evaluated in uniaxial confined compression).

Results

Quantification of ECM biosynthesis and accumulation was evaluated during 4 weeks of culture. Radiolabel incorporation and total GAG accumulation (Fig. 2A) in chondrocyte seeded peptide and agarose hydrogels cultured in FBS were similar at all timepoints. Values in agarose were comparable to previously reported data showing increasing GAG accumulation and decreasing biosynthetic rates with time in culture [6]. Total GAG accumulation was significantly lower in ITS medium during the first 2 weeks in culture. However, by day 21 there was no significant difference between the ITS and 10% FBS peptide gels. Toluidine blue staining of chondrocyte seeded hydrogels on day 15 (Fig. 2B) showed GAG deposition throughout the gel, with highest intensity in the pericellular regions. Consistent with the accumulation of a continuous GAG matrix, the equilibrium confined compression modulus increased to ~27 kPa on day 28, a ~30-fold increase over day 0 (~0.6 kPa). Immunostaining for collagen II at day 15 showed strong positive staining throughout the gel, forming a continuous matrix as seen with GAG deposition (Fig. 2C). Collagen I staining was light background only. Electrophoresis of extracted collagens showed the presence of type II and IX collagen, and no α2-chain band characteristic of type I collagen expression by de-differentiated chondrocytes (Fig. 2D). Chondrocytes encapsulated in peptide hydrogels were found to proliferate significantly over 9 days of culture. Viable cell density on day 9 increased ~80% relative to day 2 values. BrdU incorporation was consistent with MTS results, showing significant populations of proliferating cells on day 3 and 7. In contrast, viable cell densities in chondrocyte-seeded agarose hydrogels increased by only ~20% on day 7 relative to day 2, consistent with previously reported data [6].

Discussion

The data of Fig 2, along with the observed increase in compressive stiffness, demonstrate the potential of a self-assembling peptide gel for hosting a chondrogenic repair response with seeded primary chondrocytes *in vitro*. The ultimate utility of the scaffold will be determined by long-term development of functional, integrated repair tissue *in vivo*. However, *in vitro* conditioning may also be utilized to stimulate the development of optimal cell/scaffold/ECM constructs prior to implantation. Therefore, in addition to the free swelling *in vitro* cultures typified by the data of Figure 2, we also investigated the effects of dynamic mechanical compression on ECM biosynthesis by primary chondrocytes seeded in peptide gels. Groups of six 12-mm diameter seeded peptide disks were placed into the six wells of a custom chamber, one disk per well, for use with an incubator-housed loading system [9]. Compression was applied to seeded peptide samples with porous platens. A center mounted spring was used to create a 400–800 µm gap between the platens and gel disks when the lid was unloaded. In this manner, samples may alternate between periods of cyclic compression and "free-swell" culture. Seeded peptide gels were subjected to sinusoidal dynamic compression of 2.5% strain amplitude superimposed on 5% static offset strain at a frequency of 1.0 Hz starting on day 22 of culture. Loading was applied for 45 min, followed by 5.25 h of free-swell culture. Loading was applied 4×/day, every other day. Samples were radiolabeled with ^{35}S-sulfate on days 27 (non-loading period), 32 (loading period), and 33 (non-loading), and the results were normalized to the incorporation in samples maintained in static, free-swelling culture. ^{35}S-sulfate incorporation (measured to be ~98% macromolecular in all samples) in loaded samples was significantly higher than that in free-swelling culture at all timepoints. ^{35}S-Sulfate incorporation was higher on day 33 than day 32, indicating proteoglycan synthesis was greater during periods of loading. Total GAG content was measured on day 32 and 33. In both cases, GAG content was 9% higher in loaded samples. This represents a ~20–30% increase in GAG content during the 12 and 13 day loading periods. Therefore, mechanical loading may be used to accelerate proteoglycan accumulation during *in vitro* culture prior to implantation.

Flexibility in peptide design may also be advantageous to stimulating a complete repair response. Peptide sequences may potentially be designed for cell attachment to guide tissue regeneration, and for enzymatic degradation to enable spatially and temporally controlled biodegradability. Peptide sequences may also allow for tethering of growth factors to peptides for direct delivery to encapsulated cells. Such flexibility may be combined with *in vitro* conditioning, such as mechanical loading or ITS-supplemented medium, to generate multiple approaches towards complete tissue regeneration.

Acknowledgements
Funded by NIH Grant AR33236 and the DuPont-MIT Alliance.

References

1. Kisiday J, Jin M, Kurz B, Hung HH, Semino C, Zhang S, Grodzinsky AJ (2002) Self-assembling peptide hydrogel fosters chondrocyte extracellular matrix production and cell division: Implications for cartilage tissue repair. *Proc Natl Acad Sci USA* 99(15): 9996–10001
2. Zhang S, Holmes T, Lockshin C, Rich A (1993) Spontaneous assembly of a self-complementary oligopeptide to form a stable macroscopic membrane. *Proc Natl Acad Sci USA* 90 (8): 3334–3338
3. Zhang S, Holmes TC, DiPersio CM, Hynes RO, Su X, Rich A (1995) Self-complementary oligopeptide matrices support mammalian cell attachment. *Biomaterials* 16(18): 1385–1393
4. Holmes TC, de Lacalle S, Su X, Liu G, Rich A, Zhang S (2000) Extensive neurite outgrowth and active synapse formation on self-assembling peptide scaffolds. *Proc Natl Acad Sci USA* 97(12): 6728–6733
5. Marini DM, Hwang W, Lauffenburger DA, Zhang S, Kamm RD (2002) Left-handed helican ribbon intermediates in the self-assembly of a β-sheet peptide. *Nano Letters* 2: 295–299
6. Benya PD, Shaffer JD (1982) Dedifferentiated chondrocytes reexpress the differentiated collagen phenotype when cultured in agarose gels. *Cell* 30(1): 215–224
7. Buschmann MD, Gluzband YA, Grodzinsky AJ, Kimura JH, Hunziker EB (1992) Chondrocytes in agarose culture synthesize a mechanically functional extracellular matrix. *J Orthop Res* 10(6): 745–758
8. Kim YJ, Sah RL, Doong JY, Grodzinsky AJ (1988) Fluorometric assay of DNA in cartilage explants using Hoechst 33258. *Anal Biochem* 174(1): 168–176.
9. Frank EH, Jin M, Loening AM, Levenston ME, Grodzinsky AJ (2000) A versatile shear and compression apparatus for mechanical stimulation of tissue culture explants. *J Biomech* 33(11): 1523–1527

Differential effects of IL-1β on human knee and ankle chondrocytes

Matthias E. Aurich[1,2], Jürgen A. Mollenhauer[1,2], Klaus E. Kuettner[2] and Ada A. Cole[2]

[1]Department of Orthopedics, University of Jena, Waldkrankenhaus "Rudolf-Elle", Klosterlausnitzer Str. 81, 07607 Eisenberg, Germany; [2]Department of Biochemistry, Rush Medical College at Rush-Presbyterian-St.Luke's Medical Center, 1653 West Congress Parkway, Chicago, IL 60612, USA

Several studies have shown less prevalence of degenerative changes and osteoarthritis in ankle than in knee joints [1–3]. This cannot be explained exclusively with differences in anatomy and biomechanical properties [4], especially since the ankle surface is exposed to higher loads per unit surface area than the knee. Rather, the contribution of structural and biochemical differences can be assumed [5]. Moreover, knee and ankle cartilage explants show quantitatively different metabolic responses when stimulated with fibronectin fragments [6] or interleukin-1β (IL-1β) [7].

Human articular cartilage from the weight bearing areas of the knee (femoral condyles) and ankle (talar dome) of the same limb of a single donor (matched pairs) that showed no signs of cartilage tissue degeneration were received from eight organ donors (age 32 to 61 years) within 24 h of death through the Regional Organ Bank of Illinois with institutional approval. The cartilage was shaved from the subchondral bone and enzymatically digested to release the chondrocytes. To distinguish the secondary effects of the extracellular matrix from cellular differentiation, chondrocytes were seeded in alginate beads (40 000 cells/bead) and incubated for 3 weeks using standard feeding medium containing 10% fetal bovine serum (FBS). The alginate bead culture system has some unique properties with regard to the level of matrix turnover. With the monolayer culture, it shares the property of culturing isolated cells, but with a more stable phenotype. It also possesses an elevated level of steady-state metabolism compared to the explant culture system. It is appropriate to investigate effects of IL-1β on chondroytes at an elevated metabolic level since the anabolic activity in alginate bead cultures may reflect certain aspects of tissue regeneration thus providing a model of repair metabolism.

At the end of 3 weeks of initial culture, the chondrocytes were treated with IL-1β (1 to 1000 pg/ml) for 3 days as a model of a restricted inflammatory episode. The cells were then pulse-labeled for 18 h with ^{35}S-sulfate (20 µCi/ml) followed by a chase for 24, 48 and 72 h under continuing exposure to IL-1β. Chondrocytes with their cell-associated matrix (CAM) were separated from the alginate matrix (AM) and the medium (MED). ^{35}S-labeled glycosaminoglycans (GAGs) in CAM, AM and MED were quantified by liquid scintillation counting following alcian blue precipi-

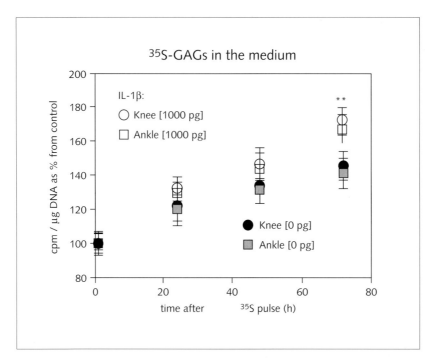

Figure 1
PG degradation expressed as release of ^{35}S-GAGs into the medium measured as counts per minute (cpm) per µg DNA and plotted as % of the 0 h timepoint (control) at 0 and 1000 pg IL-1β treatment (average from eight donors in triplicates).

tation and rapid filtration [8]. The degradation of proteoglycans (PGs) was defined by the half-life ($t_{1/2}$) of ^{35}S-pulse-labeled GAGs and the inhibition of PG synthesis was calculated by the IC50 (the concentration of IL-1β that inhibits the GAG synthesis to 50% of the untreated control).

Under control conditions (no IL-1β chase), there was no statistically significant difference between knee and ankle regarding the distribution of ^{35}S-labeled GAGs in the different compartments (CAM, AM and MED) after 18 h of ^{35}S-sulfate pulse. The ^{35}S-labeled GAGs were initially degraded in the CAM as the primary metabolic compartment and finally accumulate in the MED as the "deep sink" compartment. Consequently, PG-degradation could be measured as the half-life of ^{35}S-labelled GAGs in the CAM or alternatively as accumulated ^{35}S-labelled GAGs in the MED. The average $t_{1/2}$ of ^{35}S-labeled GAGs in the CAM was 68.6 ± 9.2 h (knee) and 70.0 ± 10.4 h (ankle). There was no statistically significant difference in the PG half-life ($t_{1/2}$) in both knee and ankle chondrocytes regardless of the IL-1β treatment, with the exception of a slight but insignificant tendency towards an

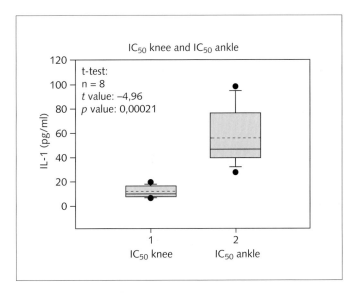

Figure 2
Tissue specific inhibition of proteoglycan biosynthesis, depicted as the 50% inhibition of ^{35}S-sulfate incorporation (IC_{50}) in knee and ankle chondrocytes in the presence of increasing IL-1β concentrations (1 to 1000 pg/ml). The IC_{50} has been calculated by regression analysis using an exponential decay function for every single matched pair. The box diagram shows the mean (dotted line), median (full line), the 95% confidence interval (box height), and the standard deviation of the mean.

decrease in $t_{1/2}$ at 1000 pg IL-1β. However, there were significantly more ^{35}S-labeled GAGs in the MED after 72 h of treatment with 1000 pg/ml in both knee ($p = 0.04$) and ankle ($p = 0.044$) (Fig. 1). When combined, these data suggest the onset of degeneration in the cellular matrix and a flux of degraded components into the culture medium. This is in analogy to an inflamed joint when degradation within the cartilage matrix causes elevated synovial fluid and serum markers of cartilage metabolism.

Both knee and ankle chondrocytes from the same donor responded strongly to IL-1β dose-dependent decrease of ^{35}S-sulfate incorporation into GAGs. However, this decrease was significantly steeper for knee chondrocytes than for ankle chondrocytes (Fig. 2). The IC_{50} for the knee ranged from 6.8 to 19 (average: 11.8 ± 4.6) pg IL-1β/ml, whereas for the ankle it ranged from 29 to 96.7 (average: 56.1 ± 24.9) pg/ml. In other words, ankle chondrocytes were four- to five-times more resistant than knee chondrocytes with respect to the downregulation of PG production by IL-1β.

In conclusion, IL-1β expresses differential effects on human knee and ankle chondrocytes even when these cells were cultured in alginate beads and therefore independent of the normally present matrix. This result strongly suggests significant differences in the state of differentiation between knee and ankle chondrocytes that is independent of the original extracellular matrix.

Acknowledgements

This work was supported in part by NIH grants AR 39239 (M.A., J.M., H.K., A.M., K.K., A.C.), a grant from the Max Kade Foundation (M.A.), and a grant from The Dr. Scholl Foundation (A.C.). We wish to thank ROBI and the donor families for the access to human cartilage.

References

1. Huch K, Kuettner KE, Dieppe P (1997) Osteoarthritis in ankle and knee joints. *Sem Arthritis Rheum* 26: 667–674
2. Muehleman C, Bareither D, Huch K, Cole AA, Kuettner KE (1997) Prevalence of degenerative morphological changes in the joints of the lower extremity. *Osteoarthritis Cartilage* 5: 23–37
3. Koepp HE, Eger W, Muehleman C, Valdellon A, Buckwalter J, Kuettner KE, Cole AA (1999) Prevalence of articular cartilage degeneration in the ankle and knee joints of human organ donors. *J Orthop Sci* 4: 407–412
4. Kempson GE (1991) Age-related changes in the tensile properties of human articular cartilage: a comparative study between the femoral head of the hip joint and the talus of the ankle joint. *Biochim Biophys Acta* 1075 (3): 223–230
5. Treppo S, Koepp H, Quan EC, Cole AA, Kuettner KE, Grodzinsky AJ (2000) Comparison of biomechanical and biochemical properties of cartilage from human knee and ankle pairs. *J Orthop Res* 18: 739–748
6. Kang Y, Koepp H, Cole AA, Kuettner KE, Homandberg GA (1998) Cultured human ankle and knee cartilage differ in susceptibility to damage mediated by fibronectin fragments. *J Orthop Res* 16: 551–556
7. Eger W, Schumacher BL, Mollenhauer J, Kuettner KE, Cole AA (1999) Human knee and ankle cartilage explants: catabolic differences. *Trans Orthop Res Soc* 24: 446
8. Masuda K, Shirota H, Thonar EJ-MA (1994) Quantification of 35S-labeled proteoglycans complexed to alcian blue by rapid filtration in multiwell plates. *Anal Biochem* 217: 167–175

Discussion

Fixed charge density and cartilage biomechanics

Q: *Juergen Mollenhauer*: I'm particularly grateful, Jill, for your comment on ion concentrations in culture conditions because I believe that some of our baseline data reflect poor conditions rather than standard metabolic states of the chondrocytes. When I look at your graphs, the pH-dependence, the lactate concentration and so forth, what does the regular chondrocyte experience? Is it always at the top of the incorporation rate curve that you can observe or somewhere down the line?

A: *Jill Urban*: I think basically the cells adapt within limits to prevailing conditions. When we prepare chondrocyte cultures, in general we are taking cells from a whole range of different osmolarities, pHs and so on, pooling them and putting them into culture medium. We know that the cells from different regions are different but we culture them under the same conditions. I don't think they like being placed in standard culture medium, which is around 300 mOsm for instance. They seem to perform better if you bump the osmolarity to around 400 mOsm which is about average for cartilage. Basically though, there is no one correct environment for all chondrocytes because the original environment varies over such a range. Whatever though, I don't think we should culture them in standard medium. It's not difficult to change the osmolarity for instance; you can just add some sodium chloride.

Q: *Ted Oegema*: If you want the cells to stay high producers, what would you do?

A: *Jill Urban*: The higher the prevailing osmolarity within limits, the higher the rate of sulfate incorporation.

Q: *Ted Oegema*: Down-regulation of the osmolarity for adaptation shifts the incorporation curve. Do the cells stay at high levels?

A: *Jill Urban*: Over the short-term however you change the osmolarity, the cells down-regulate. But they will adapt if you put them under hyperosmotic conditions for something like 12 h or more, which is the time they take to insert these transporters. Then, they should incorporate more sulfate than they did originally.

Q: *Ted Oegema*: Would you have to adjust other ions proportionally?

A: *Jill Urban*: You should really put in potassium proportionally, as well. Potassium in standard tissue culture medium is about 5 mM, but should be about 8–10 mM for chondrocytes.

Q: *Alan Grodzinsky*: Alice Maroudas and you have shown the exquisite sensitivity of cartilage and chondrocytes to their ionic environment. Invariably at a meeting every year somebody will ask what is the local pH near the cell. In one of your slides you could see dramatically the gradient in pH near the cell membrane and also as a function of depth macroscopically, which I thought was really exciting. Does the fluorescent dye that you used actually bind? Or is it just a very quick snapshot?

A: *Jill Urban*: It doesn't bind; it's a quick snapshot. I don't want to give any figures because we don't know them. We are having a lot more difficulty calibrating than we thought because even though the dye really sees hydrogen ions, there are so many more sodium ions around that they do actually affect the reading. And because the sodium ion concentration is changing around the cell and with depth, that makes the results a bit tricky to interpret.

Q: *Alan Grodzinsky*: But even forgetting about the absolute value, is it possible to estimate the gradient change?

A: *Jill Urban*: It's certainly more acidic around the cell, that's for sure.

Influence of tissue shear deformation on chondrocyte biosynthesis and matrix nano-electromechanisms

Q: *Vince Hascall*: I'm fascinated by the difference that you show between the chemical composition of calf vs. adult cartilage and the difference in the modulus, and in the way you are attempting to explain that in terms of the architecture of the tissue. I wonder whether or not we need to think about calf cartilage as more like a growth cartilage as opposed to an articular cartilage, which is a mature tissue. Is there anything in the organization of those two tissues that would explain a difference in the modulus or are we going to have to look at more detail in terms of the chemistry of the components? For example, the proteoglycans in those two tissues actually have quite different properties.

A: *Ernst Hunziker*: I agree that one explanation for this is the architectural differences, and I'm aware also that the proteoglycan population and composition are quite different. But the cell population of calf articular cartilage is very stable near the surface, and in these areas there is not much proliferation activity present. That occurs only deep down in the expanding zone and secondary ossification centers. This suggests that at least to some degree calf and mature articular cartilage are comparable.

Q: *Eugene Thonar*: In the compression experiment the superficial and deep layers did not deform equally. You did not discuss, however, if there are differences in this regard between matrix in the immediate vicinity of the cell and in the interterritorial matrix. Do you have any information about that? Is the cell protected by this ring of collagen fibrils and the structure of the immediate pericellular matrix?

A: *Ernst Hunziker*: There are two things that we should mention here. First of all these compressions are not completely physiological. They don't go so high under physiological conditions. It is for experimental purposes to understand the mechanics that such compressions are used. I cannot tell you if the cells are deformed or not, but if you look at the cell density of adult human articular cartilage, it can be as low as 1.5%. That is less than 2,000 cells per microliter, and that's really very low. So you have 98.5% just matrix, which should be enough protection for the cells. It's basically the superficial zone that the cell density is about 10 times higher. So there I would assume that deformation occurs. I would expect that the cell deformation generally occurs also differentially throughout the zones, if it does at all.

Alan Grodzinsky: I might just add that Eugene was referring to an experiment that showed the biosynthetic response to compression. In another separate pulse-chase experiment, ^{35}S-sulfate incorporation into the matrix was used as a fiducial marker to actually measure ECM strain at the cell level. When static compression was effective in shutting down new synthesis and deposition in the axial direction of compression, that in fact was the direction in which the maximum pericellular matrix compression occurred as well. That's just what Eugene was getting at, that there may be some local effects, not just macroscopic compression, but right down at the cell/matrix level.

The biomechanical faces of articular cartilage in growth, aging, and osteoarthritis

Q: *Vince Hascall*: I'm interested in the relation between tensile strength at failure and the collagen structure. It seems to me that when you're looking in the mid zone when most of the collagen fibers are vertical to the direction of the strain, you're suggesting that you actually get collagen reorientation to allow it to stretch further before it breaks.

A: *Robert Sah*: Possibly, or it could be that molecules connecting the collagen stretch a great deal before failure.

Q: *Vince Hascall*: The question is have you ever taken it almost to that failure point and then fixed it and looked at it morphologically?

A: *Robert Sah*: No, we haven't. That would be interesting to do.

Q: *Jill Urban*: Do the differences you see depend on the age of the cartilage? Are you going to see the same stiffening if you start with calf cartilage, as when you start with fetal cartilage?

A: *Robert Sah*: We've only examined fetal, 3 week old newborn and young adult bovine cartilage. We haven't looked at older bovine animals.

Q: *Vince Hascall*: Is there any way where one could measure strain in the perpendicular direction? I know you don't have much to get hold of, but have you ever thought of ways to do that?

A: *Robert Sah*: We have used video microscopy to look from both the transverse directions during a tensile test. This allows one to measure the three-dimensional deformation of cartilage. The pattern of deformation is very different in the superficial region and the deeper regions. It looks like the superficial region of cartilage necks down much more in both the lateral and the thickness directions than do the deeper regions. This pattern of deformation seems to be consistent with the organization of the meshwork of collagen in the superficial and the deep regions.

Q: *Vince Hascall*: Is there any way to measure in a compression test the strain in the superficial layer in the perpendicular direction along the lines of the collagen arcades where you don't have much depth?

A: *Robert Sah*: We have only looked at deformation and strain in a radially-confined compression configuration. Lori Setton and Elizabeth Myers have done some nice experiments where they look at how cartilage deforms in response to changes in the osmotic environment.

Alice Maroudas: If one subjects cartilage to instantaneous compressive deformation and just looks at the moduli after about a second, then we showed many years ago that you observe a very similar picture to what you are getting in tensile properties. That is, the surface extends very little, whereas the deeper zones do so more. But we find great sensitivity to the direction of prick lines. The actual direction of superficial collagen persisted on what we saw almost down to the deep zone. So I think the idea of pure perpendicular direction for collagen in the middle zone is probably not quite accurate. It's also got a relationship to the direction in the superficial zone.

Cartilage tissue engineering using new self-assembling peptide gel scaffold

Q: *Vince Hascall*: You mentioned that you were using both a 10% serum medium and an ITS medium. Were the data you presented with the serum medium?

A: *John Kisiday*: In the first part of the study before the mechanical loading, the default was with the fetal bovine serum medium. The mechanical loading was actually done with the ITS medium.

Q: *Vince Hascall*: Then how do the two culture conditions compare in terms of proliferation or the properties that you mentioned?

A: *John Kisiday*: The proliferation is actually similar with the two media. I didn't show those data although I just did that experiment recently. The ITS medium doesn't cause any greater biosynthesis rates, but it does cut down on aspects such as dedifferentiation on the surface. With serum, cultures in agarose or peptide or other scaffolds show dedifferentiation and proliferation on the surface. This is not nearly as prevalent when the ITS medium is used.

Q: *Jill Urban*: You seem to have been very successful with the mechanical loading protocol for your cultures in increasing synthesis. Can you explain why you chose that particular protocol? Was it accidental?

A: *John Kisiday*: No, it's not accidental. I spent some time working with agarose cultures, which were a little easier and cheaper to make for routine loading. I thought from the start that there would have to be both a loaded and then an unloaded period, just assuming that 24 h of loading over a period of weeks might not be good for the chondrocytes. That actually did turn out to be the case. So the current protocol is a result of working down from an hour on, an hour off every day to a point where the breaks are sufficient to allow the chondrocytes to recover.

Q: *Vince Hascall*: I actually thought those 2 days of breaks were a weekend. What is known about the biocompatibility of the synthetic peptides? Are they antigenic or not?

A: *John Kisiday*: In one study where the peptide solution was injected into rats, there was no detectable toxic reaction.

Q: *Vince Hascall*: But were there antibodies developed, for example?

A: *John Kisiday*: We don't know yet. I think this needs to be rigorously investigated.

Differential effects of IL-1β on human knee and ankle chondrocytes

Q: *Chris Handley*: Have you got any idea what happened to the chemical level of proteoglycans?

A: *Matthias Aurich*: I've only done radiolabeling experiments, so there are questions of what size of proteoglycans are synthesized, or what the actual isotope gradient from the medium is, or what size of proteoglycans remain in the matrix, and so on. Since these issues cannot be answered just with pure radiolabeling, I don't have the answer for you.

Q: *Chris Handley*: No, I'm just asking about the bulk chemical levels, not sizes. Do you get a change in the chemical level of the proteoglycans? Some of them are lost into the medium. There is a possibility that you have regional distributions of radiolabeled proteoglycans that differ compared with chemical levels.

A: *Matthias Aurich*: Yes, but I have no information on that.

Q: *Vince Hascall*: I want to try and clarify that again. I think what Chris is trying to get at is that if you measured the actual mass of glycosaminoglycans, say in the cell-associated matrix, does it stay constant with time? Are you looking at a dilution of the radiolabeled material as it passes through and leaves that compartment, or in the presence of the IL-1, are you also depleting the actual mass of the proteoglycan in the cell associated matrix as well?

A: *Matthias Aurich*: I went over the data for this experiment a bit too quickly perhaps. Under control conditions there is no loss of radiolabeled proteoglycans over time, but with IL-1 stimulation there is a clear loss of radiolabeled proteoglycans. I don't know if this is true for the total amount of proteoglycans.

Biomechanics, motor control and osteoarthritis

Introduction

Mark Grabiner

School of Kinesiology, University of Illinois at Chicago, 901 West Roosevelt Road, Chicago, IL 60608, USA

In the way that the Central and Union Pacific Railroads converged at Promontory Summit, Utah on May 10, 1869, the scientific rails of cell biology, molecular biology, biomechanics and technology converge on osteoarthritis at the level of systems biology/integrative physiology. This level of study is marked by the mantra that the properties of a system cannot be entirely understood as a function of the properties of its isolated components. This session was organized to highlight linkages between scientific rails. The following derailed excerpt from the "Ask the Pharmacist" column appeared in the 3 March 2002 *Chicago Tribune* (page 5 Section 13): "Question: I've been in pain for 12 years with osteoarthritis... I'm willing to try anything... I'm especially interested in natural products... what do you suggest? Answer: A good brand of glucosamine sulfate can, in fact, put the shock absorbing quality back in your joints. However, this miracle cure may take up to four months." (A disclaimer prominently displayed at the end of the column states explicitly that the information in the column is not intended to treat the condition.)

That is quite an answer, one that results from studies of the clinical benefits of glucosamine may not support. Increased loading of articular cartilage is associated with the pathogenesis of osteoarthritis, and damping impulsive loads such as those arising during locomotion is indeed a crucial musculoskeletal function. However, anyone who has experienced the teeth-rattling load of inaccurately timing a step down from even a short step can attest that the shock absorbing quality of one's joints is rather small.

The primary means by which impulsive loads are attenuated by the musculoskeletal system is skeletal muscle contraction. Successful shock absorption by skeletal muscle, and therefore, the reduction of the impulsive loads transmitted by articular cartilage, requires precise regulation of muscle contraction force and timing by the central nervous system (CNS). CNS regulation of skeletal muscle contraction requires the integration of signals arising from voluntary, supraspinal and spinal processes with signals arising from a complex network of peripherally located mechanoreceptors, especially those that generate signals reflecting the static and dynamic states of muscle, tendon and ligament.

Figure 1 presents a simple conceptual model that relates various factors and between-factor interactions to the progression of osteoarthritis. An implied feature of the model is its complex nonlinear behavior arising from feedback mechanisms

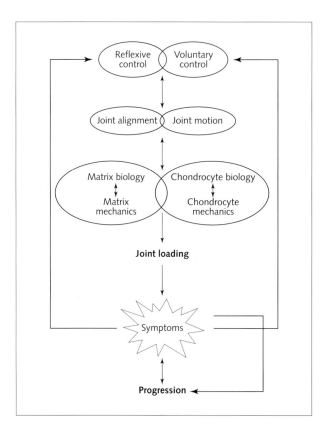

between interacting levels and components. It is this type of behavior that is necessary to characterize at the organism level. For example, the skeletal muscle contraction forces directly influence joint load and are mediated by resultant joint position and motion states. In the case of the symptomatic patient, joint pain generates independent feedback that converges upon and affects mechanisms of reflexive and voluntary motor control.

A key feature of the model is that it directly links organization levels ranging from molecule to organism. The themes in this session are similarly linked and include:

- relating the progression of osteoarthritis to loading patterns, derived using advanced imaging and motion capture and analysis techniques
- central nervous system plasticity by which some individuals adjust to the knee joint instability arising from anterior cruciate ligament rupture and maintain high levels of physical activity
- combining biological and biomechanical markers to predict joint space narrowing.

Dynamic function and imaging in the analysis of osteoarthritis at the knee

Thomas P. Andriacchi

Stanford University, Department of Mechanical Engineering, Department of Orthopaedics/Sports Medicine, Division of Biomechanical Engineering, Stanford, CA 94305-3030, USA

Introduction

Degenerative changes to the articular cartilage at the knee are caused by the complex interactions between the biology, structural properties and the mechanical environment of the articular cartilage. Understanding the causes of degenerative changes to the articular cartilage is confounded by the capacity of biological systems to adapt to early pathological changes. These adaptive changes occur over a broad range of scales from micro levels (e.g. alterations in cell metabolism) to macro levels (e.g changes in the pattern of locomotion). The nature of these adaptive changes provides a potential explanation for the large differences in the rate of progression of osteoarthritis (OA) at the knee joint.

Clearly the biology of articular cartilage is fundamental to the maintenance of its normal function. The interaction of cartilage cells with their environment is an important consideration. A change in the *in vivo* mechanical environment of the joint can be a stimulus for a biological response of the cells in the articular cartilage. While living cells occupy approximately 10% of cartilage by volume, their metabolic activity is fundamental to the synthesis, assembly and maintenance of the extra cellular matrix [1]. For example, loss of the anterior cruciate ligament produces a number of biological and structural changes in the articular cartilage [2, 3]. Often these changes precede the clinical symptoms of osteoarthritis.

A change in metabolic activity of the cells can cause changes in the material and structural characteristics of the articular cartilage. There are a number of important *in vitro* studies describing the material and structural characteristics of articular cartilage [4]. However, the *in vivo* mechanical, electrical and biological requirements must be considered relative to the *in vivo* mechanical loading at the joint, since there are likely specific modes that will stimulate a biological response.

Isolating the specific modes of loading and motion on the articular surface that stimulate the biological response represents a critical challenge. While loading has often been considered an important factor, alteration in joint kinematics is clearly an equally important factor that can influence the progression of OA at the knee.

The neuromuscular system is an important component [4] in the adaptive response to injury or disease. The rate of progression of degenerative processes in a canine model of OA can be modulated by altering the neuromuscular system's ability to adapt [5]. Adaptive changes in loading have also been demonstrated by changes in bone mineral density in a canine model [7] following transection of the ACL. Clearly, altered patterns of loading and/or kinematics at the knee follow ACL transection. These adaptive changes in the biology, structure and the mechanical environment of the joint precede symptomatic changes [2, 3, 6]. Understanding these adaptive responses in structure and function is fundamental to understanding the factors causing the progression of OA at the knee.

The purpose of this discussion is to examine the interaction between structural information obtained from imaging and dynamic functional measurements obtained from *in vivo* studies of human movement as a means to study the progression of osteoarthritis at the knee.

In vivo mechanical environment of the knee joint

Forces at the knee joint have been a primary consideration in the factors influencing the progression of osteoarthritis at the knee joint [8]. Surgical procedures such as high tibial osteotomy (HTO), designed to treat varus gonarthosis, are based on the assumption that altering loads on the damaged medial compartment will slow the rate of progression of the disease as well as produce relief of symptoms [9]. Treatment planning for HTO is often based on correcting the varus malalignment of the joint. However, outcome of high tibial osteotomy has been quite variable, and studies of *in vivo* function [10, 11] have demonstrated that the dynamic loads during walking are strongly correlated with surgical outcome following high tibial osteotomy.

Dynamic loading at the knee in patients with high tibial osteotomy suggests that the potential for dynamic adaptations to modify the loading at the knee joint can have a profound effect on the rate of progression of the disease as well as clinical symptoms [11]. Patients who adapted a gait pattern that lowered the adduction moment at the knee had a substantially improved clinical outcome relative to patients who did not develop dynamic adaptive changes in the patterns of locomotion. These clinical studies [10, 11] suggest that an increased adduction moment causes larger compressive forces on the medial compartment relative to the lateral component. The results of these clinical studies [10, 11] support the notion that the adduction motion during walking is a valid predictor of the distribution of the compressive force across the lateral and medial compartments of the knee joint. In addition, recent studies have demonstrated that the adduction moment during walking is predictive of bone mineral distribution between the medial and lateral sides of the tibia [13]. Subjects with higher bone mineral density on the medial side of the tibia

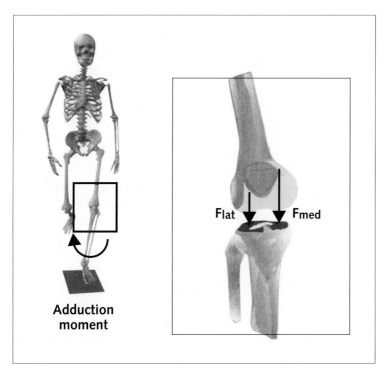

Figure 1
The dynamic adduction moment at the knee has been related to the distribution of medial and lateral forces across the knee joint. During normal gait the load on the medial compartment of the knee is larger than on the lateral compartment. This larger medial force has been related to the progression of osteoarthritis of the medial compartment.

walked with a higher peak adduction moment. Thus, the adduction moment is a useful and convenient predictor of the load on the medial compartment of the knee joint during gait.

Pain associated with medial compartment osteoarthritis of the knee has also been related to the magnitude of the adduction moment at the knee during walking [14]. Patients with mild to moderate osteoarthritis of the knee were reported to increase the magnitude of the adduction moment when pain was reduced by the administration of an analgesic in a prospective study [14]. Again, these data suggest that pain may be a stimulus for an adaptive response to change loading at the knee joint and potentially alter the rate of progression of osteoarthritis at the knee. Similarly, the radiographic severity of the disease was related to the adduction moment during gait [15].

As described above, both clinical and basic studies lead to the conclusion that the dynamic adduction moment during walking can be related to the compressive load of the medial compartment of the knee, and this component of loading is a dominant factor influencing the progression of medial compartment osteoarthritis of the knee joint. This conclusion however, is confounded by the natural history of the anterior cruciate ligament deficient knee [16]. Typically, patients with ACL injury show an increased rate of degenerative changes to the articular cartilage of the knee joint [16]. Similarly, animal models [6] use transection of the anterior cruciate ligament to produce degenerative changes to the articular cartilage at the knee. It is not likely that the loss of an anterior cruciate ligament produces increased compressive force across the medial portion of the knee joint. Therefore, either an increased laxity or altered kinematics of the joint is likely a consideration in the factors affecting the progression of osteoarthritis at the knee joint. Loss of the ACL can affect changes in contact location at the articulating surface, since the ACL can shift the neutral position of the joint. Imaging information provides a useful tool in the examination of the influence of altered contact motion on articular cartilage. In particular, variation of cartilage thickness across the joint can help to demonstrate the role of altered kinematics on progression of osteoarthritis of the knee joint.

Imaging and *in vivo* motion

Information obtained from imaging articular cartilage has been useful in isolating conditions that influence progression of osteoarthritis at the knee. Three dimensional reconstruction of articular cartilage [17, 18] from MRI images has been used to examine and monitor cartilage loss in patients with degenerative joint disease [19, 20]. Volumetric quantification [20] of the entire cartilage surface has been used by several investigators. These methods involve segmenting cartilage from the surrounding tissue using a signal intensity threshold applied to an MRI sequence that is sensitive to articular cartilage. While these methods demonstrate good intraobserver reproducibility [18], they are limited in the context of the analysis of the progression of OA, due to their lack of specificity of particular regions where mechanical loading may vary over the surface of the articular cartilage. In addition, the assessment of defects is often masked by the relationship between the volume of the defect and the overall volume of the unaffected cartilage. Images from MR have been used to demonstrate that the rate of progression is dependent on the location of the initial lesion [21]. Lesions located in the central region of the medial compartment were significantly more likely to progress to a more advanced cartilage pathology than lesions in the anterior and posterior regions of the medial femorotibial compartment. Similarly, lesions on the medial compartment were more likely to progress than lesions located on the lateral compartment of the knee joint. In fact, lesions located in the anterior region of the lateral compartment did not show any degradation over the follow-up period. The results of these studies are consistent with the previously described studies demonstrating the relationship between the adduction moment, medial compartment load, and the rate of progression of osteoarthritis. However, these studies also provide more specificity relative to the location of the defect and can provide important new information in assessing factors like the role of ACL loss and meniscal loss in progression of osteoarthritis.

Imaging and dynamic functional loading

Combining images derived from MR dynamic functional measurements can help to identify the role of loading and motion on the progression of degenerative changes. Thickness variations in femoral cartilage are perhaps related to differences in load across the articular surface of the joint [22]. A recent study [23] examined the thickness of the femoral cartilage for eight different regions distributed along the medial and lateral articulating surface of the femur (Fig. 2). These regions were selected to correspond to the locations of contact at peak loading during walking (0 to 20 degrees), stair-climbing (60 degrees), and deep-flexion activities (90 degrees), such as squatting. The thicknesses were normalized to the anterior-posterior (A-P) depth of the cartilage and expressed as a percentage to facilitate comparisons across subjects. Overall the cartilage thickness was greatest at 20 degrees of flexion on both the medial and lateral condyles, while the thickness did not differ significantly between the medial and lateral surfaces. The posterior surfaces of both condyles were significantly thinner than the cartilage at 20 degrees. The greatest number of cycles of loading (walking) take place in the range between 0 and 20 degrees of flexion, while the highest loads on the knee occur during activities of daily living such as stair climbing with the knee flexed between 45 and 90 degrees. These results indicate that cartilage thickness could be related to the number of cycles that occur in a certain contact region, as well as the magnitude of load seen in that region. Thus, a knowledge of regions of cartilage in contact during certain activities is potentially an important consideration when evaluating factors that influence progression of osteoarthritis at the knee joint.

A study [24] of patients following ACL injury provides useful insight into the need to consider regional variations in the contact location at the knee as a factor influencing long-term progression of osteoarthritis. Typically, clinical evaluations of patients with ACL injury have demonstrated an increased AP laxity relative to the contralateral knee. However, increased passive clinical laxity measurements have not correlated with clinical outcome [25]. Yet, alterations in translational and rotational motion of the knee during dynamic activities could have a profound effect on the articular cartilage as well as secondary restraints such as the medial meniscus. Recent developments [26] have made it possible to measure, *in vivo*, six degrees of

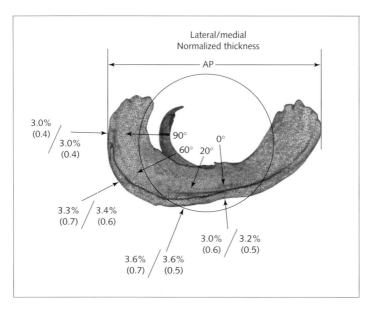

Figure 2
An illustration of the three dimensional geometry of the articular cartilage from the surface of the femur. This three dimensional geometry was created from segmented magnetic resonance images. The average thickness values were obtained from the three-dimensional models created for each subject.

freedom motion at the knee joint. This method [26] has been used to measure anterior-posterior (AP) and internal-external (IE) components of motion of the tibia relative to the femur during walking. Note, in this study, anterior-posterior motion was measured from the midpoint of the tibial surface to a point fixed at the midpoint of the transepicondylar axis on the femur.

The temporal pattern of AP and IE motion of the tibia relative to the femur was common to all knees. Interestingly, the overall AP motion of the ACL deficient knee (3.8 ± 0.9 cm) was significantly less than the control or contralateral knee (3.2 ± 0.7 cm). The overall dynamic range of IE motion was not different between the ACL deficient and normal knees. However, the temporal IE patterns, while similar, were offset between the ACL deficient group and normals. The general characteristics (Fig. 3), common to all subjects, have the tibia (at heel strike) externally rotated and displaced anteriorly. From here, the tibia internally rotates and moves posteriorly through stance phase. During swing phase, the tibia rotates externally as the knee extends towards heel strike. The average offset of the ACL deficient knee was significantly more internally rotated than the contralateral knee.

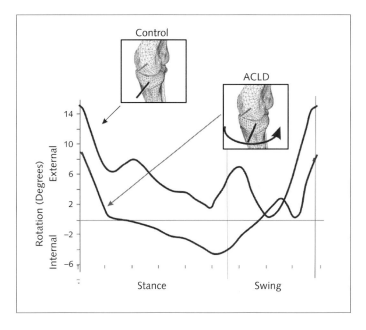

Figure 3
The tibia on the ACL deficient knee was internally rotated relative to the normal knee throughout stance phase.

A surprising result of this study was that the AP displacement of the normal knee is greater than the ACL deficient knee during walking. The difference in AP motion appears to be coupled with a more internally rotated knee than the contralateral knee throughout the walking cycle. The internally rotated offset of the tibia for ACL deficient knee occurs at heel strike and appears to be caused by a loss of the normal screw-home movement of the knee (external tibial rotation with extension) during swing phase.

A knee that functions with the tibia more internally rotated will likely produce higher strains in the secondary structures such as the menisci and could be a possible cause for the degenerative changes that are often reported following ACL loss. The rotational offset identified in this study could be an important consideration in the evaluation of the ACL deficient knee. Perhaps surgical reconstruction and clinical evaluation should consider the rotational function of the ACL as well as AP laxity. Note that model predictions [27] and clinical studies [28] have suggested similar offsets in the contact positions at the knee joint as a possible cause for degenerative changes following loss of the anterior cruciate ligament.

Conclusions

The progression and perhaps initiation of degenerative changes at the knee joint can be influenced by functional adaptations. These adaptations can produce changes in load, motion or contact position of the joint during activities such as walking. Information derived from imaging studies are important since regional variations in cartilage thickness and structure can be detected using these methods. Functional changes that alter contact location or produce morphological changes in the cartilage often precede degenerative changes. Most adaptations involve re-programming of the locomotor process. Therefore, these adaptations can only be detected during activities performed during unencumbered motion [29]. Unfortunately, present technology does not permit simultaneous imaging and unencumbered *in vivo* motion. However, with the appropriate registration of images with laboratory gait analysis systems, it is possible that eventually images obtained from either CT or MRI can be combined with motion measurements from the laboratory. The combination of these methods can provide new insights into the mechanical and structural factors that influence the progression of osteoarthritis [30].

References

1. Smith RL, Lin J, Trindade MC, Shida J, Kajiyama G, Vu T, Hoffman AR, van der Meulen MC, Goodman SB, Schurman DJ, Carter DR (2000) Time-dependent effects of intermittent hydrostatic pressure on articular chondrocyte type II collagen and aggrecan mRNA expression. *J Orthop Res* 14(1): 53–60
2. Lohmander LS, Yoshihara Y, Roos H, Kobayashi T, Yamada H, Shinm (1996) Procollagen II C-propeptide in joint fluid: changes in concentration with age, time after knee injury, and osteoarthritis. *J Rheumatol* 23(10): 1765–1769
3. Lohmander LS, Ionescu M, Jugessur H, Poole AR (1999) Changes in joint cartilage aggrecan after knee injury and in osteoarthritis. *Arthritis Rheum* 42(3): 534–544
4. Mow VC, Ratcliffe A, Poole AR (1992) Cartilage and diarthrodial joints as paradigms for hierarchical materials and structures. *Biomaterials* 13(2): 67–97
5. O'Connor BL, Brandt KD (1993) Neurogenic factors in the etiopathogenesis of osteoarthritis. *Rheumatic Disease Clinics of North America* 19(3): 581–605
6. O'Connor BL, Visco DM, Brandt KD, Albrecht M, O'Connor AB (1993) Sensory nerves only temporarily protect the unstable canine knee joint from osteoarthritis. *Arthritis Rheum* 36(8): 1154–1163
7. Boyd SK, Matyas JR, Wohl GR, Kantzas A, Zernicke RF (2000) Early regional adaption of periarticular bone mineral density after anterior cruciate ligament injury. *J Appl Physiol* 89(6): 2359–2364
8. Andriacchi TP 8 1994) Dynamics of knee malalignment. *Orthopedic Clin N Am* 25(3): 395–403

9   Nizard RS (1998) Role of tibial osteotomy in the treatment of medial femorotibial osteoarthritis. *Rev Rheum Engl Ed* 65(7–9): 443–446
10 Prodromos CC, Andriacchi TP, Galante JO (1985) A relationship between knee joint loads and clinical changes following high tibial osteotomy. *J Bone Joint Surg* 67A(8) 1188–1194
11 Wang J-W, Kuo KN, Andriacchi TP, Galante JO (1990) The influence of walking mechanics and time on the results of proximal tibial osteotomy. *J Bone Joint Surg* 72A(6): 905–913
12 Schipplein OD, Andriacchi T (1991) Interaction between active and passive knee stabilizers during level walking. *J Orthop Res* 9: 113–119
13 Hurwitz DE, Sumner, DR, Andriacchi, TP, Sugar DA (1998) Dynamic knee loads during gait predict proximal tibial bone distribution. *J Biomechanics* 31: 423–430
14 Schnitzer TJ, Popovich JM, Andersson GB, Andriacchi TP (1993) Effect of piroxicam on gait in patients with osteoarthritis of the knee, *Arthritis Rheum* 36(9): 1207–1213
15 Sharma L, Hurwitz DE, Thonar EJMA, Sum JA, Lenz ME, Dunlop DD, Schnitzer TJ, Kirwan-Mellis G, Andriacchi TP (1998) Knee adduction moment, serum hyaluronic acid level, and disease severity in medial tibiofemoral osteoarthritis. *Arthritis Rheum* 41 (7): 1233–1240
16 Daniel DM, Stone ML, Dobson BE, Fithian DC, Rossman DJ, Kaufman KR (1994) Fate of the ACL-injured patient: a prospective outcome study. *Am J Sports Med* 22(5): 632–644
17 Peterfy CG, Dijke CF, van Janzen DL, Gluer CC, Namba R, Majumdar S, Lang P, Genant HK (1994) Quantification of articular cartilage in the knee with pulsed saturation transfer subtraction and fat-suppressed MR imaging: optimization and validation. *Radiology* 192: 485–491
18 Stammberger T, Eckstein F, Michaelis M, Englmeier KH, Reiser M (1999) Interobserver reproducibility of quantitative cartilage measurements: comparison of B-spline snakes and manual segmentation. *Magn Reson Imaging* 17: 1033–1042
19 Peterfy CG, van Dijke CF, Janzen DL et al (1994) Quantification of articular cartilage in the knee with pulsed saturation transfer subtraction and fat-suppressed MR imaging: optimization and validation. *Radiology* 192: 485–491
20 Piplani MA, Disler DG, McCauley TR, Holmes TJ, Cousins JP (1996) Articular cartilage volume in the knee: semiautomated determination from three-dimensional reformations of MR images. *Radiology* 198: 855–859
21 Biswal S, Hastie T, Andriacchi, T, Bergman G, Dillingham M, Lang. P. The rate of progressive cartilage loss at the knee is dependent on the location of the lesion: a longitudinal MRI study in 43 patients. *Arthritis Rheum*; accepted for publication
22 Ateshian GA, Soslowsky LJ, Mow VC (1991) Quantitation of articular surface topography and cartilage thickness in knee joints using stereophotogrammetry. *J Biomechanics* 24(8): 761–76
23 Alexander EJ, Rinderknecht D, Andriacchi TP. Regional variations in femoral cartilage thickness relative to functional load. *Personal communication*

24. Andriacchi TP, Dyrby C, Dillingham M (2002) ACL injury causes rotational abnormalities at the knee during walking. *Trans 48th Ann Meeting, Orthop Res Soc, Poster Number 0922*
25. Snyder-Mackler L, Kelley Fitzgerald G, Bartolozzi III A, Ciccotti M (1997) The relationship between passive joint laxity and functional outcome after anterior cruciate ligament injury. *Am J Sports Med* 25(2): 191–195
26. Andriacchi TP, Alexander EJ, Toney MK, Dyrby CO, Sum J (1998) A point cluster method for *in vivo* motion analysis: applied to a study of knee kinematics. *J Biomech Eng* 120(12): 743–749
27. Wu JZ, Herzog W, Epstein M (2000) Joint contact mechanics in the early stages of osteoarthritis. *Med Eng Phys* 22(1): 1–12
28. Kvist J, Gillquist J (2001) Anterior positioning of tibia during motion after anterior cruciate ligament injury. *Med Sci Sports & Exerc* 33(7): 1063–1072
29. Andriacchi TP (1990) Dynamics of pathological motion: Applied to the anterior cruciate deficient knee. *J Biomech* 23 (Suppl 1): 99–105
30. Andriacchi TP, Lang PK, Alexander EJ, Hurwitz DE (2000) Methods for evaluating the progression of osteoarthritis. *J Rehab Res Dev* 37(2): 163–170

Linking the biology of osteoarthritis to locomotion mechanics

Eugene J-M.A. Thonar[1], Debra E. Hurwitz[1], Thomas P. Andriacchi[2], Mary Ellen Lenz[1] and Leena Sharma[3]

[1]Rush Medical College at Rush-Presbyterian-St. Luke's Medical Center, 1653 West Congress Parkway, Chicago, IL 60612, USA; [2]Stanford University, Mechanical Engineering Department, Durand Building, Stanford, CA 94305-3030, USA; [3]Leena Sharma, Northwestern University, Ward Building 3-315, 303 East Chicago Avenue, Chicago, IL 60611, USA

Introduction

Osteoarthritis (OA) may be described as a syndrome in which biochemical and biomechanical abnormalities combine to cause pain and functional deterioration associated with structural degradation of the joint. The end of the 20th century saw the emergence of assays capable of measuring molecules whose concentrations in body fluids (joint fluid, blood and urine) provide information about specific metabolic processes occurring *in vivo* in the joint tissues of patients with OA and other joint diseases [1]. The ability to quantify with precision these molecules, referred to below as metabolic markers, is helping scientists and clinicians address questions they had not been able to ask before. While the field of research on body fluid markers of joint metabolism is still in its infancy, there already is clear evidence that the measurement of some of these markers is going to shed more light on the pathogenesis of both the pre-clinical and clinical stages of joint disease, including OA; that will prove helpful in developing effective therapeutic treatments. For example, it is now relatively simple to rapidly determine if a drug, with documented effects upon the metabolism of aggrecan in cartilage or chondrocytes cultured *in vitro*, has the same effect *in vivo* [2]. The ability to measure, in a small volume of blood, metabolic markers that will help monitor or prognosticate OA-related changes remains a long-term goal of investigators in the field. There is good agreement that, over the shorter term, this may be achieved by measuring a panel of a few markers. Indeed, recent observations that the quantification of the level of hyaluronan (HA) [3], cartilage oligomeric matrix protein (COMP) [4] and C-reactive protein [5] in serum may have prognostic value in OA suggest that sensitive assays capable of measuring markers of specific metabolic processes may soon be used in the clinical assessment of patients with this condition.

In light of the obvious relevance of physiological loading to the onset and progression of OA, it seems logical to develop noninvasive means of assessing these dynamic loads in both normal and OA subjects. Gait testing provides a reliable indirect method of assessing the loading that occurs dynamically (i.e., during normal ambulation) in weight-bearing joints. These external loading parameters measured during gait are related in a predictable manner to internal joint loads [6]. These measurable external loads include moments that act at the knee during gait. One such moment is the external adduction moment, a surrogate marker of loading of the medial compartment of the knee relative to the lateral compartment.

The purpose of this presentation is to present and discuss recent preliminary evidence that measurement of the serum level of HA and the adduction moment are independent predictors of the outcome variance in medial knee OA [7], presumably representing one or more aspects of the biochemical and biomechanical components of OA, respectively.

Metabolic markers of joint metabolism

While measurement of metabolic markers in synovial fluid provides very useful information about the metabolic state of tissue in the joint from which the fluid was drawn, some investigators have chosen to quantify markers in blood or urine. The advantages, as well as disadvantages, of analyzing blood or urine rather than joint fluid have been described elsewhere [1]. It must be noted that most serum markers used to assess joint tissue metabolism originate not only from joints but also from many tissues outside the joints. For this reason, measurement of the serum level of cartilage-derived molecules usually are best used to identify systemic metabolic abnormalities involving all or most cartilaginous structures in the body [2]. In an attempt to begin to distinguish between different types of metabolic markers, those that potentially provide information about the metabolism of cartilage were recently subdivided into two classes: direct and indirect markers. According to this classification, direct markers of cartilage metabolism have the following properties. First, they are molecules that originate principally from cartilage structures in the body. Second, they provide a measure of the responses of cells or of changes that occur in the cartilaginous tissues from which they originate. A good example of a direct marker is antigenic keratan sulfate (AgKS), a molecule found almost exclusively in aggrecan molecules within cartilaginous tissues (including nucleus pulposus and annulus fibrosus of intervertebral discs) [2]. In those tissues, AgKS is present within large proteoglycan aggregates firmly immobilized in the extracellular matrix. When the aggrecan molecules are cleaved by proteolytic enzymes, the AgKS-bearing fragments rapidly diffuse out of the tissues and appear in body fluids where they can be measured [8].

The indirect markers of cartilage metabolism, on the other hand, are found in many tissues and are produced by many cell types. Many of the indirect markers are molecules with the potential to influence the metabolism not only of chondrocytes but also of synovial cells and/or other cells in joints. They include but are not restricted to: (i) proteolytic enzymes (e.g., metalloproteinases [MMPs], "aggrecanase", etc.) capable of degrading one or more molecules in the matrix, and their inhibitors, such as tissue inhibitors of MMPs (TIMPs) [9, 10]; (ii) growth factors (e.g., insulin-like growth factor-1) that can stimulate biosynthetic processes [11], (iii) proinflammatory cytokines (e.g., interleukin-1, interleukin-6, tumor necrosis factor-α, etc.) capable of mediating directly or indirectly the metabolism of the chondrocytes [1, 12] and (iv) other molecules, such as HA [4] and C-reactive protein [5], that originate in part or principally from non-joint sources and whose concentrations in body fluids provide clues as to the state of health of joints [13].

Even when they are quantified in joint fluid, these indirect markers do not always provide a useful measure of what may be happening in articular cartilage. For example, an elevated level of active MMP-3 in joint fluid does not necessarily provide a good measure of degradative changes occurring in articular cartilage within that joint since the action of this enzyme can be inhibited by several TIMPs [10]. It also must be remembered that as several of the indirect markers can mediate, directly or indirectly, the same metabolic process in many connective tissues, extreme caution should be exercised when attempting to relate their levels in blood to the state of health of cartilaginous and other structures in an OA joint.

Serum HA – a metabolic marker with prognostic value in OA

The serum level of HA, a fascinating indirect marker with key functions in the metabolism of articular cartilage and apparent therapeutic properties in synovial joints, was recently shown to have prognostic value in OA. The evidence comes from a 5-year longitudinal study of the relationship between changes in knee radiography and the baseline and 5-year levels of HA in 46 subjects with knee OA [3]. Progression was defined as a decrease ≥ 2 mm in either tibiofemoral compartment or the subject's having undergone total knee replacement [14]. Results are shown in Figure 1 (arbitrary natural log units). At baseline, subjects who progressed had a significantly higher serum HA level ($p = 0.022$, t-test) than those who did not progress. This difference persisted at 5 years ($p = 0.001$). Moreover, the level (mean ± SD) of HA decreased significantly (−0.69 ± 0.88 units, 95% CI −1.01, −0.37) in the group who did not progress, insignificantly (−0.06 ± 0.96, 95% CI −0.59, 0.47) in the group that did progress. This study further showed that HA levels were significantly related to disease duration ($p = 0.036$) and minimum joint space width ($p = 0.049$). The results were interpreted to suggest that the baseline level and changes in levels of HA were predictors of radiographic progression of OA.

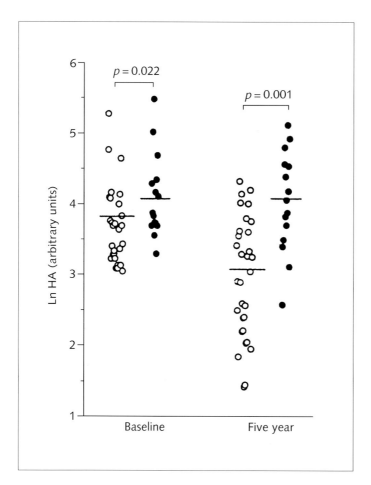

Figure 1
Serum hyaluronan (HA) levels in osteoarthritis patients at baseline and at 5 years, according to the presence (●) or absence (○) of disease progression at 5 years. Horizontal bars indicate the group means.

The usefulness of measuring the adduction moment in knee OA

Chondrocytes are responsive to biomechanical forces through altered metabolism of matrix components. Excessive load is clearly deleterious both to chondrocytes and to the cartilage matrix [1, 15, 16], although the magnitude and nature of "normal" vs. "excessive" loading *in vivo* remain ill-defined [17, 18]. In the intact knee, medial tibiofemoral cartilage contact pressure is increased in the presence of varus deformity [19], presumably because of greater loading [20]; this may exacerbate the degenerative processes. In light of the importance of physiological loading to the onset and progression of OA, it would be useful to have noninvasive means of assessing these dynamic loads in both normal and OA subjects. Measurements reflective of static loading, such as the determination of the mechanical axis by long radiography, are convenient to perform, but are insufficient to predict the true dynamic forces and rotational torques (termed "moments") present in the knee and elsewhere [21]. Direct methods of measuring dynamic joint loading, such as the use of implanted transducers, are impractical for large-scale use in humans, particularly in studies of long duration [22].

Gait testing provides a reliable indirect method of assessing the loading that occurs dynamically (i.e., during normal ambulation) at the weight-bearing joints. These external loading parameters measured during gait are related in a predictable manner to internal joint loads [6, 23]. These measurable external loads include moments that act at the knee during gait. One such moment is the external adduction moment, a surrogate marker of loading of the medial compartment of the knee relative to the lateral compartment. The adduction moment has been extensively analyzed both through mathematical modeling [6, 23] and through clinical gait testing [7, 24–26]. These studies suggest that the magnitude of the adduction moment is directly related to the radiographic and clinical severity of medial knee OA [7, 27] and to poorer surgical outcomes [27, 28]. Due to its pathophysiological significance in relating increased loading to clinically significant OA, the external adduction moment has become an important parameter in clinical OA [29, 30].

Relationship between the adduction moment and the serum level of HA in knee OA

In a study to determine the predictors of disease severity in medial tibiofemoral knee OA, the serum HA levels and the sum of the minimum medial joint space widths for each knee [31] were correlated in 23 subjects with symptomatic knee OA [7]. The serum HA level was 41.9 ± 21.9 ng/ml (mean $\pm$ SD) (range 18–100). The correlation with the summed minimum joint space width was significant ($p = 0.007$, $r = -0.55$). The correlation of the serum HA with the summed KL grade approached statistical significance ($p = 0.06$, $r = 0.41$). This result was interpreted as showing that even in a small, cross-sectional study, the random serum HA level, a marker of a systemic inflammatory metabolic response, was related to structural changes in knee OA. Although there was a significant relationship between the magnitude of the adduction moment and the radiographic outcome in OA [7], this study failed to demonstrate a correlation between the magnitude of the external adduction moment and the serum level of HA. This was interpreted as suggesting that the HA level and adduction moment are independent predictors of the outcome variance in medial knee OA, presumably representing the biochemical and biomechanical components of OA, respectively. A longitudinal study is currently underway to further test this contention.

Acknowledgements

This work was supported by NIH/NIAMS grant 2-P50 AR-39329, NIH/NIA grant AG-04736, and a grant from the Rush Arthritis and Orthopedics Institute.

References

1 Thonar EJ-MA, Manicourt D-H (2001) Noninvasive markers in osteoarthritis. In: R Moskowitz (ed): *Osteoarthritis*, 3rd ed. Harcourt, Philadelphia, 293–313
2 Thonar EJ-MA, Manicourt D-H, Williams JM, Fukuda K, Campion GV, Sweet MBE, Lenz ME, Schnitzer TJ, Kuettner KE. (1992) Serum keratan sulfate: a measure of cartilage proteoglycan metabolism. In: KE Kuettner, R Schleyerbach, JG Peyron, VC Hascall (eds): *Articular cartilage and osteoarthritis*. Raven Press, New York, 429–445
3 Sharif M, George E, Shepstone L, Knudson W, Thonar EJ-MA, Cushnaghan J, Dieppe P (1995) Serum hyaluronic acid level as a predictor of disease progression in osteoarthritis of the knee. *Arthritis Rheum* 38: 760–767
4 Sharif M, Saxne T, Shepstone L, Kirwan JR, Elson CJ, Heinegard D, Dieppe PA (1995) Relationship between serum cartilage oligomeric matrix protein levels and disease progression in osteoarthritis of the knee joint. *Br J Rheumatol* 34(4): 306–310
5 Spector TD, Hart DJ, Nandra D, Doyle DV, Mackillop N, Gallimore JR, Pepys MB (1997) Low-level increases in serum C-reactive protein are present in early osteoarthritis of the knee and predict progressive disease. *Arthritis Rheum* 40: 723–727
6 Schipplein OD, Andriacchi TP (1991) Interaction between active and passive knee stabilizers during level walking. *J Orthop Res* 9(1): 113–119
7 Sharma L, Hurwitz DE, Thonar EJ, Sum JA, Lenz ME, Dunlop DD Schnitzer TJ, Kirwan-Mellis G, Andriacchi TP (1998) Knee adduction moment, serum hyaluronan level, and disease severity in medial tibiofemoral osteoarthritis. *Arthritis Rheum* 41(7): 1233–1240
8 Thonar EJ-MA, Kuettner KE, Williams JM (1994) Markers of articular cartilage injury and healing. In: JT Andrish (ed): *Sports medicine and arthroscopy review: Chondral injuries*, vol. 2. Raven Press, New York, 13–28
9 Lohmander LS, Hoerrner LA, Lark MW (1993) Metalloproteinases, tissue inhibitor, and proteoglycan fragments in knee synovial fluid in human osteoarthritis. *Arthritis Rheum* 36: 181–189
10 Manicourt D-H, Fujimoto N, Obata K, Thonar EJ-MA (1995) Levels of circulating collagenase, stromelysin-1, and tissue inhibitor of matrix metalloproteinases 1 in patients

with rheumatoid arthritis. Relationship to serum levels of antigenic keratan sulfate and systemic parameters of inflammation. *Arthritis Rheum* 38: 1031–1039

11 Morales TI (1994) Transforming growth factor-beta and insulin-like growth factor-1 restore proteoglycan metabolism of bovine articular cartilage after depletion by retinoic acid. *Arch Biochem Biophys* 315: 190–198

12 Hauselmann HJ, Flechtenmacher J, Michal L, Thonar EJ-MA, Shinmei M, Kuettner KE, Aydelotte MB (1996) The superficial layer of human articular cartilage is more susceptible to interleukin-1-induced damage than the deeper layers. *Arthritis Rheum* 39: 478–488

13 Thonar EJ-MA, Lenz ME, Masuda K, Manicourt D-H (1999) Body fluid markers of cartilage metabolism. In: MJ Seibel, SP Robins, JP Bilezikian (eds): *Dynamics of bone and cartilage metabolism*. Academic Press, San Diego, 453–464

14 Dieppe P, Cushnaghan J, Young P, Kirwan J (1993) Prediction of the progression of joint space narrowing in osteoarthritis of the knee by bone scintigraphy. *Ann Rheum Dis* 52(8): 557–563

15 Urban JP (1994) The chondrocyte: a cell under pressure. *Br J Rheumatol* 33(10): 901–908 (Review)

16 Arokoski JP, Jurvelin JS, Vaatainen U, Helminen HJ (2000) Normal and pathological adaptations of articular cartilage to joint loading. *Scand J Med Sci Sports* 10(4): 186–198 (Review)

17 Radin EL, Burr DB, Caterson B, Fyhrie D, Brown TD, Boyd RD (1991) Mechanical determinants of osteoarthrosis. *Sem in Arthritis Rheum* 21(3 Suppl 2): 12–21 (Review)

18 Chen CT, Burton-Wurster N, Lust G, Bank RA, Tekoppele JM (1999) Compositional and metabolic changes in damaged cartilage are peak-stress, stress-rate, and loading-duration dependent. *J Orthop Res* 17(6): 870–879

19 Riegger-Krugh C, Gerhart TN, Powers WR, Hayes WC (1998) Tibiofemoral contact pressures in degenerative joint disease. *Clin Orthop* 348: 233–245

20 Fukubayashi T, Kurosawa H (1980) The contact area and pressure distribution pattern of the knee. A study of normal and osteoarthrotic knee joints. *Acta Orthop Scand* 51(6): 871–879

21 Andriacchi TP (1994) Dynamics of knee malalignment. *Orthop Clin North Am* 25(3): 395–403 (Review)

22 Hodge WA, Carlson KL, Fijan RS, Burgess RG, Riley PO, Harris WH, Mann RW (1989) Contact pressures from an instrumented hip endoprosthesis. *J Bone Joint Surg Am* 71(9): 1378–1386

23 Harrington IJ (1983) Static and dynamic loading patterns in knee joints with deformities. *J Bone Joint Surg Am* 65(2): 247–259

24 Andrews M, Noyes FR, Hewett TE, Andriacchi TP (1996) Lower limb alignment and foot angle are related to stance phase knee adduction in normal subjects: a critical analysis of the reliability of gait analysis data. *J Orthop Res* 14(2): 289–295

25 Kowalk DL, Duncan JA, Vaughan CL (1996) Abduction-adduction moments at the knee during stair ascent and descent. *J Biomech* 29(3): 383–388

26 Weidenhielm L, Svensson OK, Brostrom LA, Mattsson E (1994) Adduction moment of the knee compared to radiological and clinical parameters in moderate medical osteoarthrosis of the knee. *Ann Chir Gynaecol* 83(3): 236–242
27 Prodromos CC, Andriacchi TP, Galante JO (1985) A relationship between gait and clinical changes following high tibial osteotomy. *J Bone Joint Surg Am* 67(8): 1188–1194
28 Wang JW, Kuo KN, Andriacchi TP, Galante JO (1990) The influence of walking mechanics and time on the results of proximal tibial osteotomy. *J Bone Joint Surg Am* 72(6): 905–909
29 Hurwitz DE, Ryals AR, Block JA, Sharma L, Schnitzer TJ, Andriacchi TP (2000) Knee pain and joint loading in subjects with knee osteoarthritis. *J Orthop Res* 18(4): 572–579
30 Hurwitz DE, Ryals AR, Case JP, Block JA, Andriacchi TP (2002) The knee adduction moment during gait in subjects with knee osteoarthritis is more closely correlated with static alignment than radiographic disease severity, toe out angle and pain. *J Orthop Res* 20: 101–107
31 Lequesne M (1995) Quantitative measurements of joint space during progression of osteoarthrtitis: "chondrometry". In: KE Kuettner, VM Goldberg (eds): *Osteoarthritic disorders*. American Academy of Orthopedic Surgeons, Rosemont, IL, 427–444

Movement patterns of individuals with good potential to dynamically stabilize their knees after acute ACL rupture

Terese L. Chmielewski, Katherine S. Rudolph, Michael J. Axe, G. Kelley Fitzgerald and Lynn Snyder-Mackler

Department of Physical Therapy and Interdisciplinary Graduate Program in Biomechanics and Movement Sciences, University of Delaware, Newark, DE, USA

Previous work in our lab has demonstrated that a screening tool can be beneficial in selecting patients with anterior cruciate ligament (ACL) rupture for return to high-level activities through non-operative management [1]. The screening tool uses clinical tests to differentiate those who have the potential to dynamically stabilize their knee from those who cannot. Patients selected by the screening tool (potential copers) have greater success in returning to high-level activities without knee instability compared to either those who do not meet the criteria, or uncategorized subjects [1, 5, 6].

Rudolph et al. [2, 3] found that classifying patients with ACL rupture by functional level allows for the differentiation of movement patterns during walking and jogging and gives insight into dynamic knee stabilization strategies. Those who returned to high-level activities for at least 1 year without knee instability were classified as copers; those who participated in high-level activities prior to injury, but could not after injury because of knee instability, were classified as non-copers [2–4]. Copers had knee angles and internal knee extensor moments that were similar to uninjured subjects; whereas non-copers had a decreased peak knee angle and reduced internal knee extensor moments [2, 3]. Non-copers also had higher co-contraction between muscles that cross the knee joint [3]. The authors stated that the non-copers appeared to use a "stiffening" strategy to stabilize their knee, which could be detrimental to articular cartilage [3].

The purpose of our present work was to compare movement patterns during walking and jogging between potential copers, those selected by the screening examination, and uninjured subjects. Our intent was to reveal whether the movement patterns of potential copers were more like uninjured subjects than non-copers. We hypothesized that potential copers would move more like uninjured subjects than non-copers since clinical outcomes suggest that potential copers have superior dynamic knee stabilization strategies compared to non-copers.

Eleven potential copers participated in this investigation. These subjects had an acute, unilateral ACL tear, met the criteria of the screening examination [1] and were involved regularly in high-level activities prior to injury. The potential copers

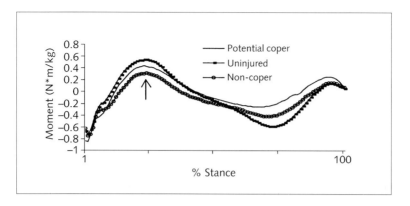

Figure 1
Internal knee extensor moment during walking. Potential copers' moment at peak knee flexion (indicated by ↑) is intermediate between uninjured and non-coper subjects.

were compared to ten uninjured subjects studied previously [3]. All subjects performed five walking and jogging trials at a self-selected speed. Kinematic and kinetic data were collected over stance. The variables for comparison were those in which non-copers differed from uninjured subjects or their own uninjured leg.

During walking, potential copers did not flex their knees as much as uninjured subjects or their uninjured knee. Potential copers' internal knee extensor moment at peak knee flexion was not different from uninjured subjects, and in fact, was intermediate between uninjured subjects and non-copers (Fig. 1). Potential copers had a decreased vertical ground reaction force during loading response, a characteristic similar to both copers and non-copers [3]. Potential copers also had altered joint support moments; the knee support moment was decreased, and the ankle support moment was increased. This is in contrast to non-copers whose decreased knee support moment is accompanied by an increased hip support moment [3]. It appears that transferring the support moment to the ankle is a more successful strategy, possibly due to the ability of the gastrocnemius and soleus muscles to assist with tibial stabilization.

During jogging, potential copers had a more extended knee angle at initial contact compared to uninjured subjects. The peak knee angle did not differ from uninjured subjects, but was decreased compared to the uninjured side. Potential copers generated an internal knee extensor moment, a peak vertical ground reaction force, and joint support moments that were comparable to uninjured subjects.

The results of our study give biomechanical evidence to support clinical findings that potential copers have better dynamic knee stability than non-copers. Potential copers have altered knee kinematics similar to non-copers, but do not show the non-copers' characteristic reduced internal knee extensor moment. Although potential copers had a decreased vertical ground reaction force during walking, it was not decreased during jogging, and this pattern is similar to copers. Potential copers movement patterns are thus a hybrid between copers and non-copers.

Our findings indicate that the dynamic knee stabilization strategies of potential copers, though superior to non-copers, are not completely developed and could benefit from further rehabilitation

References

1 Fitzgerald GK, Axe MJ, Snyder-Mackler L (2000) A decision-making scheme for returning patients to high-level activity with nonoperative treatment after anterior cruciate ligament rupture. *Knee Surg Sports Traumatol Arthrosc* 8: 76–82
2 Rudolph KS, Eastlack ME, Axe MJ, Snyder-Mackler L (1998) Movement patterns after anterior cruciate ligament injury. *J Electromyogr Kinesiol* 6: 349–362
3 Rudolph KS, Axe MJ, Buchanan TS, Scholz JP, Snyder-Mackler L (2001) Dynamic stability in the anterior cruciate ligament deficient knee. *Knee Surg Sports Traumatol Arthrosc* 2: 62–71
4 Eastlack ME, Axe MJ, Snyder-Mackler L (1999) Laxity, instability, and functional outcome after ACL injury: copers *versus* non-copers. *Med Sci Sports Exerc* 31: 210–215
5 Shelton WR, Barrett GR, Dukes A (1997) Early season anterior cruciate ligament tears: A treatment dilemma. *Am J Sports Med* 25: 656–658
6 Barrack RL, Bruckner JD, Kneist J, Inman WS, Alexander AH (1990) The outcome of nonoperatively treated complete tears of the anterior cruciate ligament in active young adults. *Clin Orthop* 259: 192–199

Neuromuscular control of the ACL deficient knee: Implications for the development of osteoarthritis

Lynn Snyder-Mackler and Terese Chmielewski

Department of Physical Therapy and Interdisciplinary Graduate Program in Biomechanics and Movement Sciences, University of Delaware, Newark, DE, USA

Many of the estimated 100,000–200,000 individuals who annually sustain anterior cruciate ligament (ACL) injuries in the USA experience substantial knee joint instability that could lead to joint deterioration over time [1–3]. Those who do not compensate well for ACL rupture (non-copers) comprise the majority of those who sustain this injury and the majority of those who undergo reconstructive surgery in the early post-injury period [4, 5]. ACL deficiency is associated with episodes of "giving way" of the knee that likely lead to excessive shear forces in the joint and uneven joint loading [6]. Shear forces are particularly detrimental to articular cartilage [7]. We have shown that joint instability and quadriceps femoris weakness, ubiquitous in non-copers, precipitate movement compensations that include reduced knee motion, reduced knee moment and excessive muscle co-contraction. Recently, Ramsey and colleagues, using markers attached to bone *via* intercortical pins, extended this finding to measurements of anterior translation during jumping in four ACL deficient non-copers ([8] and DK Ramsey, personal communication 1/24/01). The non-copers in their study over-constrained the knee, keeping the tibial translation below even that of their uninjured knees. This stiffening strategy, which persists over a wide range of activities, reflects an unsophisticated adaptation to the ACL rupture for which appropriate muscle activation strategies have not yet developed [9, 10]. This compensation strategy, which may reduce the anterior tibial translation at the knee, may also lead to less shock absorption and excessive joint compression that exacerbate joint destruction [4, 9, 10]. Compression and shear forces contribute significantly to the biochemical and metabolic changes that characterize degeneration of articular cartilage [2]. Most individuals with chronic ACL deficiency develop knee OA years after injury [11].

Some individuals undergo surgical reconstruction of the ligament to restore knee joint stability in hopes of minimizing future joint destruction. While most recover well from the surgery, long-term outcome is not uniformly good [4, 13]. Many never return to pre-injury activity levels, and those who undergo ACL reconstruction can expect a ten-fold increase in the incidence of knee OA over that of age-matched individuals [11]. Movement patterns of non-copers are not predictably normalized after

ACL reconstruction [12–15]. While kinematics eventually return to normal, kinetics and muscle activity alterations persist [12, 13, 16, 17]. It is likely that the maladaptive strategies seen prior to surgery that persist after surgery contribute to the less than optimal functional outcome for patients. Current rehabilitation programs for ACL rupture and surgical reconstruction are not uniformly successful in restoring functional outcome or mitigating the joint destruction over time [3, 4, 18].

Proprioceptive deficits, which are present in patients after ACL rupture and reconstruction, may also contribute to the movement strategies adopted by these persons [2, 19, 20]. The movement pattern abnormalities are exacerbated when quadriceps strength is not restored [9, 13–15]. Other investigators have demonstrated that quadriceps weakness and loss of proprioception occur in those with knee OA even without ACL rupture [19, 21]. More importantly, proprioceptive deficits and quadriceps weakness have been implicated in the development of osteoarthritis. Sharma et al. [20] found proprioceptive loss in people with unilateral OA, however the deficits were observed in both knees. This suggests that proprioceptive loss is not the result of OA but may predispose individuals to develop OA. Altered efferent input would also affect the reflexive activation of the quadriceps and could prevent appropriate muscle responses to joint loads. Individuals with ACL deficiency have similar proprioceptive loss and quadriceps femoris weakness exacerbated by joint instability thus increasing the risk for the development of OA.

The association of weakness and OA was demonstrated by Slemenda et al. [21], who studied the effect of quadriceps femoris muscle weakness on the development of OA in elderly, community dwelling men and women. As would be expected, people with pain and radiographic evidence of OA had greater quadriceps strength deficits and functional limitations than those without pain or evidence of OA. Individuals with radiographic evidence of OA and no pain, however, also had significant quadriceps strength deficits. This suggested that quadriceps weakness might precede the development of OA in the knee. Even relatively small increases in quadriceps femoris strength predicted a 20-30% decrease in the odds for having OA. Quadriceps weakness is a consistent sequelae of ACL rupture and persists after reconstructive surgery. Our own work and a recent study by Bush-Joseph and colleagues showed that after ACL reconstruction, persistent aberrant movements patterns occur even in those who are satisfied with the outcome of surgery, and that these patterns are strongly correlated with quadriceps strength [13, 16].

In addition to weakness and proprioceptive losses, the disruption of the ACL may influence the neuromuscular responses in the knee. Mechanoreceptors in the intact ACL can help to trigger reflexive muscle contraction in response to high stresses in the knee [22–24]. When the ACL is ruptured, however, other motor control mechanisms must be necessary to induce the appropriate protective motor responses to maintain the dynamic knee joint stability seen in people who can cope well with ACL rupture (copers). The gamma-system, which controls muscle stiffness and is mediated by spinal pathways and higher CNS centers, is one proposed neuromuscular control mechanism for maintaining dynamic knee joint stability [25]. Through movement experiences, the gamma motor system appears to alter the excitability of muscle spindles, causing a higher state of readiness of muscles to respond to perturbing forces [25]. Coordinated, compensatory muscle activity patterns that maintain dynamic knee stability in patients with ACL rupture involve several muscle groups in the lower extremity, which could be mediated by force-dependent reflex inhibitory neural pathways. Nichols [26] proposed a force-feedback hypothesis. When a perturbing force is applied to a joint, muscles are stretched and become activated to resist the perturbation, while there is reflex inhibition of muscles that would tend to act in the same direction as the perturbation. The net result is a coordinated co-activation of muscles affected by the perturbation to stiffen the joint and maintain dynamic stability. This reflex inhibitory mechanism appears to be force-dependent [26–29]. These proposed mechanisms for neuromuscular control of dynamic joint stability have several implications for design of treatment programs to enhance dynamic stability. The force-dependent nature of the mechanisms suggests that exposing the joint to potentially destabilizing loads during training may be the necessary stimulus to encourage the development of effective neuromuscular stabilization responses after injury.

Recent studies suggest that patients who achieve higher levels of functional recovery after ACL rupture, alter muscle activity patterns in a manner that improves dynamic stability of the knee [9, 10, 30–32]. Alterations in onset and duration of the quadriceps femoris, hamstrings, gastrocnemius, and tibialis anterior muscle activity during various functional tasks have been associated with successful compensation in patients with anterior cruciate ligament deficiency [9, 10, 30, 31, 33–36]. Alterations in the activity patterns of these muscles appear to control the anterior translational forces on the tibio-femoral joint, resulting in improved dynamic knee stability. The success of rehabilitation for patients who wish to return to rigorous physical activity should improve if treatment techniques that induce appropriate compensatory alterations in muscle activity patterns are incorporated into treatment programs. Although there is consistency across studies with respect to the muscles that are altered to improve dynamic knee stability, the patterns of altered activity themselves are remarkably varied [9, 10]. Our most recent work [9] demonstrates that there is no single "good compensation" pattern adopted by the copers; individuals adopt idiosyncratic compensation patterns that are related to rate of muscle activation and unrelated to quadriceps strength.

The lack of a "Gold Standard" compensatory muscle activity pattern creates a complex problem for designing treatments that are meant to enhance dynamic stability in patients with anterior cruciate ligament injuries. The solution is, therefore, more complex than training a single muscle group. Electromyographic biofeedback training paradigms are impractical since several muscle groups must be controlled and controlled differently for different tasks. Successful training strategies have to provide the opportunity for development of individualized compensatory alterations in activity of several lower extremity muscles during many different activities. Treatment techniques that attempt to promote the development of protective reflexes are needed to encourage involuntary muscular responses to destabilizing forces. Refinement of these protective neuromuscular responses may be possible through repeated movement experiences because spinal mediated pathways are influenced by input from higher motor control centers in the central nervous system. Treatment programs should not only allow for repeated practice of muscular responses to destabilizing forces, but should also provide practice in the context of functional and sport specific tasks [7, 37].

Interventions that target the neuromuscular contributions to strength and joint laxity may provide a more targeted and comprehensive approach to rehabilitation than traditional approaches. How people manage altered neuromuscular responses, quadriceps weakness and knee joint laxity may provide insight into the progression of OA over time. Recently, investigators and clinicians have recommended that "proprioception training" techniques be incorporated into anterior cruciate ligament rehabilitation programs [33, 38, 39]. The proprioceptive function of the knee joint becomes diminished after anterior cruciate ligament injury [22, 25, 40–42], in turn, reducing the ability of the neuromuscular system to maintain dynamic knee stability. Balance and agility training techniques are commonly used as "proprioceptive training" and are thought to improve balance and coordination skills necessary for dynamic joint stability. Although these techniques may assist patients in developing compensatory muscle activation patterns, limited success of these rehabilitation programs suggests that other treatment paradigms may be needed to improve treatment effectiveness.

Treatment techniques that involve perturbing support surfaces allow varying forces and torques to be applied to the lower extremity in multiple directions in a controlled manner to induce individualized dynamic stabilization patterns [30]. We have recently demonstrated in a randomized trial that a rehabilitation program that includes purposeful perturbation of support surfaces results in superior return to functional activity in high level athletes with good rehabilitation potential compared to a standard training program [43]. Our data and those of others [44, 45] support our contention that a successful knee stabilization strategy in the ACL deficient knee involves rapid motor responses. Traditional non-operative management of individuals with ACL rupture focuses on minimizing impairments: attain full active knee extension, which requires quadriceps activation and full passive knee extension, and eliminate joint effusion [46, 47]. The persistent altered movement patterns even after surgery and the high incidence of knee OA over time in these patients suggest that current rehabilitation programs are insufficient. The incorporation of a perturbation training intervention paradigm into the treatment regimen will potentially allow for the cycle of altered movement pattern-quadriceps weakness-proprioceptive deficit-osteoarthritis to be broken. More importantly, if the perturbation training program's ability to induce dynamic knee stability actually results in joint protection, its application to others at risk for the development knee OA may help reduce the incidence of this disabling clinical condition.

References

1   Buckland-Wright JC, Lynch JA, Dave B (2000) Early radiographic features in patients with anterior cruciate ligament rupture. *Ann Rheum Dis* 59(8): 641–646
2   Felson DT, Lawrence RC, Dieppe PA, Hirsch R, Helmick CG, Jordan JM, Kington RS, Lane NE, Nevitt MC, Zhang Y et al (2000) Osteoarthritis: new insights. Part 1: The disease and its risk factors. *Ann Intern Med* 133(8): 635–646
3   Shelbourne KD, Gray T (2000) Results of anterior cruciate ligament reconstruction based on meniscus and articular cartilage status at the time of surgery. Five- to fifteen-year evaluations. *Am J Sports Med* 28(4): 446–452
4   Daniel DM, Stone ML, Dobson BE, Fithian DC, Rossman DJ, Kaufman KR (1994) Fate of the anterior cruciate ligament-injured patient. A prospective outcome study. *Am J Sports Med* 22(5): 632–644
5   Fitzgerald GK, Axe MJ, Snyder-Mackler L (2000) A decision-making scheme for returning patients to high-level activity with nonoperative treatment after anterior cruciate ligament rupture. *Knee Surgery Sports Traumatology and Arthroscopy* 8(2): 76–82
6   Wu JZ, Herzog W, Epstein M (2000) Joint contact in the early stages of osteoarthritis. *Medical Engineering & Physics* 22: 1–12
7   Schmidt RA, Lee TD (1999) *Motor control and learning. A behavioral emphasis* (3 ed). Human Kinetics, Champaign, IL, 285–321
8   Ramsey DK, Lamontagne M, Wretenberg PF, Valentin A, Engstrom B, Nemeth G (2001) Assessment of functional knee bracing: an *in vivo* three-dimensional kinematic analysis of the anterior cruciate deficient knee. *Clin Biomech* 16(1): 61–70
9   Rudolph KS, Axe MJ, Buchanan TS, Scholz JP, Snyder-Mackler L (2001) Dynamic stability in the ACL deficient knee. *Knee Surg Sports Traumatol Arthrosc* 9: 62–71
10 Rudolph KS, Axe MJ, Snyder-Mackler L (2000) Dynamic stability in the ACL deficient knee: Who can hop? *Knee Surg Sports Traumatol Arthrosc* 8: 262–269
11 Gillquist J, Messner K (1999) Anterior cruciate ligament reconstruction and the long-term incidence of gonarthrosis. *Sports Med* 27(3): 143–156
12 DeVita P, Hortobagyi T, Barrier J (1998) Gait biomechanics are not normal after anterior cruciate ligament reconstruction and accelerated rehabilitation. *Med Sci Sports Exerc* 30(10): 1481–1488
13 Lewek M, Rudolph KS, Axe MJ, Snyder-Mackler L (2002) The effect of insufficient quadriceps strength on gait after anterior cruciate ligament reconstruction. *Clin Biomech* 17: 56–63
14 Snyder-Mackler L, Delitto A, Bailey SL, Stralka SW (1995) Strength of the quadriceps

femoris muscle and functional recovery after reconstruction of the anterior cruciate ligament. *J Bone Joint Surg* 77-A: 1166–1173

15 Snyder-Mackler L, Ladin Z, Schepsis AA, Young JC (1991) Electrical stimulation of the thigh muscles after reconstruction of the anterior cruciate ligament. *J Bone Joint Surg* 73-A: 1025–1036

16 Bush-Joseph CA, Hurwitz DE, Patel RR, Bahrani Y, Garretson R, Bach BR, Andriacchi TP (2001) Dynamic Function After ACL Reconstruction with Autologous Patellar Tendon. *Am J Sports Med* 29: 36–41

17 Timoney JM, Inman WS, Quesada PM, Sharkey PF, Barrack RL, Skinner HB, Alexander AH (1993) Return of normal gait patterns after anterior cruciate ligament reconstruction. *Am J Sports Med* 21(6): 887–889

18 Jarvela T, Paakkala T, Kannus P, Jarvinen M (2001) The incidence of patellofemoral OA and associated findings 7 years after ACL reconstruction with a bone-patellar tendon-bone autograft. *Am J Sports Med* 29: 18–24

19 Sharma L (1999) Proprioceptive impairment in knee osteoarthritis. *Rheum Dis Clin North Am* 25(2): 299–314

20 Sharma L, Lou C, Felson DT, Dunlop DD, Kirwan-Mellis G, Hayes KW, Weinrach D, Buchanan TS (1999) Laxity in healthy and osteoarthritic knees. *Arthritis Rheum* 42(5): 861–870

21 Slemenda C, Brandt KD, Heilman DK, Mazzuca S, Braunstein EM, Katz BP, Wolinsky FD (1997) Quadriceps weakness and osteoarthritis of the knee. *Ann Intern Med* 127(2): 97–104

22 Barrack RL, Skinner HB, Buckley SL (1989) Proprioception in the anterior cruciate deficient knee. *Am J Sports Med* 17: 1–8

23 Solomonow M, Baratta R, D'Ambrosia R (1989) The role of the hamstrings in the rehabilitation of the anterior cruciate ligament deficient knee in athletes. *Sports Med* 7: 42–48

24 Solomonow M, Baratta R, Zhou BH, Shoji H, Bose W, Beck C, D'Ambrosia R (1987) The synergistic action of the anterior cruciate ligament and thigh muscles in maintaining joint stability. *Am J Sports Med* 15: 207–213

25 Johansson H, Sjölander P (1993) Neurophysiology of ioints. In: V Wright, EL Radin (eds): *Mechanics of human joints: Physiology, pathophysiology, and treatment*. Marcel Dekker, New York, 243–289

26 Nichols TR (1993) A biomechanical perspective of the integrative actions of spinal reflex circuitry controlling the cat hind limb. *IUPS Abstr* 170/10

27 Bonsera SJ, Nichols TR (1994) Mechanical actions of heterogenic reflexes linking long toe flexors and extensors of the knee and ankle in the cat. *J Neurophysiol* 71: 1096–1110

28 Nichols TR (1994) A biomechanical perspective on spinal mechanisms of coordinated muscular action: An architecture principle. *Acta Anatomica* 151: 1–13

29 Nichols TR (1989) The organization of heterogenic reflexes among muscles crossing the ankle joint in the decerebrate cat. *J Physiol* 410: 463–477

30. Chmielewski TL, Rudolph KS, Snyder-Mackler L (2002) Development of dynamic knee stablility after acute ACL injury. *J Electromyogr Kinesiol* 12: 267–274
31. Ciccotti MG, Kerlan RK, Perry J, Pink M (1994) An electromyographic analysis of the knee during functional activities. II. The anterior cruciate ligament-deficient and reconstructed knee. *Am J Sports Med* 22(5): 651–658
32. Rudolph KS, Eastlack ME, Axe MJ, Snyder-Mackler L (1998) Movement patterns after anterior cruciate ligament injury: a comparison of patients who compensate well for the injury and those who require operative stabilization. *J Electromyogr Kinesiol* 8(6): 349–362
33. Friden T, Zatterstrom R, Lindstrand A, Moritz U (1991) Anterior cruciate insufficient knees treated with physiotherapy: A three year follow-up study of patients with late diagnosis. *Clinical Orthopaedics and Related Research* 263: 190–199
34. Lass P, Kaalund S, leFevre L, Sinkjaer T, Simonsen O (1991) Muscle coordination following rupture of the anterior cruciate ligament Electromyographic studies of 14 patients. *Acta Orthop Scand* 62(2): 9–14
35. Shiavi R, Zhang L, Limbird T, Edmonstone MA (1992) Pattern analysis of electromyographic linear envelopes exhibited by subjects with uninjured and injured knees during free and fast speed walking. *J Orthop Res* 10: 226–236
36. Sinkjaer T, Arendt-Nielsen L (1991) Knee stability and muscle coordination in patients with anterior cruciate ligament injuries: an electromyographic approach. *J Electromyogr Kinesiol* 1: 209–217
37. Horak FB, Nashner LM (1986) Central programming of postural movements: Adaptation to altered support-surface configurations. *J Neurophysiol* 55(6): 1369–1381
38. Irrgang JJ (1993) Modern trends in anterior cruciate ligament rehabilitation: nonoperative and postoperative management. *Clinics in Sports Medicine* 12(4): 797–813
39. Wilk KE, Andrews JR (1992) Current concepts in the treatment of anterior cruciate ligament disruption. *J Orthop Sports Phys Ther* 15: 279–293
40. DiFabio RP et al (1992) Effect of knee joint laxity on long-looped postural reflexes: Evidence for a human capsular-hamstring reflex. *Exp Brain Res* 90: 189–200
41. Wojtys EM, Huston LJ (1994) Neuromuscular performance in normal and anterior cruciate ligament-deficient lower extremities. *Am J Sports Med* 22: 89–104
42. Wojtys EM, Huston LJ (2000) Longitudinal effects of ACL injury and patellar tendon autograft reconstruction on euromuscular performance. *Am J Sports Med* 28: 336–344
43. Fitzgerald GK, Axe MJ, Snyder-Mackler L (2000) The efficacy of perturbation training in nonoperative anterior cruciate ligament rehabilitation programs for physical active individuals. *Phys Ther* 80(2): 128–140
44. Beard DJ, Dodd CAF, Trundle HR, Simpson AHRW (1994) Proprioception enhancement for anterior cruciate ligament deficiency. A prospective randomized trial of two physiotherapy regimens. *J Bone Joint Surg* 76-B: 654–659
45. Ihara H, Nakayama A (1986) Dynamic joint control training for knee ligament injuries. *Am J Sports Med* 22(5): 309–315
46. Shelbourne KD, Foulk DA (1995) Timing of surgery in acute anterior cruciate ligament

tears on the return of quadriceps muscle strength after reconstruction using an autogenous patellar tendon graft. *Am J Sports Med* 23(6): 686–689
47 Shelbourne KD, Urch SE (2000) Primary anterior cruciate ligament reconstruction using the contralateral autogenous patellar tendon. *Am J Sports Med* 28(5): 651–658

Discussion

Dynamic function and imaging in the analysis of osteoarthritis at the knee

Q: *Paul Dieppe*: Did you factor in quadriceps strength into the latter experiments? There is some evidence that quadriceps weakness might be a risk factor to OA, and it seems that it is a main controller of the sliding motion. Perhaps people with pure quads weakness but without ACL deficiency should be investigated.

A: *Tom Andriacchi*: We did do strength testing on these patients. The patients are definitely weaker and, as you know, after ACL injury there is a gait adaptation in which the quadriceps use is modified. One explanation may be that as the knee extends during the swing phase, activation of the quadriceps generates an anteriorly directed force on the tibia. If, as Radin has suggested, OA is a neuromuscular disease, perhaps the failure of the neuromuscular system to limit the anteriorly directed force contributes to the disease. Individual differences in the progression of the disease could reflect, in part, differences in the neuromuscular adaptation to the instability.

Q: *Wolfgang Eger*: We are doing between 350 and 400 tibial osteotomies per year and 90% of our patients are doing well after 7 years. But, that leaves 10% who do not do well. Have you tried to improve the tools used to predict who will do well and who will not do well based on the adduction moment? Can the adduction moment during gait be statistically related to the level of instability we see clinically? Is there any chance to correlate the adduction moment during gait to the lower extremity alignment during static weight bearing that we can measure using an x-ray?

A: *Tom Andriacchi*: We have been trying to establish a relationship between the adduction moment during gait and the pre-operative static weight bearing x-ray to predict the post-operative alignment given a specific correction. We have found that the operative results are improved in patients with a large adduction moment by incorporating gait analysis with predicted correction. Other factors are related to instability. Lateral laxity combined with a high adduction moment is a bad clinical sign due to the possibility that the load will be distributed on the medial side of the knee joint.

Q: *Ted Oegema*: Does the way that anterior cruciate repair is presently done correct the motions? One would suspect the motions are not corrected.

A: *Tom Andriacchi*: That is a big question to which an answer is needed. We have not yet done a study of post ACL reconstruction patients using this methodology.

Q: *Stefan Lohmander*: As a follow-up comment to Ted's question, there is no evidence in the literature that the currently used surgical techniques for ACL reconstruction prevent osteoarthritis. If anything, the literature suggests that surgery increases the likelihood of developing OA.

A: *Tom Andriacchi*: I think this may be getting at the issues related to the bigger picture. We may be treating the wrong thing if we are simply restoring statically measured anterior/posterior stability. When patients are walking, passive laxity does not seem to be a factor.

Q: *Klaus Kuettner*: The question from my point of view is whether compensation in the ankle joint is observed in the early phases?

A: *Tom Andriacchi*: Actually, the foot and the ankle are the driving forces that change the adduction moment. One way that patients change their adduction moment when walking is to walk more toed-out. I like to compare this to Charlie Chaplin who, based on his stage gait, probably never had medial compartment arthritis. The line of action seems to fall through the center of the knee joint and, therefore, the adduction moment is zero. At Stanford we have looked at a group of patients with flat foot deformity. These patients tend to walk like that, and they have very low adduction moments. Foot and ankle biomechanics are critical to the adaptive mechanism.

Linking the biology of osteoarthritis to locomotion mechanics

Q: *Phil Osdoby*: You may be 100% correct about your calcitonin studies, but there are also data and indications that calcitonin has added properties as well. So you may want to consider a control in your study where you add analgesics that don't have effects on bone to see what your outcome is.

A: *Eugene Thonar*: The big question in my mind is what happens to individuals who have OA and are receiving one or other drug treatment of osteoporosis. If bone changes drive the OA processes one may expect some of these drugs to affect the rate of progression of OA in these individuals. It should be noted that the fact that the serum level of antigenic KS in dogs who are administered calcitonin decreases markedly suggests that the effect was systemic.

Q: *Dick Heinegard*: Concerning the levels of keratan sulfate vs. hyaluronan, the same cells in the draining lymph nodes and the same cells in the liver that will remove hyaluronan very efficiently will also remove the larger fragments glycosaminoglycans. Could you actually be measuring the levels of hyaluronan and the hyaluronan turnover since when you have too much load on the system, you'll actually leave the keratan sulfate in the circulation?

A: *Eugene Thonar*: I don't think so but it is relatively easy to test.

Q: *Ted Oegema*: If the keratan sulfate is sialylated, the asialoglycoprotein receptor won't pick it up. If the neuraminidase in the serum removes the sialic acid, it would be. So there is actually a couple of mechanisms for getting rid of the keratan sulfate.

A: *Eugene Thonar*: You are correct. First, it should be noted that KS molecules that return to the blood circulation from the lymphatic system, *via* the thoracic duct, have lost their antigenicy. We do not know whether this is because the backbone of the chain has been partially or fully desulfated or because the chain has been more fully degraded in the lymphatic system. The point is that we are no longer able to detect the epitope. Second, antigenic KS chains on large peptidoglycans are, on average, removed from the circulation much more rapidly than single antigenic KS chains. At this time, we do not know if large and small fragments are cleared in different organs. I agree that the absence of a terminal sialic acid group on a KS chain is the major reason that chain is very rapidly removed from the circulation. We have evidence that sialylated KS chains have much longer half lives than their unsialylated counterparts in blood.

Q: *Robin Poole*: In rheumatoid arthritis it has been established in a number of studies that there is a significant relationship between joint involvement, as demonstrated clinically, and circulating levels of hyaluronan. Thus, would one suspect that there might be at least some relationship in OA? I think it is important to bear in mind that although you are dealing clinically with a single joint that has been identified as exhibiting pathology, when we are using biomarkers we may well be picking up preclinical changes. Therefore, changes in HA or CRP may be caused by upregulation of TGFbeta or cytokines such that the HA and the CRP are an indirect readout of cytokine generation. This change may not be restricted to one single joint. It could well be involving other joints.

A: *Eugene Thonar:* I agree. The excess HA circulating in blood is probably coming not only from the disease joints but also from unaffected synovial joints and connective tissues such as skin. I always have believed that serum HA in RA and, to a lesser extent, in OA may have prognostic value because this molecule is a surrogate marker for the presence of cytokines, i.e. interleukin-1, that are most effective in suppressing cartilage repair.

Movement patterns of individuals with good potential to dynamically stabilize their knee after acute ACL rupture

Q: *Mark Grabiner:* Terry, do you still test strength on these subjects?

A: *Terese Chmielewski:* We administer a screening examination, and a selection criterion is that the isometric strength of the involved leg is at least 70% of the uninvolved side. The people that we studied actually had approximately 85% strength. In my experience of the past 3 years, the people who passed this screening examination don't always pass it with a quadriceps strength that is really high. Some people are 71%, so the quadriceps strength doesn't really factor into who ends up passing.

Q: *Mark Grabiner:* The observation that Lynn indicated before – persistent weakness – now is many years old. I think someone asked the question about mechanisms – it doesn't seem to be related to coactivation, it doesn't seem to be related to inability to activate quadriceps because you have identified that it goes up. In fact, many of your patients are capable of full activation when based against electrical stimulation, so the question of the mechanisms of this persistent weakness remains and would seem to be important. The relationship between the components of the vastus lateralis and medialis may play a key role because even subtle changes in the lateralization and medialization of the patella could profoundly affect knee extension strength.

Q: *Ted Oegema:* You said these potential copers are on the way to becoming copers. Does your training program help them or not?

A: *Terese Chmielewski:* The subjects we have identified as potential copers and who have completed the support surface training are five times more likely to return. Their success rate was 92%. After the training, the quadriceps activity seems to be coupled to that of the gastrocnemius, soleus and the hamstrings. We think this is a mechanism underlying increased knee stability.

Q: *Tom Andriacchi*: My question relates to the long-term relationship between the activity levels of the copers, that is, the knee joint buckling that may occur during running, cutting and pivoting activities, and the progression of arthritis. Take an individual with ACL injury and think about him 20 years from now. What is the possibility that he's going to have arthritis based on identifying early on whether he's a coper or non-coper?

A: *Terese Chmielewski*: We just don't know. That will be a focus of follow-up investigations. We are thinking that perhaps our training will be associated with improved post-operative movement patterns. In particular, we would like to think that we are having a beneficial effect on the knee joint loading patterns during activity that may be related to the progression of osteoarthritis.

Q: *Tom Andriacchi*: It's possible that there is not a connection between high performance tests and the mechanism of osteoarthritis. In my own experience, patients I tested 20 years ago are now getting arthritis. I've recently retested two patients. Initially they tested normally for quadriceps moments. Now, at about 45 years of age and just prior to having a high tibial osteotomy for medial compartment arthritis, they still test normally for quadriceps moment.

A: *Terese Chmielewski*: What we are able to measure using the high performance tests is the level to which subjects can function without experiencing instability. We know that instability is associated with the progression of OA: But, I agree with you, there are other factors that need to be looked at.

Q: *Juergen Mollenhauer*: Of course in this acute phase of trauma, pain is a major issue. Have you looked at whether copers or non-copers are different in individual pain management? It could well be that this gait behavior and biomechanical response are related to the pain.

A: *Terese Chmielewski*: Exclusion factors that are screened for include reduced motion, effusion and pain during unilateral hopping. The people I have reported on didn't have pain.

Stefan Lohmander: This whole discussion ignores the fact that it takes considerable force to tear an ACL in a human. The forces involved in an ACL injury may very well be directly injurious to the cartilage itself. The knee joint forces, at the instant when the ACL ruptures, may be killing cartilage cells and damaging the matrix. It is quite possible that whatever we do clinically after that instant may be irrelevant in many cases.

Neuromuscular control of the ACL deficient knee. Implications for the development of osteoarthritis

Q: *Joel Block*: I wonder whether the age at which ACL injury occurs influences whether a patient is a coper or a non-coper.

A: *Lynn Snyder-Mackler*: We presently don't have enough data to really know, but age at the time of injury does not appear to be a factor. Most people will tear their ACLs in late adolescence and during a period of high activity. Although older individuals do tear their ACLs, these patients are fundamentally different from our patients who had participated in level 1 and level 2 sports, which are jumping, pivoting, cutting, racquet sports. All of the patients in our study had to have returned to that amount of activity and at their previous level of participation without episodes of giving way to be considered copers. Our subject population was fairly homogenous with respect to activity.

Q: *Paul Dieppe*: You have dichotomized your subjects into copers and non-copers. Is this really a fair representation, or are you actually looking at the ends of a continuum? The second question relates to the fate of these two distinct patient groups. What happens to them down the line, or in the long term?

A: *Lynn Snyder-Mackler*: The variable that we use to dichotomize the groups is the presence of episodes in which their knee gives way, or buckles, during activity. So, the dichotomy is made only on a single characteristic. These two groups may represent a continuum using other variables that contribute to these episodes.

Q: *Paul Dieppe*: But are there people who are in-between?

A: *Lynn Snyder-Mackler*: There are people who are in-between the copers and non-copers. The majority of people, at least in the US, who don't have surgery for the ACL are those who are willing to alter their activities. We call them adaptors. These are the people who are willing to discontinue playing tennis, or whatever activity in which they were engaged. So, we don't really know about that group even though it's a group of interest. Regarding the fate of these subjects, we know a little about the copers. For example, a person may present clinically, and we learn that they are ACL-deficient. Interestingly, each of the subjects we have evaluated using x-rays and arthroscopy has reasonably healthy knees. We think that it relates to the knee joint stability that arises from the neuromuscular adaptation to the injury. Going back to a comment made after Tom Andriacchi's presentation, in our experience, most of the individuals who undergo surgical reconstruction are non-copers and have these bad patterns, weak quads, and recurrent instability. These factors may be key for long-term degenerative processes regardless of the decision for or against surgery. A question that we are interested in is whether we can make these non-copers even potentially better operative candidates by having them develop better dynamic stability before surgery.

Stefan Lohmander: Your comment about the majority of the copers being men is interesting in view of the higher incidence of knee injuries and ACL injuries, in particular, in women compared to men. We have a fairly large cohort of women division level soccer players. Looking at these athletes 12 years down the line, more than 30% of them have definite radiological osteoarthritis of the knee, and they have significant functional and subjective problems and impairments. There are about 100,000 ACL reconstructions per year in the US, so this is not an insignificant problem. And by the way, there was no difference in the rate of OA between those operated on or not.

Photos from the conference

Photo series 1
Top left: *Roger Mason, Klaus Kuettner;* top right, standing: *Thomas Andriacchi, Richard Loeser;* seated: *Stefan Lohmander, Sue Eyre, Roger Mason, Inger Lohmander;* middle: *Jürgen Mollenhauer, Gabriella Cs-Szabo;* bottom left: *Linda Sandell, Ernst Hunziker, Jenny Tyler, Alan Grodzinsky;* bottom right: *Vince Hascall*

Photo series 2
Top left: *Vince Hascall, Klaus Kuettner, Elizabeth Dieppe;* top right: *Kathy Vukovich, Marliese Annefeld, Hari Reddi, Anu Reddi;* middle left: *Joel Block, Richard Loeser, Warren Knudson, Cheryl Knudson;* middle right: *Inger Lohmander, Stefan Lohmander;* bottom left: *Elizabeth Dieppe, Paul Dieppe;* bottom right: *Tom White, Carol Muehleman, Klaus Kuettner, Julia Freeman, Susan Chubinskaya*

Photo series 3
Top left: *Paul Dieppe, Robin Poole;* top right, seated: *Wolfgang Eger, Holger Koepp, Stefan Söder;* standing: *Johannes Flechtenmacher, Bernd Rolauffs, Monika Schulze, Klaus Kuettner, Matthias Aurich, Hans Häuselmann;* middle left: *Margaret Aydelotte, Klaus Kuettner;* middle right: *Ada Cole, Allan Valdellon, Arkady Margulis;* bottom left: *Mark Grabiner, Rick Sumner;* bottom right: *Warren Knudson, Ted Oegema*

Photo series 4
Top left: *Klaus Kuettner, Verhonda Hearon Eggleston, Matthias Aurich, Erzsebet Kuettner, Jürgen Mollenhauer;* top right: *Susan Chubinskaya, Klaus Kuettner, Carol Muehleman;* middle left: *David Howell, Fred Woessner, Vince Hascall, Teresa Morales, Klaus Kuettner;* middle right: *Fred Woessner, Nina Woessner, Klaus Kuettner, Eugene Thonar;* bottom left: *Erzsebet Kuettner, Vince Hascall, Klaus Kuettner, Hari Reddi, Inger Lohmander;* bottom right: *Vanja Hascall, Hari Reddi, Vince Hascall*

Photo series 5
Top left: *"The Scoriers" (no names needed)*; top right: *Klaus Kuettner*; middle left: *Gene Homandberg, Eugene Thonar*; bottom left: *Jill Urban, Alice Maroudas*; bottom right: *Robert Sah, Cheryl Knudson, Teresa Morales*

Index

actin 221
activin receptor-like kinase (ALK) 72, 83
ADAM 280
ADAMTS 280
adduction moment 457
adult human normal cartilage 82
adult human articular cartilage 82, 415
aggrecan 141, 147, 155, 207, 294, 387
aggrecan solution properties, comparison with hyaluronan 141
aging 248, 249, 409, 417
AgKS, direct metabolic marker of joint metabolism 454
alginate 77
alginate bead 429
ALK-1 72
ALK-2 83
ALK-3 83
ALK-6 83
alkali pH, effects on hyaluronan solution properties 136
alkaline phosphatase 254
alpha1(II) collagen (COL2A1) 72
amelogenin peptides, alternative premRNA splice products 103
analgesic effect of hyaluronan 196
analyzer crystal 352
animal model 45, 49
ankle 27, 353, 369, 429
ankle joint 27, 353
ankle (talocrural) joint 27
anterior cruciate ligament (ACL) 461, 465

apoptosis 256
applied stress 249
arthritic pain 197
articular cartilage 45, 67, 82, 279, 351
articular cartilage collagen 279
atrophic destructive OA 357, 360
autologous chondrocyte transplantation (ACT) 67, 68
automated electron tomography (AET) 124
autoradiography, quantitative 400

basic calcium phosphate (BCP) 357, 359
biglycan 294, 369, 370
bikunin 207, 208
bikunin-gene knockout mouse 209
biocompatible scaffold 423
biomechanical properties, cartilage 409
biomechanics 239, 409, 441
blood circulating system 211
blood vessel invasion 255
BMP/TGFβ superfamily 214
BMP-2 261, 262
BMP-7/OP-1 77, 178, 220
body mass index 33
bone collagen 277
bone dust 357
bone marrow 68, 73
bone matrix 276
bone mineral density 359
bone morphogenetic protein (BMP) 5, 7, 63, 64, 77, 81, 82, 101, 119, 178, 214, 220, 227, 261, 262

bone morphogenetic protein (BMP) family 81
bone morphogenetic protein (BMP) antagonist 7
bone morphogenetic protein (BMP) receptor 82
bone remodeling 46, 49
bone resorption 275
bone scintigraphy 358
brachypodism (bp) 214
brevican 147

calcified cartilage, zone of 254
calcium pyrophosphate dihydrate (CPPD) 356, 357
calcium pyrophosphate dihydrate crystal 356
calcium, effects on hyaluronan solution properties 134
camptodactyly-arthropathy-coxa vara-pericarditis syndrome 22
camptodactyly-arthropathy-coxa vara-pericarditis protein (CACP) 159
cementum attachment protein (CAP) 104
cartilage 46, 49, 63, 64, 77, 81, 82, 151, 154, 163, 192, 214, 239, 251, 261, 263, 278, 285–287, 309, 353, 369, 402, 403, 409, 415, 423, 453
cartilage biomechanics 409
cartilage collagen 278
cartilage degeneration 353, 369
cartilage degradation 49
cartilage destruction 261
cartilage development 214
cartilage differentiation 309
cartilage marker 285
cartilage matrix protein (CMP) 151
cartilage oligomeric matrix protein (COMP) 46, 163, 286, 287, 453
cartilage oligomeric matrix protein, assays 287
cartilage regeneration 81
cartilage repair 81, 263, 369, 423
cartilage-derived morphogenetic protein (CDMP-1) 63, 64
cartilage-specific gene 309
casein kinase 223

cathepsin K 276
CD44 178, 219, 255
CD44-HA interaction, modification 211
CD44, short-tail 225
cDNA-array technology 293
cell culture model 163
cell expansion 67
cellularity 416
cementum attachment protein (CAP) 104
Cereberus, *Xenopus* gene 64
chondrocyte 67, 77, 178, 392
chondroitin sulfate chain, bikunin 207
chondrophyte 358
chordin 64
chromosome 16p 43, 44
chrondrogenesis 91
chrondrogenesis, bone induction 101
COL2A1 72
collagen 45, 117, 119, 121, 155, 249, 250, 275–279, 294, 413, 414
collagen II, III, IX, XI 278, 279
collagen catabolism 277
collagen cross-links 45, 275
collagen early fibril (CEF) 121
collagen fibril 278
collagen fibrillogenesis 119
collagen network 249, 413
collagen network stiffness 247, 250
collagen network, tension 248
collagen network Pc, tension stiffness 250
collagen tension 249
collagenase 279, 296
Collins grade 370
complex genetic disease 18, 21, 22
conditional gene targeting 216
confocal fluorescence recovery after photo-bleaching (confocal-FRAP) 131
congenital hip dysplasia 411
continuum model 416
copers 461
CPPD 359
C-reactive protein 453

cross-linked telopeptide 276
cross-linking interaction 279
crystals, markers of disease 359
CTx immunoassay (cross-linked C-telopeptides of type I collagen) 275
C-type lectin 147
cumulus oophorus, expansion of 209
cytoskeleton 221, 222, 226

decorin 294, 369, 370
de-differentiation 68
deformation, of cells and matrix 383
degenerative cartilage 82
dentin matrix protein (DMP2) 103
deoxypyridinoline 275
diagnosis of OA 299, 409
differential gene trapping 310
differentiation 213
Diffraction Enhanced Imaging (DEI) 351
distal femur 413
dual-photon X-ray absorptiometry 365
dynamic compression 398
dynamic joint loading 457
dynamic knee stabilization 461
dynamic shear modulus 404
dysplastic joint 359

electrokinetic behavior, of cartilage 402
electron microscopy 121
ELISA 210
embryogenesis 5, 213
endocardial cushion tissue 149
endochondral ossification 149
energy metabolism 391
explant culture 429
extracellular matrix (ECM) 5, 65, 117, 149, 213, 147
extracellular matrix, organization of 149

fatigue 416
FCD 248
femoral condyle 412

femoral head 358, 412
fibril 117
fibroblast growth factor receptor-3 (FGFR3) 72
fibronectin fragment 78
fibronectin 78, 192, 194, 294
fibulin-1 147
fibulin-2 147
fixed charge density 383, 387, 388
fixed charge 387, 413
fluid flow 383
fractal signature analysis 358
fragment of biglycan 370
fragment of decorin 371

gait testing 457
gelatinase A 244
gene targeting 155
GlcNAc, of HA 208
glycation, non-enzymatic 249, 414
glycosaminoglycan (GAG) 387, 388, 397, 402, 414, 429
glycosaminoglycan, electrostatic interactions 402
gremlin, *Xenopus* gene 64
growth factor 261
growth law 416
growth plate 254
growth/differentiation factor 5 (Gdf5) 178
growth/differentiation factor 9 (Gdf9) 214

heart development 149
heavy chain 207
heparan sulfate, binding of matrilysin 242
hip 39–44, 303, 363, 367
hip OA 303, 363, 367
hip OA, genetic component/background 39–44
histopathological indices 414
human articular cartilage 27, 429
Human Genome Project 17
humerus, distal 413
hyaline cartilage 355–360
hyaline cartilage defect 357
hyaline cartilage degradation 355, 358

hyaline cartilage disease 360
hyalocyte 193
hyaluronan (HA) 131, 132, 134, 136–138, 141, 147, 177–179, 189, 195, 196, 201, 207, 209, 210, 213, 219, 453, 455
hyaluronan levels, in sera of RA patients 209, 210, 455
hyaluronan-rich matrix 207
hyaluronan synthase (HAS) 178, 213
hyaluronan synthase 2 (HAS2) 178
hyaluronan synthesis 195
hyaluronan, self diffusion 132
hyaluronan/aggrecan aggregates, cross-linking 147
hyaluronan-binding chondroitin sulfate proteoglycan 207
hyaluronan, therapeutic use 201
hydration 251, 387, 388
hydraulic permeability 180, 411
hydrodynamic radius (R_H), of hyaluronan 138
hydrogen ion 391
hydrostatic pressure 383
hydroxylysylpyridinoline (HP) 46
hylan A 201
hylan B 201
hypertrophic bone response 356
hypertrophic chondrocyte 254
hypertrophic hip OA 303
hypertrophic OA 356, 357
hypertrophic osteophytosis 358
hypertrophic pathway 257
hypertrophy 254

IL-1 64, 178, 220
IL-1β 78, 86, 429
IL-17 64
imaging 329, 351, 442
immobilized joint 195
immuno-histochemistry 45, 85
in situ hybridization 45, 71, 83
in vitro expansion 72
India ink 414

instantaneous deformation (ID) 250
instantaneous deformation, increase 251
inter-α-trypsin inhibitor (ITI) 178
inter-α-trypsin inhibitor (ITI) family molecule 207
inter-α-trypsin inhibitor (ITI) family molecule, structure 208
inter-α-trypsin inhibitor (ITI) family molecule, intracellular assembly 208
intracellular composition 392
intracellular pH 392, 393
ionic environment 387
ionic strength 387, 388

joint degradation 303
joint development 214
joint space width (JSW) 363–366
joint space width, rate of decline 364

knee 27, 369, 429, 456, 465
knee OA 456
knee (tibiofemoral) joint 27

lactic acid 391
lamina splendens 192
limb development 214
locomotion mechanics 453
lubricating function 412
lubrication 160
Lubricin 22, 159
lysylpyridinoline (LP) 46

macrophage 262
magnetic resonance imaging (MRI) 339, 357, 358
MAP kinase pathway 384
marked risk 43
marker, joint metabolism 453
marker, osteoarthritis 297
marker practicality 298
marrow oedema 357
matrilysin 242, 282

matrix accretion 411
matrix fragmentation 416
matrix metalloproteinase (MMP) 7, 241, 280, 281, 295
matrix metalloproteinase 3 281
matrix metalloproteinase 7 (matrilysin) 282
matrix nano-electromechanics 397
mechanical compression 427
mechanical integrity 416
mechanical properties, of cartilage 403
mechanical stimuli 383
mechanoreceptor 199
mechanotransduction, cellular 383
megakaryocyte stimulating factor (MSF) 159
membrane patches, from oocyte 199
membrane transporter 384
Mendelian disease 17
mesenchymal stem cell 73
microenvironment 416
microfibril 124
migration 213
Milwaukee shoulder 357
mineralization 256
mitogen activated protein (MAP) kinase 388
molecular marker 72, 285
molecular polarity reversal 122
molecular switch region 123
monolayer culture 429
morphogenesis 214
MRI 339, 357, 358
multiple epiphyseal dysplasia 156
multiple markers, use of 300

nanoscale electrostatic repulsive interaction 384
national genealogy database 39
national registry 39
negative staining EM 147
neurocan 147
NH_2-propeptide 5
NHE ($Na^+ \times H^+$ exchange), chondrocytes 392
noggin 64

non-copers 461, 465
NTx immunoassay (cross-linked N-telopeptide fragments of type I collagen) 275

OA cartilage 82, 251
oligosaccharide 222, 226
OP-1 (BMP7) 77, 178, 220
osmolality 387, 388
osmolyte 393
osmotic pressure 383
osmotic pressure, GuHCl extracts 248
osmotic pressure, proteoglycan (PG) 247, 402
osmotic swelling pressure, proteoglycan 402
osmotical compression 413
osteoarthritic samples 251
osteoclast 276
osteogenic protein-1 81
osteonectin 294
osteophyte 46, 261, 262, 357
osteophytosis 358
overweight 35

Paget's disease of bone 278
patella 412
periosteum 68, 73
permanent cartilage 254
permeability 414
PG 46, 71, 77, 78, 147, 207, 247, 250, 369, 402, 430
PG synthesis 78
PG-M/versican 207
pH 387, 390-393
phenotypic stability 68
physical activity 33
physical forces and flows 383
physical regulation, of cartilage metabolism 398
pointed tips 119
poroelasticity 402
preovulatory follicles 209
pressure 387, 393, 411
pressure gradient 411
protease 239

proteoglycan (PG) 46, 71, 77, 78, 147, 207, 247, 250, 369, 402, 430
proteolytic processing 152
pseudoachondroplasia (PSACH) 165
pseudogout syndrome 356, 357
PTH 257
PTHrp 257
pycnodysostosis 276
pyridinoline 275
pyrophosphate arthropathy 356

quadriceps 466
quantitative autoradiography 400

RA progression, degree 209
radiographic imaging 353
radiography 351
radiology 331
rapidly destructive hip OA 303
regeneration 411
regenerative medicine 65
remodeling 417
repair 411, 417
risk factors, development of lower limb osteoarthritis 31
risk factors, human osteoarthritis 4
RUNX2 254

scanning electron micrograph 415
scanning electron microscopy 45
scintigraphic subsets 358
scintigraphic subsets, extended pattern 358
scintigraphic subsets, tramline pattern 358
self-assembling peptide hydrogel 423
semi-quantitative histology 45
semi-quantitative RT-PCR 83
sera of OA patients 209
serum cartilage oligomeric matrix protein 288, 289
serum concentration, SHAP-HA complex 209, 210
serum factor 207, 208

serum factor, from blood 208
serum HA concentration 209, 210, 455
serum-derived hyaluronan-associated protein (SHAP) 178, 207–210
SHAP-HA complex, levels in RA sera 209
SHAP-HA complex formation, possible pathway 208
shear modulus of cartilage 403
skeletal development 5
skeletal scintigraphy 358
small proteoglycan (PG) 369
stem cell 65
STIR sequence 357, 360
strain 409, 411
streaming potentials and currents 383
stress 249, 409, 411
subchondral bone 253, 356–360
subchondral bone, blood flow 360
subchondral bone, density 358
subchondral bone response 357
subchondral bone response, crystals 356
subchondral cyst 356
subchondral marrow 357
subchondral sclerosis 358
superficial layer, of cartilage 192
superficial tangential zone (STZ) 417
superficial zone chondrocytes 159
superficial zone protein (SZP) 22, 159
superficial zone 251, 412
surrogate marker, hip OA 367
surrogate parameter 49
swelling 387, 388, 413, 416
swelling pressure 387
synchrotron 351
synovial cavity of the articular joint 207
synovial cell 159
synovial fluid 159, 177, 179, 190, 208
synovial interstitial matrix 180
synovial membrane 207
synovial tissue 193
synoviocyte 177, 180, 193
synovium 73, 177, 189

talus 412
telopeptide 275, 280, 281
telopeptidase 280
telopeptide domain epitope 281
tenascin-R 147
tensile 411, 412
tensile modulus 411
tensile strength 412
TGFβ 72, 214, 257, 261-264
TGFβ superfamily 72, 214
TGFβ type II receptor 263
tidemark 253, 255, 256
TIMP-3, binding to matrix 243
tip-to-tip fusion 123
tissue engineering 65, 411, 423
tissue engineering scaffold 423
tissue expansion 411
tissue layer 410
tissue repair 5
tracer diffusion, in hyaluronan solutions 137
transcription factor 91
transport pathway 392

transporter 389, 392, 393
trauma 33
TSG-6 gene 209
type 2A procollagen 3
type II collagen 46, 281
type II procollagen 64
type II receptor 83
type IIA procollagen 5
type IV collagen 64
type X collagen 254

urea, effects on hyaluronan solution properties 139
urinary hydroxyproline 278
urinary trypsin inhibitor (UTI) 208

versican 147, 207, 294
viscoelastic properties 177
viscosupplement 177
viscosupplementation 201

Xenopus genes 64
x-ray 334, 351